Deploying Wireless Networks

Do you want your wireless network to be profitable? Wireless operators will find this practical, hands-on guide to network deployment invaluable. Based on their own extensive experience, the authors describe an end-to-end network planning process to deliver the guaranteed QoS that enables today's wireless IP services such as VoIP, WWW and streaming video. The trade-off between enhanced user experience and operator cost is explored in the context of an example business model, and shown to be a key driver of project value in the planning process.

Comprehensive examples are provided for:

- GSM/GPRS/EDGE
- WCDMA-UMTS/HSDPA
- OFDM-WiMAX/LTE
- mesh WiFi
- packet backhaul

Topics addressed include:

- capacity/peak data rates
- service latency
- link budgets
- lifecycle costs
- network optimisation

With a focus on practical design, the book is ideal for radio and core network planners, designers, optimisers and business development staff at operators and network equipment manufacturers. Extensive references also make it suitable for graduate and postgraduate students.

Andy Wilton is Senior Director and Chief Technologist of Motorola's GSM Systems Division (GSD) and is responsible for technology management of GSM and UMTS products worldwide. His major activities are product innovation, technology sourcing, standards strategy and technical oversight of all GSD programmes.

Tim Charity is Head of Applied Technology at Motorola's Global Networks Services. He has led the creation of a service portfolio for GSM, CDMA, UMTS and WiMAX networks, and holds several patents relating to cellular technologies.

Deploying Wireless Networks

ANDY WILTON
Motorola Ltd

TIM CHARITY
Motorola Ltd

Shaftesbury Road, Cambridge CB2 8EA, United Kingdom

One Liberty Plaza, 20th Floor, New York, NY 10006, USA

477 Williamstown Road, Port Melbourne, VIC 3207, Australia

314–321, 3rd Floor, Plot 3, Splendor Forum, Jasola District Centre, New Delhi – 110025, India

103 Penang Road, #05–06/07, Visioncrest Commercial, Singapore 238467

Cambridge University Press is part of Cambridge University Press & Assessment, a department of the University of Cambridge.

We share the University's mission to contribute to society through the pursuit of education, learning and research at the highest international levels of excellence.

www.cambridge.org
Information on this title: www.cambridge.org/9780521874212

First published 2008

A catalogue record for this publication is available from the British Library

Library of Congress Cataloging-in-Publication data
Wilton, Andy, 1951–
Deploying wireless networks / Andy Wilton, Tim Charity.
p. cm.
Includes bibliographical references and index.
ISBN 978-0-521-87421-2 (hardback)
1. Cellular telephone systems. 2. Personal communication service systems. I. Charity, Tim. II. Title.

TK5105.78.W54 2008
621.384 – dc22 2008020088

ISBN 978-0-521-87421-2 Hardback

To my son Nicholas and to my wife Jenny for making this and so many other things possible.

Andy

To my family.

Tim

Contents

Foreword

Telecommunications networks have always fascinated me. My interest was sparked when, as an engineering student in the late sixties, I was told that the telephone network was the biggest machine on earth yet it was constructed from a few basic building blocks replicated many, many times over. It seemed to me that a machine with those characteristics was both already a remarkable engineering feat and a perfect platform for the rapid development of more sophisticated services. So I decided upon a career in telecommunications engineering.

I soon discovered that there was nothing basic about either the building blocks or the architecture of those networks. They were already engineeringly sophisticated at every layer and in every enabling technology. That sophistication was to lead to the continuing development of telecoms networks at a far greater pace than any of us working in the field three or four decades ago could possibly have imagined.

From voice to data; analogue to digital; terrestrial to satellite; tethered to untethered, the progress has been remarkable. Yet undoubtedly the most remarkable development of all has been in wireless networks. Nearly half of the world's population take it for granted that the purpose of telecoms networks is to connect people, not places. An increasing proportion of them use those connections for exchanging text and images as readily as voice. The transformational effect on national economies, education, health and many other factors that bear upon the quality of life is apparent. Wireless networks are helping us to improve our world in all sorts of ways.

It seems an age ago that my network engineering adventures began. Indeed it was an age ago. It is likely that for those who read this book, the adventure is just beginning. They will take wireless networks to the next high and, along the way, deliver true seamlessness across an increasing diversity of networks, services and devices. They will offer the benefits of that seamlessness to the second half of the world's population. They will transform lives, no less.

I envy them those opportunities and wish them every success. And I thank Andy Wilton and Tim Charity for writing a book that is destined both to become a classic in its field and to inspire all with an interest in wireless networks.

Sir David Brown, FREng
Chairman, Motorola Ltd

Preface

At the time of writing, and to an extent never seen before, there is an expectation that almost any information or service that is available through communication systems in the office or home will be available wherever the user happens to be. This is placing incredible demands on wireless communications and has been the driver for the genesis and deployment of three generations of cellular systems in the space of 20 years. In parallel with this revolution in access technology has come the recognition that any information, whether for communication, entertainment or, indeed, for other purposes as yet unenvisaged, can be stored and transported in a universal digital format. The former technology-driven distinctions of analogue storage and transport for high bandwidth signals, such as video, and digital storage for other content are no more. These changes, together with an increasing international consensus on a 'light-touch' regime for regulation to stimulate competition, have enabled the first generation of quad-play multinational companies to become established. Such companies seek to spread a strong base of content and services across what would formerly have been known as broadcast (cable, satellite, terrestrial), fixed telephony, mobile and broadband access channels. However, the ability for such companies to deliver applications and services that operate reliably and consistently, regardless of user location, is ultimately predicated on their ability to design solutions that deliver an appropriate and guaranteed quality of service (QoS) over what will certainly be a finite and potentially narrow-access data pipe. As today's CEOs know, the delivery of best-effort bit pipes is increasingly a commodity business. Bankers have known since the Battle of Waterloo that the delivery of timely information is the basis on which fortunes are made. The design and deployment of wireless networks that can deliver guaranteed QoS form the focus of this book.

The book is organised into ten chapters. Chapter 1 provides an overview of the factors currently at work in the communications marketplace worldwide; it also reacquaints the reader with *Shannon's Law*, which drives the architecture of many of the solutions discussed. Chapter 2 aims to establish an understanding of the way most cellular networks operate, using the ubiquitous GSM system as a baseline, and highlights the key differences that can be expected in networks providing fixed or 'nomadic' wireless access. The concept of *contexts* – central to the timely and efficient delivery of packet data in GSM and UMTS – is also introduced. Factors that contribute to cellular network operating expense are explored to highlight activities that significantly impact the operator's profit and loss account and, finally, the profit and loss account is used as an agenda to identify wireless network technologies that are likely to change in the future.

Chapter 3 is a key chapter for any reader new to wireless network planning. It introduces the processes to be followed when planning the wireless network segment of most cellular networks. Particular attention is paid to techniques that can be adopted to maximise return on investment through intelligent network rollout. A review of the different approaches that are adopted when planning voice and multimedia networks is provided and there is emphasis on the planning of wireless packet networks. A technique is introduced that may be used to plan such networks so that the delivery of QoS is assured. Such techniques will assume increasing importance as it becomes necessary to ensure the correct operation of the new classes of applications, which tend to have arbitrary traffic burst distributions. The chapter concludes with a methodology that may be used to assess the commercial viability of new wireless network proposals. Chapter 4 provides a more extensive discussion of propagation models and introduces a framework for capturing the requirements of multimedia traffic. Insight is provided into the sources of information that are needed together with a discussion of the key steps in the generic aspects of the RAN planning process.

Chapters 5, 6, 7 and 8 adhere to a common format. A description of the key features and principles of a particular radio access network (RAN) is followed by a discussion of the factors that influence the system sensitivity and, hence, coverage and capacity. Additional functionality that can be configured to influence performance is then reviewed. The intention of these sections is to provide the reader with a good understanding of the factors that drive performance for the wireless network in a given scenario. The second part of each of these chapters features a worked example, which addresses one of the deployment scenarios detailed in Chapter 4. The worked example template is consistent across the four chapters and serves to illustrate the strengths and weaknesses of the air interface under discussion. Chapter 5 addresses GSM and GPRS/EDGE, Chapter 6 UMTS, Chapter 7 UMTS long-term evolution (LTE) and WiMAX and Chapter 8 Wi-Fi mesh networks.

Chapter 9 addresses core network and transmission, which typically forms the other major element of a wireless network deployment. An overview of core network evolution is followed by a more comprehensive discussion of two specific core network configurations. The first of these is the Release 98 MSC-based core network with its separate circuit and packet network elements and MSC or IN-based services. This is included because it represents the overwhelming majority of network deployments to date. The second configuration is a core network comprising the Release 8 evolved packet core (EPC) augmented by an IMS-hosted application solution. Over a period of time, this architecture is likely to be widely deployed because it represents a single media-independent solution. The chapter concludes with a summary of transmission systems and a worked example, which addresses the packet transport requirements necessary to connect the SAE and PDN gateways via a shared packet transport link, whilst preserving application QoS.

Chapter 10, the concluding section of the book, aims to provide insight into the activities that take place during the initial deployment of the planned network and subsequent day-to-day operations. The importance of network optimisation in assuring the ongoing quality of the network is highlighted along with a detailed discussion of the way optimisation is carried out for GSM/GPRS and UMTS networks.

Deploying Wireless Networks was written to appeal to a broad audience. It will be an invaluable resource for system architects, radio and core network planning staff, deployment and optimisation teams and mobile network business development staff. The extensive references and worked examples mean that the text may also be attractive to graduate and undergraduate students.

Andy Wilton
Tim Charity
Swindon, UK

Acknowledgements

We would like to express our warmest thanks to Dr Paul Barton, a friend and colleague for many years, for his major contribution to this work. Not only did he provide consistently useful advice regarding the format and content of the book, but he also undertook the unenviable task of reviewing the finished work and, where necessary, authoring invariably superior contributions in lieu of the original text. It is thanks to his perspicacity that a number of errors are no longer present in this work; we claim full credit for those that remain. Contributions from Oliver Tyce (on the HSDPA worked example) and Jose Gil and Chris Murphy on GPRS and UMTS optimisation respectively are also gratefully acknowledged.

A number of our colleagues at Motorola have spent time reviewing sections of the text. Particular thanks are due to Dr Luis Lopes, who provided expert comments on Chapters 5, 6 and 7, and Dr Rashmi Misra and Gerry Foster, who carried out similar duties on Chapter 9. We would also like to thank Ian Doig who provided invaluable help in locating the more obscure specifications and documents from ETSI and 3GPP meetings.

Over and above these specific contributions, we would like to recognise the large numbers of 'corridor conversations' with our Motorola colleagues that have helped clarify various aspects of this work. It is impossible to include a complete list, but we would like to recognise Steve Aftelak, David Bhatoolaul, Paul Flynn, Amitava Ghosh, Tim Jeanes, Howard Thomas, Gary Western and Tim Wilson. Special thanks are due to our employer, Motorola, for their encouragement and agreement to use some of the illustrations in the book.

Outside of Motorola, thanks are due to Professor William Webb for his insightful comments and Dr Julie Lancashire from Cambridge University Press who has kept us on the straight and narrow!

Lastly (and by no means least) we are extremely grateful to our families for their patience and support of non-existent weekends over a protracted period and to Nicholas Wilton for his assistance in the design of the more challenging figures.

Authors' disclaimer

Note that the views and opinions presented in this book are those of the authors and not necessarily of the organisations which employ them. These views should in no way be assumed to imply any particular strategic direction or policy recommendation within the organisations thus represented.

1 Introduction

The human species is unique amongst all life forms in developing a sophisticated and rich means of communication – speech. While communication may have had its origins in the need for individuals to work co-operatively to survive, it is now deeply embedded in the human psyche and is motivated as much by social as business needs. Historically, this was met simply as individuals with similar interests and values chose to form small settlements or villages and all communication was face to face. It was not until the introduction of the telephone in the late nineteenth century that social and business networks could be sustained even when the individuals concerned did not live in the vicinity. Although the coverage and level of automation of the fixed telephony network improved dramatically over the next 100 years, the next major step, communication on the move, was only possible with the introduction of wireless networks.

The term 'wireless network' is very broad and, at various points in history, could have included everything from Marconi's first transatlantic communication in 1901, the first truly mobile (tactical) networks, in the form of the Motorola walkie-talkie in use during the Second World War, to the wide-area private mobile networks in use by the emergency services and large companies since the late 1940s. However, 'wireless networks' didn't really enter the public consciousness until the commercial deployment of cellular mobile radio in the 1980s. The planning and deployment of such networks to provide mobile voice and multimedia communications form the subject of this book.

1.1 Liberalisation of the communications industry

In the late 1960s and early 1970s the communications market was stagnant; the retail market *was* telephony, the only 'applications' being directory enquiries and the speaking clock! This was to change dramatically with the liberalisation of communications markets. The first significant instance of communications liberalisation occurred in the USA when the licence to compete for public-switched long-distance services was granted to MCI in 1969. This was followed in 1984 by the break-up of the AT&T monopoly into the long distance carrier AT&T and seven Regional Bell Operating Companies [1]. At the same time, in the UK, Mercury Communications was issued with a licence to build and operate a second network in competition with BT, the incumbent PTT. This

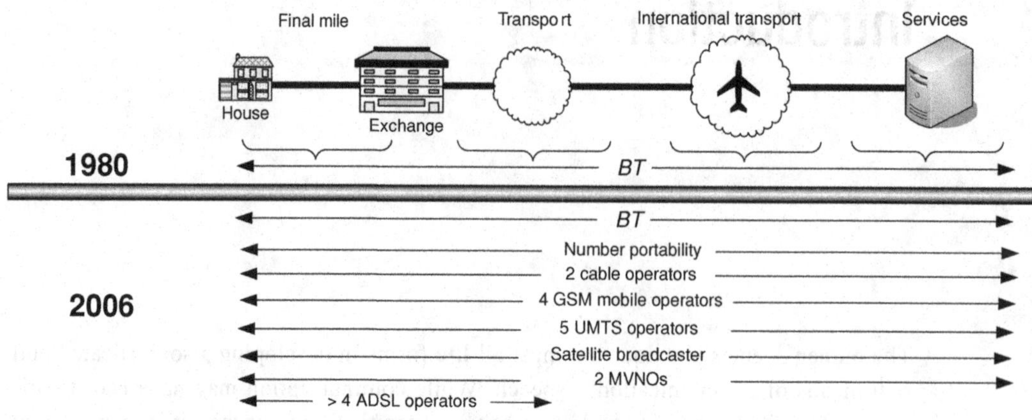

Figure 1.1 UK competitive environment – 1980 and 2006

process continued apace both in the USA and in the UK but it was not until the late 1980s that change started in the EU [2, 3], and culminated in the modification of an existing *Directive* to liberalise fully all European Union telecoms networks with effect from 1 January 1998. Since the 1998 *Directive* the EU has introduced a regulatory framework for the electronic communications sector [4], which is aimed at broadening the scope of competition policy. Figure 1.1 summarises other measures introduced in the UK during this period, which were generally adopted by mainland Europe in due course.

A key result of these changes was that current and future operators could, for the first time, provide almost any communication service; the previous silo structure of 'fixed', 'mobile', 'cable', etc., operators was eliminated. The impact of competition following liberalisation was not long in coming. For the first time, if consumers did not like the pricing or responsiveness of their current operators they could transfer to another service provider with little inconvenience. The result has been that prices for mobile voice have gone down to the point where they are little different from wired networks, and operators have had to look elsewhere to help sustain their margins.

1.2 Digitalisation of content

In parallel with the liberalisation activities of the last 30 years, technology has also moved forward dramatically. The inexorable increase of transistor density in integrated circuits, in line with Moore's Law [5], has meant that the cost of processing and storing information in a digital format has, for most applications, become trivial. Computers, of course, have been conveying information, such as words and numbers, in a digital format since the 1950s but the first modern application conveying 'content' in a digital format was probably the use of the G711 codec [6] as part of the PCM-based telephone networks deployed in the late 1970s. This was followed by the compact disc in 1982, the DVD in 1996 and digital photographs with equal or better quality than their 'film' counterparts

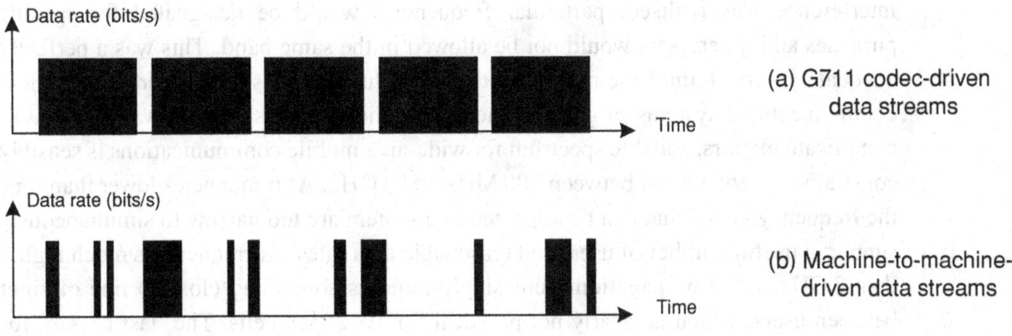

Figure 1.2 Application characteristics

in 2007. Paralleling the evolution of transmission has been the rise of new applications, such as multiplayer network gaming and web browsing. Nonetheless, for the first time, one (digital) transport solution can convey any content likely to be of interest to the user, provided the data rate, delay and bit error rate requirements can be met.

The dramatic reduction in the cost of signal processing has also enabled much more sophisticated codecs to be used, which reduce the amount of information to be transported whilst still allowing reproduction of the original content without significant degradation. The new algorithms usually reduce the information sent by a combination of *differential encoding* (only transmitting changes from the previous frame) and *tokenising* information (recognising, for instance, that the possible sequential time samples of speech are in fact constrained by the human larynx). This is very significant in that it changes the coded output from a regular sequence of constant data bursts of the type shown in Figure 1.2(a) to sequences of arbitrary size bursts of lower duty cycle shown in Figure 1.2(b). This 'bursty' traffic is also representative of 'gaming', web browsing and, indeed, most of the newer applications. Such applications are most efficiently conveyed on 'packet networks' as it is normally possible to convey the traffic from several users over the *same physical link*, which would previously have been dedicated to support one circuit-based application. This is achieved by interleaving other user traffic in the gaps between the bursts of the first user's traffic.

It is also straightforward to evaluate the sort of sustained bandwidths required. Although it is unlikely that wide area mobile systems will be used to sell and download full length DVDs, selling individual CD tracks on the move to a multimedia phone is already becoming feasible. If a transaction time of 10 seconds is desired, and assuming that a 5 minute track is about 350 Mbits, then sustained bandwidths of 35 Mbits/s are needed.

1.3 Changes in spectrum management

Until the 1980s, there had been a consistent approach to spectrum management since spectrum was first used for communications purposes and its vulnerability to unintended interference was realised: particular frequencies would be designated for specific purposes and other users would not be allowed in the same band. This was a perfectly adequate approach until the rapid adoption of cellular phones, and other new wireless communications systems in general, meant that more radio spectrum was needed. To complicate matters, suitable spectrum for wide-area mobile communications is sensibly constrained to the region between 300 MHz and 3 GHz. At frequencies lower than this, the frequency bands that can be allocated to a system are too narrow to simultaneously support a useful number of users and reasonable data rates. At frequencies much higher than 3 GHz, radio propagation increasingly requires something close to line of sight between users, which is clearly not practical for 10–20 km cells. The 'last hurrah' for this conventional spectrum policy was the EU Directive of 1986 requiring the reservation of the 900 MHz band for GSM.

The first change to spectrum management was to retain the concept of dedicated bands for specific uses but to attempt to regulate demand using market forces. In the decade from 1994, the FCC in the USA [7] and regulators in Europe [8] raised a total of about $130 billion through a series of spectrum auctions. More recently, the FCC [9], Ofcom [10, 11] and the EU [12] have all been exploring a more flexible approach to spectrum management. The objectives for these new regimes, which are being progressively introduced, are essentially common to the three regulators:

- Ensure that the needs of all users are met,
- Maximise the economic benefits of spectrum,
- Promote the use of spectrum efficient technologies,
- Seek ways of making more spectrum available.

This new regulatory environment is much more relaxed and is characterised by:

- Market-driven allocation (spectrum auctions),
- Technology neutrality (subject to interference management),
- Only broad use designation (generally in line with the ITU),
- Freedom to trade some or all of the licensed spectrum with others,
- Ability to sub-lease spectrum for secondary applications.

The new regulatory environment has implications for future wireless systems. It will encourage operators to review at regular intervals whether it is commercially attractive to replace a deployed system with a new network that is more spectrally efficient. The ability to cram more bits/second into the same spectrum allocation translates to more revenue. Similarly, if it is possible to use the same *air interface* standard but in a variety of channel bandwidths with only minor changes to the equipment, it will offer operators economies of scale with corresponding cost benefits. These considerations account for the attractiveness of the *OFDMA* air interface and *MIMO* found in WiMAX and the planned UMTS LTE standard [13], which will be discussed in later sections.

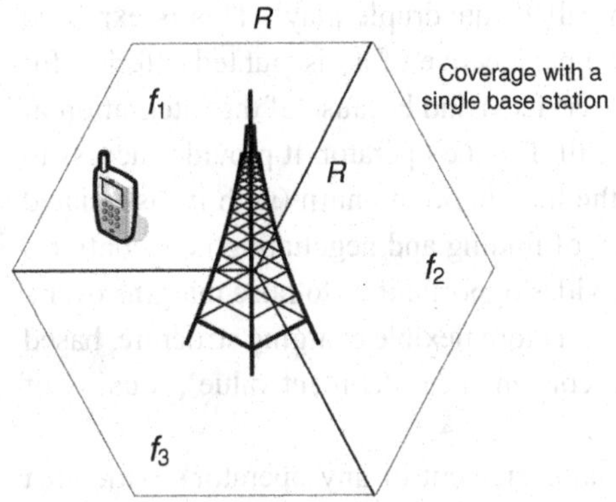

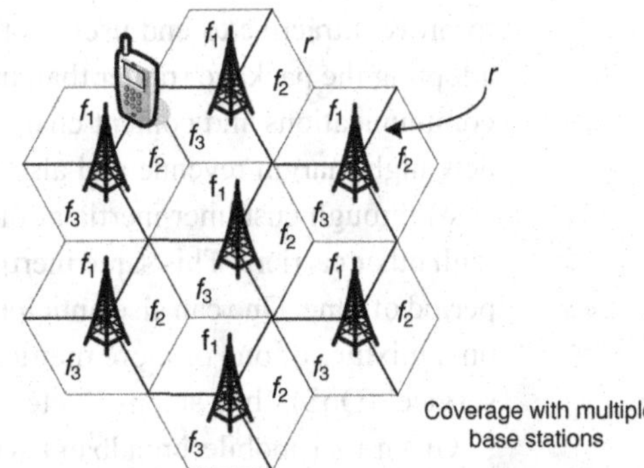

Figure 1.3 Why cellular?

1.4 Why cellular reuse?

The universal adoption of cellular solutions makes it apparent that such solutions are optimum for wide-area mobile networks. Why is this? Why not deploy one transmitter in the middle of the country, which is the configuration adopted for national time standards?

The primary consideration is coverage. Mobile radio systems need to support two-way communication, rather than unidirectional broadcasting. Whilst it is practical to achieve large coverage ranges in the downlink (from base station to the mobile) by using high transmit powers and large antennas, in the reverse direction (the uplink) the power is likely to be limited to 1 W or less if talk times of several hours are to be delivered by a small, light handheld device. A secondary consideration is that when networks mature, there is a need to address a much larger number of subscribers. Figure 1.3 shows that each deployment has three frequencies and is 'sectorised' into three cells. As the hexagonal shape factor and cell capacities will cancel out, the ratio of subscriber densities is simply the square of the ratio of the sides of the two hexagons, R/r. This will be discussed further in Chapter 3.

1.5 The drive towards broadband

Bringing together the discussion of these last few sections, it is apparent that the most efficient (and, therefore, most commercially valuable) operator would be one providing broadband packet-based access to content and applications over the transmission medium, which is most cost effective at the user's current location. It is becoming clear that the current generation of 'mobile' operators will seek to sustain their profitability in the face of the eroding premium on mobile voice by offering a 'one-stop shop' for all the users' telecommunication needs. The industry is already seeing mergers and acquisitions to enable operators to provide a common content and application base across TV,

Internet, fixed and mobile systems – the so-called 'quadruple play'. This is expected to prove attractive to end users both from the perspective of a discounted offering for adopting the package (rather than individual services) and because of the integration of communications and content charges in one bill. For the operator, it provides access to new high-margin revenue and also reduces the likelihood of churn (with its associated cost) through customer inertia at the prospect of finding and negotiating new contracts with other carriers. This same inertia also provides opportunities to raise margins over a period of time. One can also anticipate a new and more flexible charging structure, based on a mixture of one or more metrics such as 'convenience', 'content value', 'quality of service' (QoS), 'bits shipped', etc.

Given that mobile broadband will be a major element of any operator's wide-area multimedia offering, what are the key elements of such a solution? For wide-area coverage, a cellular solution is essential for the reasons already discussed. However, what determines the applications that can be supported in the cell? A key consideration is the maximum data rate that can be sustained over the radio link to the user. This is governed by a fundamental relationship known as Shannon's Law [14]. Claude Shannon was a mathematician who helped build the foundations for the modern computer and developed a statement on information theory that expresses the maximum possible data speed that can be obtained in a single data channel. Shannon's Law makes it clear that the highest obtainable error-free data speed, expressed in bits per second, is a function of the bandwidth and the signal-to-noise ratio. It can be expressed as:

$$C = B\log_2(1 + S/N), \tag{1.1}$$

where:

C is the channel capacity in bits/s,
B is the channel bandwidth in Hz,
S is the signal power in the channel,
N is the noise power in the channel.

If this result is considered in the context of any system where mobile coverage is provided by a central base station, the power density from the central station must fall off as the distance r from the base station increases. Because the noise power N is independent of the distance r and constant, it follows that the maximum data rate available to a user must decrease as the user moves away from the base site. In a cellular system, where there is interference (arising from the frequency in the serving cell being reused in surrounding cells) it is necessary to substitute 'N' with '$N + I$' where I is the interference at the mobile location. If an application is to be supported over the whole cell, then it is clear that the system must be designed to support the required application data rate at the worst-case cell-edge location. Secondly, the total data rate (or throughput) provided by the cell is the summation of the data rate supported by each mobile at its location. Thus the cell capacity and the peak data rate at the cell edge are both ultimately limited by the spectrum available to the operator. It is primarily these two considerations and the finite amount of spectrum suitable for mobile purposes that account for:

- The large fees paid in auctions for mobile spectrum,
- The continued evolution of air interface designs in efforts to get closer to the theoretical Shannon capacity limit.

1.6 Organisation of the text

Chapter 2 seeks to establish a common understanding of the way most cellular networks operate, using the ubiquitous GSM system as a baseline, and discusses the key differences that can be expected in networks providing fixed or 'nomadic' wireless access. The concept of 'contexts', central to packet services in GSM and UMTS, will be covered in some detail. Chapter 2 will also explore the factors that contribute to cellular network operating expense, to determine activities that heavily impact the operator profit and loss account. Finally, the profit and loss account is used to establish metrics that can be used to assess the relative merits of wireless network technologies addressed in later chapters.

Chapter 3 develops the principles and processes that are used to plan wireless access networks. The major focus will be on cellular networks, as these usually represent the most complex planning cases, but the corresponding process for *Wi-Fi* will also be discussed. Circuit voice networks will be examined initially, but the treatment will be extended to understand the additional considerations that come into play as first circuit data and subsequently packet data based applications are introduced. Significant space is dedicated to the way in which packet networks are planned to deliver guaranteed QoS and thus ensure satisfactory operation of arbitrary applications. With the planning sequence understood, the way in which information from such processes can be used to explore the potential profitability of networks well in advance of deployment is addressed. Choices regarding which applications are to be supported in the network and the quality of service offered are shown to have a major impact on the profitability of projects.

Chapters 4 to 8 develop a *detailed* radio access network (RAN) design process to complement the generic principles and processes discussed in Chapter 3. These chapters will address, step by step, the practical planning process.

Chapter 4 describes the steps in the detailed planning process that are essentially common regardless of the specific air interface in use. It takes cell sizes and estimates of the numbers of cell sites from the generic planning activity described in Chapter 3 and, using *specific* topographical data for the regions to be planned, selects *actual* cell locations, antenna heights, etc., for the network.

Chapters 5 to 8 deal with the multiplicity of aspects that are specific to the planning of a particular radio access network (RAN). A major part of these sections will be dedicated to providing a detailed understanding of the particular factors that influence the coverage, capacity and system latency for each of these air interfaces. It is an understanding of such factors that enables system designers to exploit fully the potential of each air interface. The following air interfaces are addressed in the four chapters:

- GSM/GPRS/EDGE,
- UMTS,
- OFDM,
- 802.11 mesh networks.

Each air interface chapter concludes with a planning example.

Chapter 9 provides an overview of the system architecture, principal network elements, transmission and service support for two distinct core networks. The first architecture reflects the networks that would have dominated deployments up to 2006/7, when the majority of traffic was circuit-based voice and, with most services, supported off the MSC. The second architecture is becoming increasingly important and will come to dominate deployments in the future. It features IP transport for all services with applications hosted by IMS. The chapter concludes with the dimensioning of IP transmission required to transport multiple applications with different but guaranteed QoS.

Chapter 10 discusses the last phases of the network life-cycle: provisioning, operation and optimisation. The activities to be performed in the provisioning and operational phases are described in a technology-agnostic manner as far as possible, whereas methods and techniques for radio network performance optimisation are explained separately for the GSM, GPRS and UMTS access technologies. In addition to identifying the sources of information and discussing analysis techniques used for optimisation, indications of the expected values of the key performance indicators (KPIs) of optimised systems are provided.

References

1 V. Mayer-Schoenberger and M. Strasser, A closer look at telecom deregulation: the European advantage, *Harvard Journal of Law and Technology*, **12** 3 (1999).

2 T. Kiessling and Y. Blondeel, The EU regulatory framework in telecommunications – a critical analysis, *Telecommunications Policy*, **22** 7 (1998) 571–592.

3 UK DTI, *Communications Liberalisation in the UK, Communications and Information Industries Directorate* (UK Department of Trade and Industry, March 2001).

4 The European Commission, *Overview of the EU Regulatory Framework for the Electronic Communications Sector*, http://europa.eu.int/information_society/policy/ecomm/todays_framework/overview/index_en.htm

5 G. Moore, Cramming more components onto integrated circuits, *Electronics*, **38** 8 (1965).

6 The ITU, G 711 codec, www.itu.int/rec/T-REC-G.711/e.

7 P. Cramton and J. A. Schwartz, Collusive bidding: lessons from the FCC spectrum auctions, *Journal of Regulatory Economics*, **17** (2000) 229–252.

8 P. Klemperer, How (not) to run auctions: the European 3G telecom auctions. *European Economic Review*, **46** 4–5 (2002) 829–845.

9 FCC Office of Engineering and Technology, *FCC Issues Guiding Principles for Spectrum Management*, Report No ET 99–6 (1999).

10 UK Ofcom and Radio Communications Agency, *A Statement on Spectrum Trading: Implementation in 2004 and Beyond* (2004).

11 UK Ofcom presentation, *The Spectrum Framework Review* (2005).

12 Communication from the Commission to the Council, *A Market-based Approach to Spectrum Management in the European Union (COM(2005) 400 final)* (2005).

13 *3GPP, UTRA-UTRAN Long Term Evolution (LTE) and 3GPP System Architecture Evolution (SAE)*, www.3gpp.org/Highlights/LTE/LTE.htm.

14 C. E. Shannon, *The Mathematical Theory of Communication* (University of Illinois Press, 1949) (reprinted 1998).

2 Wireless network systems

The discussion in Chapter 1 indicated that dividing the planned coverage area into a number of radio cells results in a more spectrally efficient solution with smaller and lighter end-user devices. In such a network, as the individual moves further away from the cell to which he or she is currently connected, the signal strength at the mobile eventually falls to a level where correct operation cannot be guaranteed and the call may 'drop'. However, because the cellular system is designed to ensure good coverage over the plan region there will be one or more other cells at this location that *can* be received at adequate signal strength, provided some mechanism is found to 'hand over' the call to one of these cells. Most of the complexity in practical cellular systems arises from the need to achieve this handover in a way that makes this process as imperceptible to the user as possible.

This chapter aims to establish a common understanding of the way most cellular networks operate, using the ubiquitous GSM system as a baseline, and highlight the key differences that can be expected in networks providing fixed or 'nomadic' wireless access. It will also explore the factors that significantly contribute to cellular network operating expense and thus determine activities that impact the operators' profit and loss account. Finally, the profit and loss account will be used as an agenda to identify wireless network technologies that are likely to change in the future.

The standards that define cellular networks, as a relatively new technology, are subject to constant evolution as vendors and operators work to improve performance and introduce new features. Even though GSM was first deployed in 1992, it is still subject to very active development and so it is necessary to define a particular version as the baseline for this discussion. The specific version of GSM known as GSM *Release 98* will be used for all discussion in this chapter; it supports circuit and packet transport along with much of the functionality found in more modern systems.

2.1 Cellular networks

At the risk of stating the obvious, the original requirement for cellular networks (and still the dominant source of revenue) was to enable people to communicate by voice whilst they were away from their 'normal' fixed-line home or office phones. The architecture of first-generation digital cellular systems, such as GSM, was, therefore, based on the then prevalent designs in fixed voice networks, with specific extensions to add mobility. Indeed the first entrants into the GSM market, such as Ericsson and Siemens, used their fixed network platforms as the basis of new mobile elements known as *mobile switching centres* (MSC) and *base station controllers* (BSC). In a 1980s fixed network (when GSM was conceived), transport capacity over medium and long distances was still a significant cost. A series of switches at the local exchanges and prior to long-distance transport was therefore used to 'concentrate' the traffic from the small percentage of telephones active at any one time (for instance within the catchment area of an exchange) on to a much reduced number of 'lines' for (expensive) long-distance transport. The architecture of GSM reflects this pedigree. Figure 2.1 shows a top level block diagram for a GSM network [1, 2] and if, for a moment, the BSC and BTS are ignored and instead an imaginary wire is used to connect the mobile station (MS) to the 'A' interface of the MSC, the same fixed-line architecture can be seen.

Section 2.1 seeks to establish an understanding of the key principles of cellular operation, which will not only be found in GSM but will also largely recur in the next generation systems described in Chapters 6 and 7. These chapters will then focus only on the key differences compared with this GSM baseline.

2.1.1 The GSM circuit switched network

The mobile switching centre (MSC) takes incoming mobile calls and switches the majority of these to its base station controllers (BSC), which handle the target mobiles. If the dialled mobile is actually in a different part of the country, it is likely to be managed by another MSC and the gateway MSC (GMSC) will route calls to the appropriate MSC. It will also, when necessary, switch traffic to other public land mobile networks (PLMN) and to fixed networks. The network elements above the BSC that ultimately connect to the public switched telephone network (PSTN) or public packet data network (PPDN) together comprise the *core network*.

The one or more BSCs connected to the MSC together with the one or more base transceiver stations (BTS) connected to each BSC form the *access network*. The access network supports the signalling between the MS, MSC and dialled MS necessary to establish the call and subsequently transports the bearer path traffic for the duration of the call. For most calls, it will also handle all aspects of mobility necessary to ensure that the connection is maintained seamlessly as the user moves from one cell to another. Occasionally, the MSC is involved to execute an inter-BSC handover, which occurs when a mobile moves from a cell managed by one BSC to another cell managed by a different controller. The core and access networks perform similar functions for packet traffic for those networks supporting GPRS (GSM packet radio system) but using the packet control unit (PCU) – a part of the BSC, the serving GPRS support node (SGSN) and gateway GSN (GGSN).

Finally, the mobile phone, which for many will be the only element with which they are familiar. Known in ETSI terminology as the mobile station or MS, it is the element central to any transaction in the mobile network. It is therefore appropriate to spend some

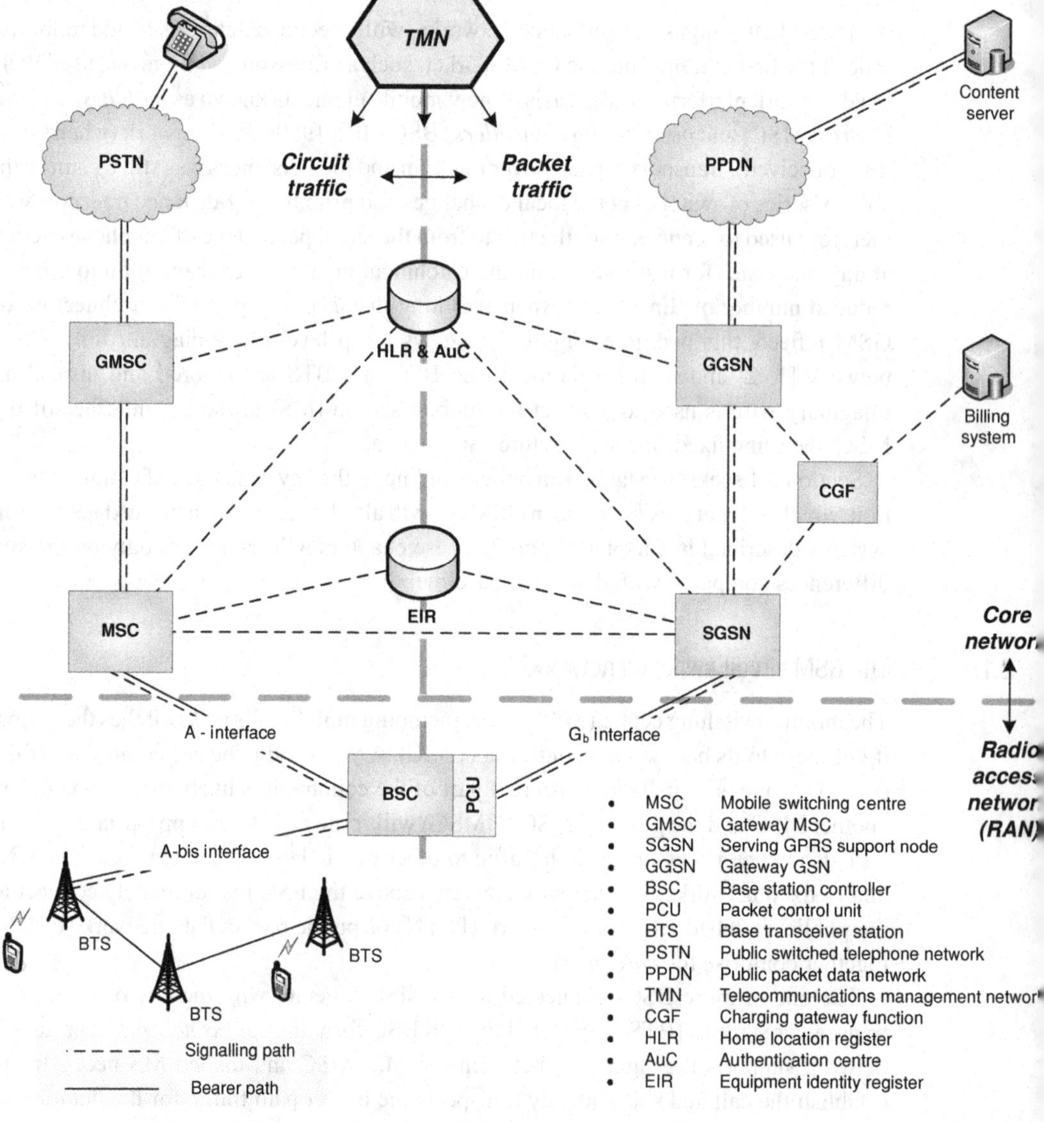

Figure 2.1 GSM Release 98 network architecture

time looking at the key attributes of these devices; the ETSI terminology will be used throughout this section. The key features and functionality of the MS are summarised in Figure 2.2. The mandatory features of mobile stations are defined in ETSI specifications [3] together with descriptions of some optional functionality, which is more fully documented in [4, 5, 6, 7]. The distinction between mandatory and optional functionality is essential to ensure inter-operability of phones from any manufacturer on any vendor's network infrastructure. Conformance testing against these mandatory features is undertaken by vendors and operators before new phones are introduced into an operator's network.

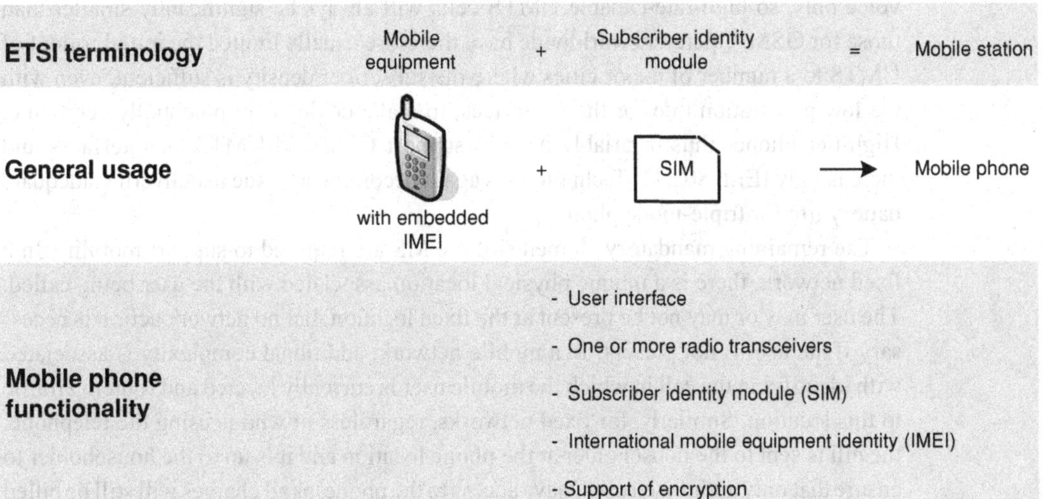

Figure 2.2 Mobile phone attributes

The user interface is the means by which all information is entered into or accessed from the network. At one level it is simply used to enter the phone number of the person who is to be called or the internet location to be addressed and subsequently to be the audio or visual interface for the resulting traffic and transactions. In practice, vendors recognise this as one of the key elements of mobile phone design and the way in which differentiation can be achieved in a standardised system. With phones now increasingly becoming multimedia devices, the number and complexity of transactions to be completed has soared and vendors devote much effort to understanding the way these facilities are used in practice so that the information needed for each particular transaction sequence is never more than one or two keystrokes away. Similarly, the quality of the display and keyboard and the way these are presented to give the largest screen and most user-friendly interface in a small and attractive package are more important than aspects such as battery life, which is generally now perceived to be sufficient for most users.

It is a measure of the rate at which radio access technology has evolved that the technical performance of the GSM radio transceiver in the MS is no longer a market differentiator. But this does not mean that radio technology in today's phones doesn't matter. Mobile devices are now used for much more than making phone calls, and so different communications standards have been developed to provide capability better matched to these new applications. The UMTS interface has been developed by the 3GPP standards group to support voice and wider bandwidth data traffic. The IEEE 802.11 standard (Wi-Fi) [8, 9] has been developed to provide even higher data rates over short ranges. An initial reaction might be that in the near future the industry will develop devices that either support only UMTS or perhaps both UMTS and IEEE 802.11 – but this is unlikely. As was discussed in Chapter 1, for a given mobile transmit power the cell radius of BTSs supporting high data rates will always be smaller than those supporting voice only, so high-rate-capable UMTS cells will always be significantly smaller than those for GSM. Operators worldwide have therefore usually limited the initial rollout of UMTS to a number of major cities where the subscriber density is sufficient, even with the low penetration rate for these services, to make deployment potentially economic. High-tier phones thus invariably have to support GSM and UMTS air interfaces and increasingly IEEE 802.11. Technology is again becoming an issue in delivering adequate battery life for triple-mode phones.

The remaining mandatory elements of the MS are required to support mobility. In a fixed network, there is a unique physical location associated with the user being called. The user may or may not be present at the fixed location, but no network action is necessary if the user is not present. In a mobile network, additional complexity is associated with identifying the cell in which the mobile user is currently located and routeing traffic to this location. Similarly, for fixed networks, regardless of who is using the telephone, the bill is sent to the householder at the phone location and it is up to the householder to ensure that only authorised users have access to the phone, as all charges will still be billed to that address. In mobile networks, unless additional measures are taken, anyone with a compatible phone could use the network for free! The subscriber identity module (SIM) addresses these problems by preventing normal mobile phone calls unless a validated SIM is present in the mobile equipment (ME). The SIM has the international mobile subscriber identity (IMSI) embedded within it, an identity which is unique worldwide and which is allocated by the operator to an individual subscriber to their network. It is this feature of the SIM that allows operators to associate call charges with individuals – whichever network operator (e.g., in another country) may be providing the service at the time. It is also this feature that provides users with the convenience of being able to buy another phone and immediately use the new phone on their existing network merely by swapping their SIM to the new ME. To prevent wrongful use of the network merely by copying valid IMSIs, the SIM card also contains a subscriber authentication key, which is used together with information from the network to validate the SIM to the network.

The international mobile equipment identity (IMEI) is embedded in the mobile equipment, not the SIM. Its purpose is to identify uniquely a particular mobile equipment and it plays no part in determining the charges due to a subscriber. The network can require the disclosure of the user's IMEI prior to allowing access. The use of the IMEI to deny access in this way can be useful for two reasons: (1) to prevent stolen mobiles accessing the network, and, more importantly, (2) to prevent mobiles with known or suspected adverse performance characteristics utilising the network to the detriment of others. The IMEI comprises four segments of information and a check digit [10]. The type approval code (TAC) can be used by operators to gather performance statistics for individual equipment models in conjunction with data made available from the operations and maintenance centre (OMC); such statistics might, for instance, be used by operators to persuade vendors of limitations found in a new mobile phone model. The remaining three segments are particularly useful to vendors and comprise a final assembly code (FAC), a serial number (SNR) and a software version number (SVN).

A few key procedures central to the operation of most mobile networks will be used to explore the role and functionality of the GSM Release 98 network.

Table 2.1. GSM channel structures

	Channel	Type	Direction	Purpose
TCH/F	Traffic channel (full rate)	User	Bidirectional	Support of circuit voice or data traffic (point to point)
TCH/H	Traffic channel (half rate)	User	Bidirectional	Support of circuit voice or data traffic (point to point)
BCCH	Broadcast channel	Control	Down	Broadcast cell specific information (point to multipoint)
RACH	Random access channel	Control	Up	Used to request allocation of a dedicated control channel (SDCCH) – (point to multipoint)
AGCH	Access grant channel	Control	Down	Used to provide details of a newly allocated SDCCH – (point to multipoint)
PCH	Paging channel	Control	Down	Used to page a specific mobile and ultimately cause the establishment of a TCH in response to an incoming call for the mobile (point to multipoint)
SACCH/ FACCH	Slow (or fast) associated control channel	Control	Bidirectional	Primarily used to convey radio measurement information to the BSS related to the TCH or SDCCH with which it is associated (point to point)
SDCCH	Stand-alone dedicated control channel	Control	Bidirectional	Assigned in response to a RACH request from the mobile and used to support signalling during call set-up. SDCCH is terminated once call set-up is complete (point to point)

2.1.1.1 GSM channel structures

At the beginning of this chapter, when the core network architecture was discussed, it was necessary to assume that the MS could signal and ultimately transport user traffic to the MSC and other core network elements. In reality, a structure has to be put in place to transport connection management information over the GSM air interface together with other signalling to assist the base station system (BSS) (comprising the base station controller (BSC) and base transceiver stations (BTS)) to support radio resource allocation (which frequencies to use, etc.) and mobility management (e.g., when to hand over to the next cell). This is achieved by adopting a *channel structure* [11, 12], the key aspects of which are summarised in Table 2.1. It is important to recognise that similar channel structures exist for all cellular systems – regardless of the modulation scheme adopted for transmission over the radio path.

When a mobile is switched on for the first time, or moves to a new network, it has no information about the channels on which BTSs may be transmitting. It therefore scans the predefined GSM frequencies looking for *broadcast channels* [13, 14]. A unique broadcast channel, or BCCH, is transmitted from every cell at a constant power and is used to make mobiles aware of frequencies within the cell that might be chosen to provide the radio link to the fixed network. The mobile selects a cell as suitable provided it can be received at adequate signal strength and is a cell which its subscription potentially allows it to access. Once selected, the mobile will 'camp' on the chosen cell. In addition to the burst for power measurement, the BCCH contains much additional information, including *cell ID*, *location area*, the number and location of *control channels* (PCH, AGCH and RACH) and the location and number of *dedicated channels* (SDCCH and SACCH), together with the *BCCH allocation*. The 'allocation' is a list of RF channels carrying the BCCH information for cells in that geographic region and which the mobile is required to monitor to provide the BSS with received power measurements (used to determine potential handover candidates).

In addition to the BCCH, there are a number of other channels that may be categorised as *traffic channels* (TCH/F and TCH/H), which are point-to-point bearers, and two forms of *control channels*: the *common control channels*, which are point-to-multipoint channels (RACH, AGCH and PCH) and the point-to-point *dedicated control channels* (SDCCH, SACCH and FACCH).

The way in which these channels are used is best illustrated by an example. If a previously registered mobile is switched on and 'camps' on the same cell that it was most recently registered in, it will immediately enter the 'idle' state. In this state, the mobile listens to the BCCH and also monitors the *paging channel* (PCH) so that it can respond to advice of incoming calls intended for it. If the mobile remains in this cell and has good coverage, it will only move from the idle state if it receives an incoming call or if the user decides to initiate a call. In response to one of these stimuli, the mobile will request allocation of an SDCCH using the RACH channel. The network will respond via the AGCH with details of the assigned channel. The SDCCH will then be used to exchange all information required by the network to establish a *traffic channel* (TCH) to the called number. Once call establishment is complete, the SDCCH will be terminated. Whilst the call is in progress, the received signal – at either the mobile or the base station – can vary, either because the user is moving the handset or because the immediate environment of the user is changing (vehicles, trees, etc.). Similarly, interference can cause the received signal quality to degrade – even though the received power is still high. It is, therefore, essential to provide the BSS with regular updates of mobile received signal strength and quality for the allocated TCH. This information is conveyed on the *slow associated control channel* (SACCH); the 'association' recognises that the same physical channel transports traffic and the control information. A SACCH is active for the entire duration of the call and is also used to provide information on the target handover channel once the network has decided a handover is required [15].

2.1.1.2 Authentication and location updating

As mentioned earlier, one of the additional complexities of a mobile network is the need for the network to 'know', physically, where the mobile is so that it can route signalling and traffic to the appropriate cell [16]. In practice, it is convenient to group numbers of cells together and label them as a 'location area'. This reduces the amount of signalling the network has to support as the mobile moves across cells that might be as small as 500 metres diameter, yet still enables the MS to be located quickly by 'paging' it within the location area to identify the specific cell within which it is located. For this reason,

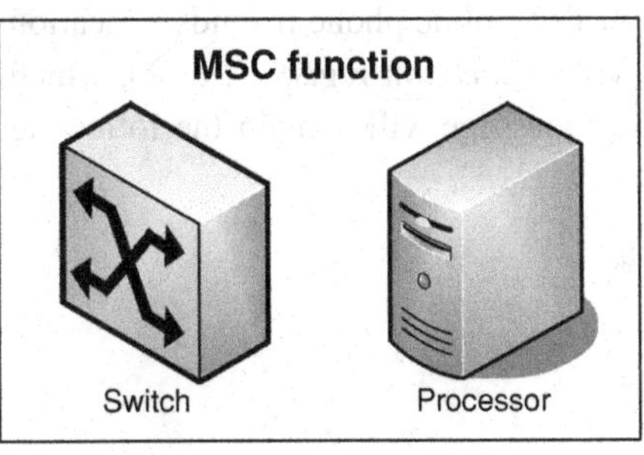

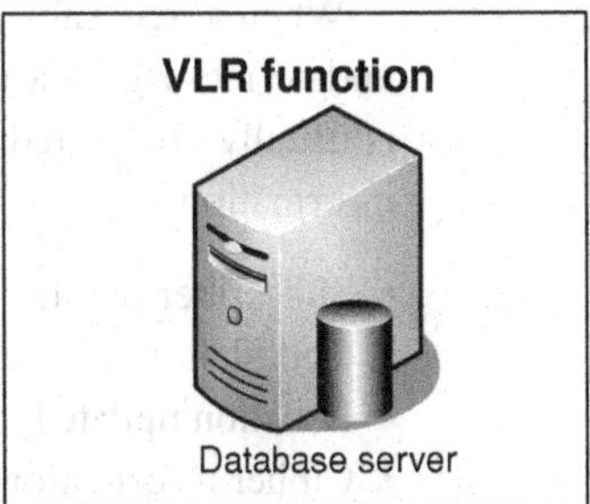

Figure 2.3 Typical MSC functionality

whenever a mobile moves from a cell in one location area to a second cell in a new location area, the MS is required to send a 'location update' message to the core network so that it knows where each MS is likely to be when it receives incoming calls. This same 'location update' process is used when a new subscriber first switches on his or her mobile, as it can exercise all of the functionality needed to potentially recognise it as a valid subscriber of the particular mobile network (PLMN). The core network elements involved in this process are shown in Figure 2.3 and Figure 2.4 along with a summary of their key functionality [2].

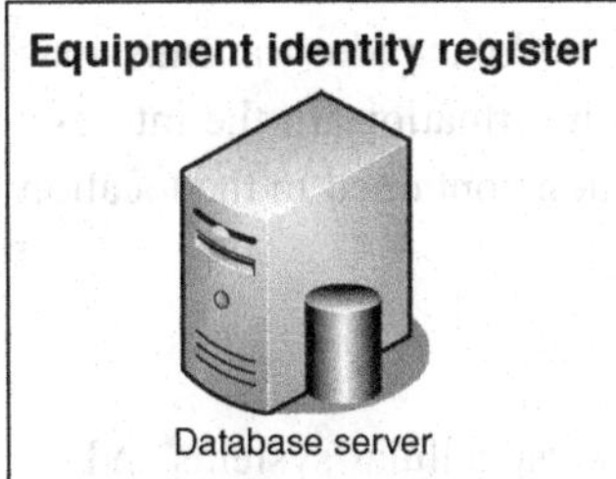

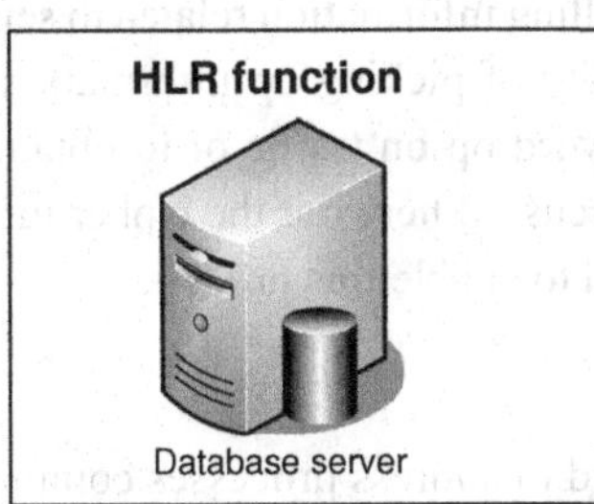

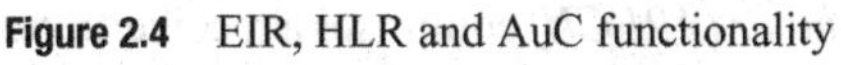

Figure 2.4 EIR, HLR and AuC functionality

When a new subscriber first switches on his or her mobile phone it sends a location update message via the access network to the *visitor location register* (VLR), which is usually co-located with an MSC [17, 18]. The message will contain the following information:

- Subscriber IMSI,
- Cell identifier,
- Location update type,
- Cipher information.

If the location update type indicates that IMSI authentication is required (as would be the case for the initial registration), the VLR causes the MSC to interrogate the *authentication centre* (AuC) using the mobile IMSI [19, 20]. For subscribers registered with the PLMN, the authentication centre will hold an authentication key that is associated with each specific IMSI; this key is the same as that embedded in the SIM with the IMSI. On interrogation, the AuC will send a random number to the mobile and ask it to return a security response computed using a predetermined algorithm, the authentication key and the random number. If the security response returned from the mobile agrees with that generated by the AuC then authentication is completed. The VLR will also request the IMEI from the mobile and check that the mobile equipment is not blacklisted. With the mobile subscription and equipment validated, the remainder of the location update process can be completed. The 'cell identifier' information is unique; it is used by the MSC to derive the new 'location area' associated with this cell and this is stored in the VLR. The VLR then contacts the *home location register* (HLR) to transfer additional subscriber-specific information needed for call handling (service subscription details, etc.). A positive outcome of the location update process is acknowledged by a return message to the MS containing the new location area, which is stored by the mobile.

It should also be realised that GSM and most other mobile systems encipher the bearer traffic and signalling information related to sensitive subscriber information. This minimises the possibility of picking up information from cellular radio transmissions that can be used to eavesdrop on traffic or to obtain user information. In the interests of space, this is not discussed here but the cipher information contained in the location update message is used to enable this process.

2.1.1.3 Physical channels

This chapter is intended to address processes common to most cellular systems and, as such, has stayed away from detailed implementation issues. It is nevertheless helpful to provide an overview of the physical resources used to convey both traffic and signalling information. For GSM, physical channels are delivered through a combination of frequency and time division multiple access.

Traffic and signalling information is modulated onto RF carriers at a rate of approximately 271 kbits/s. *Gaussian minimum shift keying (GMSK)*, a form of bi-phase modulation, is employed and allows a channel spacing of 200 kHz to be adopted whilst still delivering acceptably low adjacent channel and inter-symbol interference [22]. A GSM *absolute radio frequency channel number (ARFCN)* comprises a receive/transmit channel

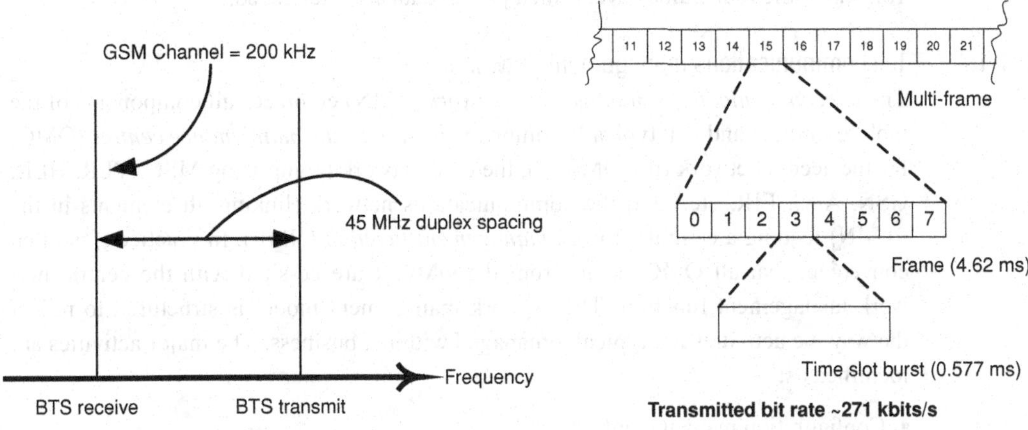

Figure 2.5 GSM physical channels

pair. In the primary GSM 900 MHz band, the pair of 200 kHz channels is separated by 45 MHz.

A GSM channel, for both the uplink and downlink, is time sliced into basic units of eight time slots called *frames*. Each time slot within a frame (known as a *normal burst*) conveys up to 116 payload bits [12]. A specific time slot within a sequence of frames of known length (26, 51, 52, etc.) is used to carry the traffic channels and dedicated signalling channels defined in Table 2.1. The associated control channels (SACCH, FACCH) steal frames or bits from these multi-frames to carry their information. Variations of the *normal burst* and multi-frames are used to convey all the remaining channel structures. The physical channel structure is summarised in Figure 2.5.

Over the radio interface, the 116 payload bits are subject to interference, both from reuse of the GSM frequencies and other impulse interference. Two approaches are employed to reduce the impact of these effects:

- Coding the information bits; this provides resilience against interference and can allow recovery of corrupted bits except in extreme cases. However this can reduce the information rate to less than half the payload rate, depending on the scheme selected.
- Interleaving a single user's information over a number of frames; this means that impulse interference, which might corrupt an entire burst for one user, can be spread amongst four, eight, or even more users, so that in speech the effects can be tolerated and for data the small data loss can be recovered by the higher layers.

The impact on speech channels is a maximum data rate of about 13 kbits/s and delay, introduced by interleaving over eight frames, of about 40 ms. For the packet data channels, four coding schemes have been introduced with data rates ranging from approximately 9 kbit/s to 21 kbit/s and a constraint on the minimum time period that a single user can share a *physical channel*, known as a *radio block*, corresponding to approximately 18.5 ms or the four frames over which packet data are interleaved.

2.1.1.4 Telecommunications management network

The *telecommunications management network* (TMN) embraces all components of the mobile solution and will typically comprise *operations and maintenance centres* (OMC) for the access network (the OMC-R), the core network (comprising MSC, VLR, HLR, GSN, AuC, EIR, etc.) and the communications network (linking all elements in the PLMN) feeding a central *network management function* (NMF). In practice it is often convenient that all OMCs apart from the OMC-R are co-sited with the central network management function. The network management model is structured to reflect the way the activities are typically managed within a business. The major activities are identified as:

- Configuration management,
- Customer administration,
- Operations,
- Maintenance.

Because of the sensitivity of customer data and the need to maintain accurate and traceable records by law, considerable attention has been paid in the TMN specifications [23] to ensure that there is a hierarchical security structure that allows tiered levels of access to TMN functions. Additionally, for certain levels of access, events are automatically logged so that records of who has made changes are available.

Configuration management includes all those functions that allow the network to be established in a known configuration or status and to log changes to the PLMN or the TMN. The integrity of the PLMN, i.e., its ability to perform consistently the functions for which it was designed in the manner expected, is related to the stability of the network. Performing any change places this integrity at risk. Functions implemented in this area are designed to minimise, if not eliminate, the risks. When configuration changes result in modifications to data values, the system allows the change to be temporary or permanent, to be brought into instant use or delayed, and to provide options to update the reference database if required. The configuration process is the vehicle used to roll out a new network and to manage the introduction of new functionality through software or hardware upgrades. The ability to build databases reflecting new configurations or functionality offline is key to trialling changes at times that will cause minimal disruption to customers if something goes wrong – e.g., early in the morning. Similarly, the reference database concept allows trial configurations to revert quickly to a known 'good' configuration.

Customer administration includes those functions that relate to management of subscribers, subscriber data, mobile equipment data, the collection of call data and other data upon which charges may be levied. Billing is discussed separately but other tasks will include the registration of new customer data in the HLR to enable access to the network, information on service subscription, roaming privileges. In the future, it is expected that this function will increasingly address over-the-air provisioning of mobiles so that they can be configured by the operator to receive new services and, eventually, software upgrades to phone operating systems.

Operations addresses the major day-to-day activity for network mangers and describes the functions that allow the PLMN operator to monitor the performance of the network and, by means of the management network, to optimise the parameters of the PLMN to improve the quality of service provided to subscribers. Functions are also provided to allow for the collection of comprehensive activity reports including call and call path information for individually identifiable subscribers. This can be used to investigate persistent problems with an individual subscriber's calls or particular mobile device types. Considerable detail is available:

- Call statistics resolved to individual cell level (blocked calls, dropped calls, etc.),
- Call quality (individual call and bulk statistics),
- Network element statistics,
- Call trace,
- Etc.

As will be discussed in later chapters, this information can be used by operators to:

- Optimise performance in newly deployed cells,
- Identify gradual degradation in equipment performance,
- Identify trends in cell usage and thus plan for upgrades,
- Gather statistics on service class usage,
- Deduce actual coverage areas without drive testing,
- And much more.

The *maintenance* process, as in its wider usage, defines the work necessary to ensure that the network continues to operate at planned performance levels. Vendors' equipment will be designed to achieve specified *mean time between failures* (MTBF) and *mean time to repair* (MTTR). Each of the network elements will include sensors that will flag fault conditions. The status of the sensors within all the elements of the particular network sub-system (access, core or communications) will be integrated in the relevant OMC and outputs from these OMCs will, in turn, be concentrated in the network management function. From the NMF it is usually possible to take a total network view and then obtain increasing levels of information on any fault conditions visible at the top level. In addition to specific fault conditions flagged by this process, the maintenance function will also access performance data available to the operations team and use this to diagnose 'soft' failures – e.g., misaligned antennas caused by storm damage giving rise to poor cell statistics.

2.1.1.5 Billing systems

This section is logically part of the TMN system but as a matter of key interest to the operator is treated separately here. This brief commentary will focus on the metrics available to drive subscriber billing and the mechanisms used to collect these data; Table 2.2 shows some of the factors that are used to drive competitive billing for circuit services.

Table 2.2. Tariff factors

Tariff factor	Call origination	Call location	Time and duration	Service	Subscription information
Example options	– Mob. originated – Mob. terminated	– Home cell – Home PLMN – Other PLM	– Working hours – Evenings – Weekends	– Circuit speech – Circuit data – Supplementary services	– Post paid – Pre paid – Special tariffs

Key to the charging structure in GSM is that the MSC/VLR knows the IMSI of any subscriber that is active in the network. It is thus able to unambiguously associate use of network resources with a specific subscriber. Particular events, such as mobile call origination, use of supplementary services, etc., together with other information that will be used to calculate the eventual charge, such as time of call and call duration, are collected in real time. Specific functionality for this purpose is included in the MSC, VLR, HLR and BSS and is used to build *call records* for each subscriber network activity [24]. The data are collected whether the mobile is in its home PLMN or in a visited PLMN. In the latter case, the call record will contain charges for the use of at least two networks.

The availability of comprehensive information in call records is a key consideration in allowing operators to comply with national policies and to innovate and gain competitive advantage. In many regions, mobile-terminated calls originated in the same country do not attract a fee. However, in general, received calls from outside the country of the mobile network are billed. Availability of *mobile-originated* (MO) or *mobile-terminated* (MT) information in call records allows complete flexibility to meet national obligations. On the other hand, with the traditional demarcations between fixed and mobile operators eroding and with broadband in most homes, mobile operators will need accurate knowledge of the location in which the call was made to create low-tariff 'home zones' if they are to compete with others providing home coverage using Wi-Fi and broadband back-haul.

2.1.2 The general packet radio system (GPRS)

So far, discussion of the GSM system has been focused on the transmission and reception of circuit traffic, primarily voice. However, as was pointed out in Chapter 1, it is expected that this mix will change to include an increasing percentage of traffic generated by applications (e.g., web browsing, network gaming, etc.) and, in the future, machine-to-machine communications. This traffic is characterised by aperiodic bursts of data separated by relatively long periods of inactivity. The basic GSM system can convey such information (see the *high speed circuit-switched data (HSCSD)* system [12]) at the cost of dedicating a GSM physical channel for the application session – even though it may be utilised 10% of the time or less. The GPRS architecture was added to the original GSM system to allow a physical channel to be shared by multiple users. This allows more 'data' users to be supported by a given amount of spectrum and a given amount

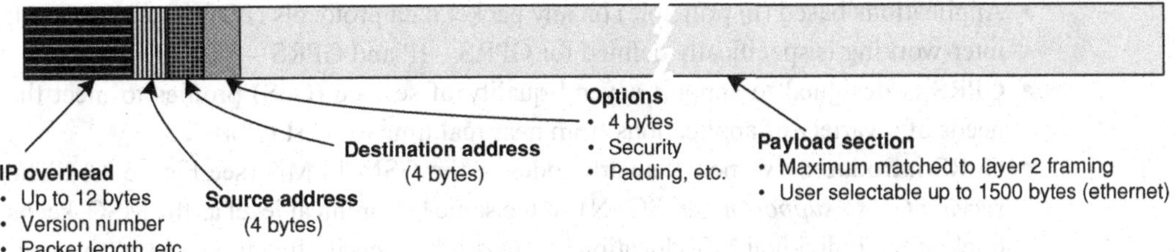

Figure 2.6 Internet protocol (IP) packet format

of infrastructure than would be the case if HSCSD was used, with the corresponding impact on profitability.

Packet protocols are increasingly the default choice in fixed networks, to multiplex traffic from many users on to a common bearer and realise an efficient transport solution. Such protocols support payloads that can be varied on a packet-by-packet basis, source and destination addressing, and means of enabling differential QoS, security, etc. When the source traffic is 'bursty', sufficient packets are sent to convey the information in each burst – but no packets are sent in the gap between bursts. These gaps can be used to convey packets from other users. This contrasts with dedicated circuit connections, where the whole connection is tied up regardless of whether there is information to be sent. Packets from different users emerging from this common transport are routed to their destination using the address contained in the header. Although many packet protocols are possible, the *Internet protocol* (IP), as defined by the *Internet Engineering Task Force* (IETF), is, far and away, the most popular and is supported by GPRS. The key elements of the protocol are illustrated in Figure 2.6.

One key difference to note between packet traffic over fixed networks and packet traffic supported over cellular systems is that, in the latter case, the communication is intrinsically point to multipoint; all mobiles are capable of communicating over the specified time slot and frequency combination that represents the physical channel. Multiplexing of packets over the air interface is thus achieved by allocating this physical channel in turn to those mobiles with queued traffic.

2.1.2.1 Overview

The functionality and key features of GPRS are summarised below [25]:

- As many as possible of the GSM elements and identities are reused. These include the AuC, EIR, IMSI and IMEI.
- New GPRS radio channels are defined. The allocation of these channels is flexible; from one to eight of the GSM radio interface time slots can be allocated to GPRS per TDMA frame. Time slots are shared by the active users, and uplink and downlink resources are allocated separately. The GSM time slots may also be shared dynamically between speech and data traffic, in sympathy with the short-term voice loading and operator preference. Various radio channel coding schemes are specified to allow bit rates from 9 kbits/s to more than 150 kbits/s per user.

- Applications based (in principle) on any packet data protocols (PDPs) are supported; inter-working is specifically defined for GPRS – IP and GPRS – X.25 networks.
- GPRS is designed to support several quality of service (QoS) profiles to meet the needs of a variety of applications from near real time to 'best efforts'.
- GPRS introduces two new network nodes in the GSM PLMN (see Figure 2.1). The *serving GPRS support node* (SGSN), at the same hierarchical level as the MSC, keeps track of the individual MS's location and performs security functions and access control. The *gateway GSN (GGSN)* provides inter-working with external packet-switched networks.
- The HLR is enhanced to support GPRS subscriber information including one or more PDP addresses for each PPDN over which it wishes to support applications. It also contains the IP address of the SGSN serving the mobile associated with the particular IMSI, along with the IP addresses of the GGSNs to which traffic should be routed when the user initiates applications supported by a particular PPDN.
- New functionality, the *packet control unit* (PCU), is added to the BSC to segment application protocol units and make them suitable for transmission over the air interface. It also manages the requests and granting of packet channels.
- There is a set of GPRS channel structures largely equivalent to those for GSM discussed in Section 2.1.1 and summarised in Table 2.1. Thus, there are the packet data traffic channel (PDTCH), packet broadcast channel (PBCCH), packet paging channel (PPCH), packet random access channel (PRACH), etc. However, no use is made of a stand-alone dedicated control channel (SDCCH). Instead, the normal packet bearers are used to convey the signalling necessary at call set-up.

A typical GPRS network will comprise one or more GGSNs, with each GGSN supporting one or more SGSNs. Interestingly, in networks containing more than one GGSN, each SGSN is also connected to each GGSN. This arises because GPRS networks are designed to support traffic from a number of different public packet data networks (PPDNs), where each network can employ a different protocol (IP, PPP, X25, etc.). A GGSN forms the gateway to each network with a different protocol and must, therefore, be linked to each of the SGSNs in the GPRS network. An incoming 'call' for a packet-capable mobile is routed by the GGSN to the SGSN 'serving' the targeted mobile. This network of GSNs linked to the elements common to both the GSM and GPRS networks (HLR, AuC, EIR, etc.) is known as the *GPRS packet core*.

The one or more BSCs connected to each SGSN, together with the one or more BTSs connected to each BSC, form the *GPRS access network*. Note that whilst the same BTSs are used for both GSM and GPRS, in the case of GPRS capable networks, the BSC functionality is, as discussed earlier, augmented by a *packet control unit (PCU)*, which manages the multiplexing of packet traffic over GSM physical channels. It is the northbound output of the PCU, the G_b interface, that connects the GPRS access network to the SGSN. The GPRS access network supports signalling between the originating mobile, the SGSN, the GGSN and the *application server* in the targeted PPDN during session setup. Subsequently, together with the SGSN and GGSN, it routes and manages the flow of traffic to the server for the duration of the application session; more of this later. For most sessions, it will also handle all aspects of mobility necessary to ensure that the session is maintained seamlessly as the user moves from one cell to another. Occasionally, the SGSN will be involved, as the mobile moves from a cell handled by one BSC to a cell handled by another controller. In addition to the subscriber data and capability already discussed in GSM, a GPRS mobile subscription includes one or more packet data protocol (PDP) addresses. Whilst GPRS can support multiple subscriptions for more than one PPDN type, in practice it will usually contain only one address (for IP networks).

Most of the detailed procedures used to manage a GSM mobile read across to support similar functionality in a GPRS capable network. Changes have only been introduced where they are necessary to share GSM physical channels efficiently:

- There are equivalent GPRS *channel* structures for most of those used by GSM to execute connection management, mobility management and radio resource management. Because GPRS is designed to share GSM physical channels amongst many mobiles, there is also new functionality to manage the state of connectivity between a mobile and the SGSN. This is known as *mobility management context* (MM context) information.
- The underlying transport bearer across the radio interface is still a GSM physical channel, even though it is now shared with a number of mobiles. A similar set of regular measurements of physical channel quality is, therefore, made by mobiles for which MM context information exists and is used by the BSC in managing handover between cells. The difference is that the measurements are only made on cells indicating they are GPRS capable by the presence of a *packet broadcast control channel* (PBCCH).
- Authentication is carried out by similar processes to those described for GSM. However, for GPRS, it is the SGSN (rather than the MSC/VLR) that works with the HLR, AuC and EIR.
- A process similar to the GSM 'location update' procedure is used by GPRS mobiles for which a MM context exists. In GPRS the updates are executed between the SGSN and HLR. A *routeing area* is defined, which is analogous to the 'Location Area' in GSM but may comprise a subset of the cells within the location area.

The remainder of this section on GPRS discusses the functionality that is necessary to support packet traffic in a mobile cellular system efficiently. This will be used in later chapters to contrast packet implementations introduced since GPRS.

2.1.2.2 GPRS 'contexts'

The requirement to support packet traffic efficiently implicitly dictates the need for additional GSM network functionality. In GSM circuit calls, a circuit comprising an air interface physical channel, circuits in the BTS, BSC, MSC and GMSC, and transmission between these elements is dedicated to the transport of voice or HSCSD traffic from call initiation to call termination. In services most efficiently supported using packet transport, the application will be enabled by establishing a route between the mobile and the server in a public network, and sending packets of data only in response to user or application events. When there are no data to send, all of the route resources can be temporarily released to their respective networks. In typical applications however, the

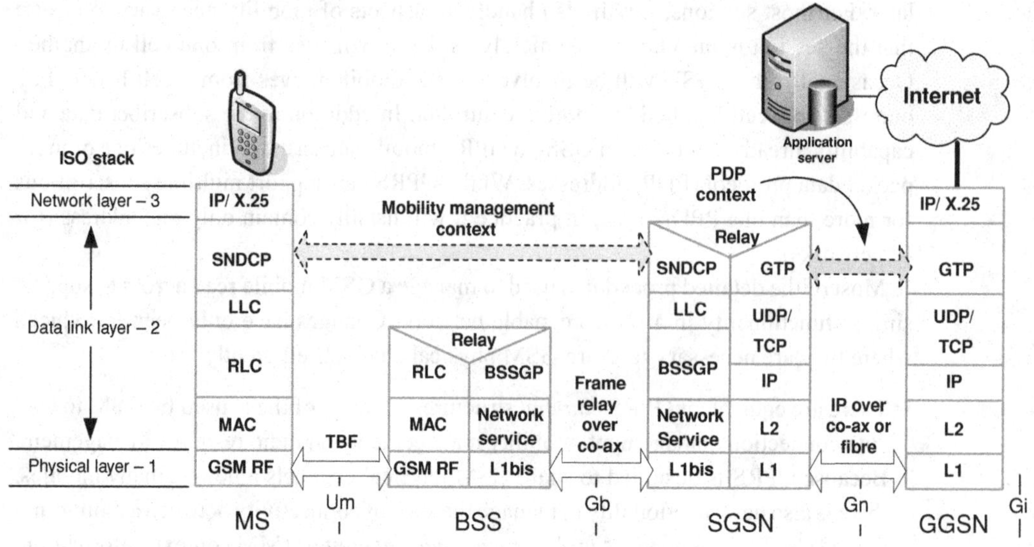

Figure 2.7 GPRS contexts

time taken to *establish* the route may be long compared with the interval between packet events. An attractive packet bearer thus needs to meet the twin goals of efficient resource utilisation *and* the ability to resume an established route quickly if an acceptable user experience in practical applications is to be delivered. These requirements lead to the concept of GPRS *contexts* between various network elements. These contexts comprise the information necessary to quickly allow the resumption of traffic flow between the mobile and the server targeted during the original call set-up. The context data, which are stored when all the queued data from the original call set-up has been sent, will typically include the mobile identity, public server address, and the routes and addresses to intermediate network nodes. This ability to resume the packet 'call' quickly when GPRS context information is available very much parallels the ability of people to resume more quickly an earlier conversation once the context of the original discussion (down the pub last Saturday with Claire and Lynn!) is provided – hence the terminology of GPRS contexts.

Figure 2.7 shows the major entities of a GPRS network, together with the physical and logical connections that exist in the network. The peer entities for any GPRS application are the mobile station and the application server and these exchange data using the network layer protocol of the PPDN hosting the server. The purpose of the GPRS network is to provide ISO layers 1 and 2 connectivity between these end points. However, as for the 'fixed' network, there are a number of different physical layers connecting the device to the server. Traffic needs to be routed and managed across these physical media, with the additional complication that the client device is mobile.

2.1.2.3 The PDP context

The PDP context is a *tunnel* (using *GPRS tunnelling protocol* (GTP) [26]) between the GGSN, acting as a gateway to the *public packet data network* (PPDN) hosting the targeted

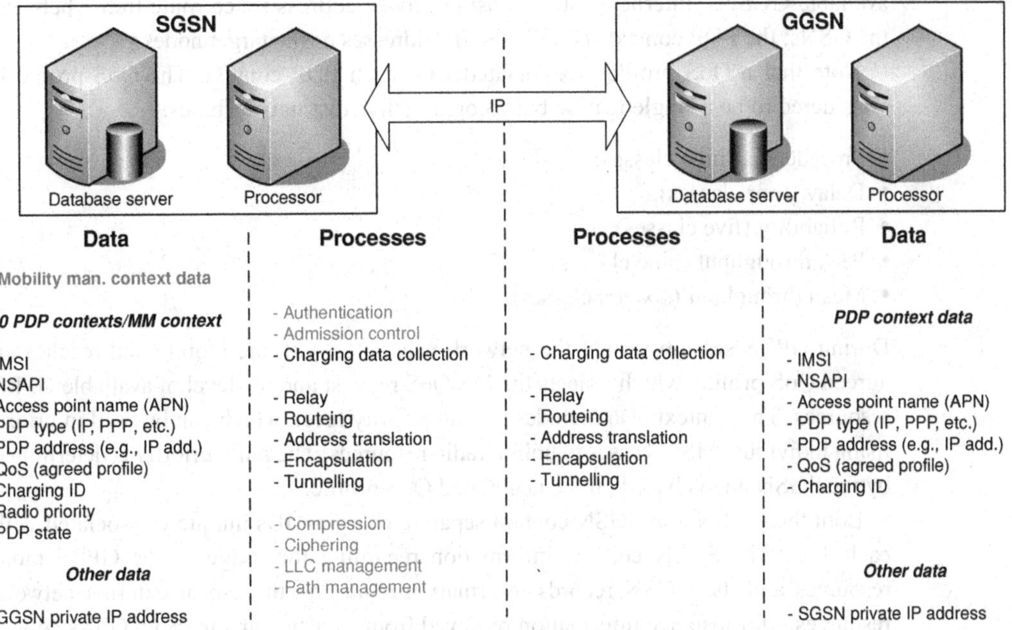

Figure 2.8 PDP context data

application server, and the SGSN serving the mobile. A GTP tunnel is defined by a PDP context (two associated data sets) in the SGSN and GGSN nodes and is identified with a tunnel ID. The make-up of these associated data sets is shown in Figure 2.8. Network *protocol data units* (PDUs), from the PPDN, are encapsulated in the *GSM tunnelling protocol* (GTP), which transports them between the GGSN and SGSN.

Key benefits afforded by the PDP context include:

- The ability to provide a static network address (IP, PPP, X25, etc.), which hides the user mobility from the PPDN.
- Releasing intra-PLMN backbone transmission resources when the PDP context is not active.
- The ability to resume packet data flow quickly as traffic appears. Once a PDP context has been established between the mobile and the host network (i.e., it is *active*) it is allocated a unique tunnel ID (TID). Figure 2.8 shows that for each TID – even when data transfer is not taking place – all information is present at each GSN to resume traffic flow quickly once packets referencing the specific TID are received.
- The ability to provide MS to PPDN connectivity independently of the underlying PPDN network protocols.
- The existence of PDP context information – even in the inactive state – means that downlink data can be buffered at the GGSN and SGSN when radio resources are not available to the mobile.

Physical connectivity is realised by an intra-PLMN network, generally comprising a mixture of co-ax and fibre transport with intermediate nodes, linking the SGSNs to all available GGSNs. Internet protocols using private-address space route traffic between the GSNs; the PDP contexts contain the IP addresses of the target nodes.

Note that a QoS profile is associated with each PDP context. The QoS profile is considered to be a single feature but recognises five distinct attributes:

- Precedence (three classes),
- Delay (four classes),
- Reliability (five classes),
- Peak throughput (nine classes),
- Mean throughput (sixteen classes).

During GPRS session set-up, the network negotiates with the mobile and reaches an agreed QoS profile, which reflects the MS QoS request and the level of available GPRS resources. The context also includes a radio priority level, which can be used in prioritising individual MS accesses to uplink radio resources. The radio priority is determined by the SGSN and is based on the negotiated QoS profile.

Both the GGSN and SGSN contain separate charging IDs uniquely associated with each TID. The SGSN collects information regarding the usage of the GPRS radio resources and the GGSN records information regarding the use of external network resources – for instance information retrieved from a particular site. Both GSNs collect information regarding the use of GPRS network resources. The SGSN is also a relay and manages data both to and from the GGSN and to and from the mobile. In Figure 2.8, data and processes associated with SGSN and mobile transactions are 'greyed' out.

2.1.2.4 The mobility management (MM) context

As has been discussed, the GPRS component of the overall GSM solution has been designed to release network resources whenever they are not required, whether these are for signalling or traffic. This requires some significant changes compared with the GSM circuit switched architecture. When a mobile is first powered up, it goes into *GPRS idle* mode. It camps on to a GPRS-capable cell using similar signal quality criteria to GSM but it does not communicate the cell identity to the network. Thus, unlike circuit GSM, in the *GPRS idle* state the mobile is not visible to the network. It can neither be paged nor immediately start a GPRS session. However, in this state, it is not required to advise the network as it moves between cells or routeing areas and so does not utilise signalling resources when it is not active on the GPRS network; this is not the case for an 'idle' GSM subscriber. One consequence of this architecture is that an MS will first make its presence known to its network peer, the SGSN, by performing a GPRS *attach*, in order to access GPRS services. This relationship is illustrated in Figure 2.7. The *attach* operation establishes a logical link between the MS and the SGSN known as an *MM context* and makes it possible for the MS to originate a GPRS session and to receive paging via the SGSN to notify it of incoming GPRS data. Authentication procedures will also be carried out as part of the *attach* process if this is the first *attach* since power-up. The concept of the MM context is perhaps easier to understand with reference to Figure 2.9, where it is possible to see that, for every MM process and data set in the MS, there is a companion process and data set in the SGSN. Thus, the compression process employed

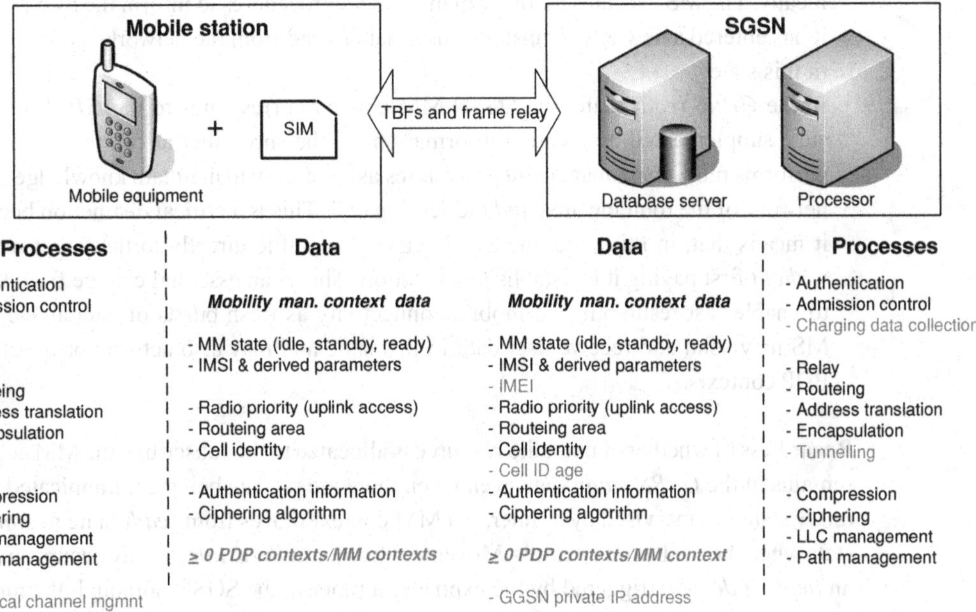

Figure 2.9 The mobility management context

in the mobile to transport user data efficiently over the air interface has a companion decompression process in the SGSN; this process is reversed for data flowing from the SGSN to the MS. The common data sets in the MS and SGSN only exist when MM is in the *GPRS standby* or *GPRS ready* states and represent the shared context.

Figure 2.9 details another important distinction. Because there is only ever one MM context active at a time for a given MS, it follows that zero or more PDP contexts are multiplexed onto a single MM context. These are distinguished by their PDP context identity at the mobile. The figure also shows (greyed out) some processes and data unrelated to the MM context.

To summarise, the MM context for a GPRS subscriber is characterised by one of three different MM context states. Each state describes a certain level of functionality and information available at the peer nodes:

- In the *GPRS idle* state, the subscriber is not attached to the GPRS mobility management. The MS and SGSN context hold no valid location or routeing information for the subscriber. Subscriber-related mobility management procedures are not performed. To establish MM contexts in the MS and the SGSN, the MS must perform the *GPRS attach* procedure.
- In the *GPRS standby* state, the subscriber is attached to GPRS mobility management. There are associated data sets for the subscriber's IMSI stored in the MS and SGSN and the network may page the MS to enable reception of network-originated data. The MS performs GPRS *routeing area* (RA) and GPRS cell selection and reselection

locally. The MS executes mobility management procedures to inform the SGSN when it has entered a new RA. Transfer of user data to and from the network is not possible in this state.

- In the *GPRS ready* state, the SGSN MM context corresponds to the *GPRS standby* state supplemented by location information for the subscriber at *cell* level. The MS performs mobility management procedures as necessary to maintain knowledge in the network of the routeing area *and* the serving cell. This is a critical distinction because it means that, in this state, the SGSN can route traffic directly to the target mobile *without* first paging it to establish its location. This is an essential change from GSM to enable fast resumption of mobile connectivity as fresh bursts of data arrive. The MS may send and receive user data in this state and may also activate or deactivate PDP contexts.

Regardless of whether or not radio resource is allocated to the subscriber, the MM context remains in the *GPRS ready* state even when there are no data being communicated. The *ready* state is supervised by a timer. An MM context moves from *ready* state to *standby* state when the ready timer expires. Movement from *standby* to *idle* usually occurs through an *implicit detach*, triggered by the expiry of a timer in the SGSN, although the mobile can request an *explicit detach*. Transition of the MM context to *idle* causes the PDP contexts in the SGSN and GGSN to return to the *inactive* state. The MM and PDP context information sets in the SGSN, MS and GGSN are then deleted.

2.1.2.5 MS-SGSN physical layer

It is appropriate to provide a short overview of the physical layers and protocols that connect the MS to the SGSN via the base station sub-system (BSS), since a major part of the operating cost of a typical cellular network is determined by the BSS.

The SGSN is a major traffic node, typically supporting many BSCs and PCUs with each BSC or PCU typically supporting 20 or more BTSs. The physical layer connection between the SGSN and each of its PCUs has historically been $N \times$ E1 links but more typically is now STM-1. Frame relay protocol [27] running over this layer creates virtual circuits to ports in the BSC connected to each of the BTS cells. Because these ports are located in the same BSC, physical-layer transmission can be efficiently shared amongst all cells, as load varies transiently as a function of the number of mobiles with active mobility management contexts in each cell.

Although not explicitly shown in Figure 2.7, it is necessary to provide point-to-point connections between the BSC and each of 20 or more BTSs. Although this is not an issue in dense urban areas, where metropolitan area networks abound, expensive E1 or other links need to be employed as sites are rolled out into rural areas; in these cases it is often necessary to use specially deployed microwave links. Normal practice is to associate a DS0 link with four *physical channels*, whether the channel is supporting voice or data.

Across the U_m or GSM air interface, data are still transported via the *physical channels* defined in Section 2.1.1.3 but *packet channel structures* are introduced to complement the GSM channel structures and enable uplink access to the radio resources through channels such as the *packet random access channel* (PRACH), the *packet access grant channel* (PAGCH) and the *packet data traffic channels* (PDTCH). A *packet data channel* (PDCH) is the name given to a GSM *physical channel* configured to support GPRS rather than GSM traffic. One or more PDCHs may be used to support traffic flows in the uplink and downlink, depending upon the mobile capability. However, these PDCHs are now shared amongst many mobiles by the introduction of the *temporary block flow* (TBF) concept [28].

A TBF is a physical connection used by the RR entities in the MS and BSC to support the unidirectional transfer of *logical link control* (LLC) PDUs across the air interface. The TBF comprises a number of *radio blocks* carrying one or more LLC PDUs; the TBF is temporary and is maintained only for the duration of the data transfer (i.e., until there are no more RLC/MAC blocks to be transmitted). Once a given TBF has completed, the same PDCHs can be used to support a new TBF for a different mobile. Depending upon the nature of the access request, the mobile priority and network congestion, the PAGCH may make a fixed allocation of blocks.

A given mobile is in a *packet idle* state when no temporary block flow exists. This may be because there are no data to transmit at a specific instance, or because no radio resources have been assigned to it and it is in a queue. A mobile is in *packet transfer* state when it is allocated radio resource providing a TBF for a physical point-to-point connection on one or more PDCHs for the unidirectional transfer of LLC PDUs between the network and the mobile station. Note that even with the MM context at *GPRS ready*, the mobile may be in either of these two states.

2.1.2.6 MS-SGSN protocols

From the earlier discussion of GPRS, recall that it is designed to support one or more concurrent applications over one or more different protocols. The multiplexing of these PDP contexts onto a single radio connection to the mobile (MM context) occurs at the SGSN (for the downlink). Also, PDUs for network-layer protocols, such as IP, likely to be employed by most applications, can have payloads of up to 1500 bytes. It is thus possible that the SGSN may receive up to 12 000 bits or even 24 000 bits from the network in a short space of time. This traffic must be delivered across a GSM *physical channel* that is only capable of about 21 kbits/s even for the highest rate coding scheme! Thus, if this channel is to be shared amongst several mobiles using the TBF structure, it is very clear that flow management and queuing in the downlink direction is essential. The role of protocols, such as *sub-network dependent convergence protocol* (SNDCP) [29], *logical link control* (LLC) [30], *base station system GPRS protocol* (BSSGP) [31] and *radio link control* or *media access control* (RLC or MAC) [28, 32], shown in Figure 2.7, is to execute these and other essential processes.

The key functions of the SNDCP protocol are:

- Multiplexing of multiple PDP contexts on to one MM context,
- Compression of control and user data,
- Segmentation of compressed data to form LLC frames.

The LLC protocol supports the following functionality:

- Multiplexing;
  - Up to five user data streams per mobile,
  - MM context signalling.
- Flow control between the SGSN and MS,
- Ciphering,
- Error correction.

The RLC/MAC protocol is concerned with:

- Multiplexing of control and data for the uplink and downlink,
- Priority handling,
- Queuing and flow control of downlink packet data across the U_m interface,
- Segmentation of LLC PDUs into RLC/MAC blocks,
- Error correction across the U_m interface.

The corresponding functions operate in the reverse manner for flows from the mobile to the SGSN.

Between the BSC and the SGSN, the BSSGP protocol provides the following functionality over the G_b interface:

- Provision of virtual circuits between the SGSN and each GPRS cell to transport LLC control and data PDUs,
- Provision of flow control and buffering for LLC PDUs between the BSC and the SGSN.

Of necessity, this discussion of the very complex protocols is brief and omits many detailed aspects necessary for effective operation and recovery from error conditions. Nevertheless, the complexity and number of transactions and transaction handshakes needed to establish and manage MM and PDP contexts can begin to be understood and accounts for the time taken to establish an initial packet call, typically in the region of 2 seconds from GPRS idle state.

2.1.2.7 GPRS operation

To conclude this review of the key aspects of the GPRS network, it is perhaps helpful to examine the sequence of events that take place once a user has decided to set up a GPRS session. A representative sequence of transactions is shown in Table 2.3.

(i) At power on, the mobile will carry out its own initialisation processes and then search the GSM frequencies for cells transmitting a packet BCCH and 'camp' on a suitable cell. It will decode the information on the PBCCH and locate the packet RACH and packet AGCH for future use.

(ii) When the user initiates a GPRS session from the handset, a request for radio resources will be made on the PRACH channel. If the network is busy the user may have to queue until TBF resources are allocated and then the necessary signalling can take place to execute GPRS attach. Note that, because this is the first time

Table 2.3. Transactions in a GPRS session

	Event	Process	GPRS status			GSN status	
			Mob.–RAN	MM context	PDP context	Transaction	Mobile location
(i)	Power on	Mob initialisation	Packet idle	GPRS idle	Inactive	None	Unknown
(ii)	Packet send	TBF queue	Packet idle	GPRS idle	Inactive	None	Unknown
		GPRS attach	Packet transfer	GPRS ready	Inactive	Signalling	Cell level
		PDP context activate	Packet transfer	GPRS ready	Active	Signalling	Cell level
(iii)	Data burst	Send data	Packet transfer	GPRS ready	Active	Data transfer	Cell level
(iv)	TBF complete	Release TBF	Packet idle	GPRS ready	Active	None	Cell level
		Ready timer expires	Packet idle	GPRS standby	Active	None	Routeing area
(v)	Data burst	TBF queue	Packet idle	GPRS standby	Active	None	Routeing area
		Cell update	Packet transfer	GPRS ready	Active	Signalling	Cell level
		Send data	Packet transfer	GPRS ready	Active	Data transfer	Cell level
(vi)	TBF complete	Release TBF	Packet idle	GPRS ready	Active	None	Cell level
		Ready timer expires	Packet idle	GPRS standby	Active	None	Routeing area
		Standby timer expires	Packet idle	GPRS idle	Active	None	Unknown
(vii)	Session end	PDP context deactivate	Packet idle	GPRS idle	Inactive	None	Unknown

(for a while) that the user has been connected to the network, authentication will take place as part of the attach process. Once this has been successfully completed, cell and routeing area information will be held in the MM context and the status will be GPRS ready. With a connection now in place between the mobile and the SGSN, the SGSN will send a PDP-context activate request to the GGSN forming the gateway to the network hosting the requested site.

(iii) With a route now established from the MS to the GGSN and target site, the initial transactions can take place.

(iv) When no further data are awaiting transmission, the mobile releases the TBF, the status of the mobile BSS link moves to GPRS idle but, for a while, the MM context remains ‘ready’. Once the timer expires, the MM context falls back to ‘standby’. This situation can persist for some time and it is possible during this period that the user may move sufficiently far away to provoke a routeing area update.

(v) Once further transactions with the server are pending, the mobile will try to carry out a ‘cell update’ to trigger the MM context move to ‘ready’. As before, it may be necessary to queue for TBF resources before these are granted. With the MM context at ‘ready’, transactions can once more take place and this sequence of transitions between 'standby' and 'ready' states enables the radio resources to be efficiently shared.

(vi) Once the session has been completed, the regression from 'ready' to 'standby' will take place and, eventually, the SGSN 'standby' timer expires, tearing down the MM context.

(vii) As a consequence of MM context deactivation, the PDP context is deleted and all radio and network resources are released back to the system.

Although some of the considerable complexity and flexibility of the GPRS system has been omitted for clarity, it is still possible to see how a fairly efficient packet transport system has been overlaid on a GSM architecture established to efficiently support *circuit* traffic. This is no mean achievement! For the operator there are, as usual, choices to be made. The dimensioning of by far the most complex element, the SGSN, can be greatly influenced by the settings of the 'ready' and 'standby' timers. Low settings on these timers mean that the number of concurrent MM and PDP contexts that must be supported at the SGSN is greatly reduced. It also increases the number of mobiles that can share the radio and network resources – two factors apparently working together to maximise revenue. The reality is that, because of the number of transactions that need to take place to establish both the MM and PDP contexts, the user experience in terms of delay to access a site can quickly deteriorate as the probability of queuing for TBFs increases, particularly for busy systems. Considerable care is therefore necessary to optimise the user experience–capacity trade-off if revenue is to be maximised for the network.

If the discussion of Section 2.1.2.3 on QoS attributes is recalled for a moment, whilst there is, in principle, a large number of QoS profiles that can be supported, in practice, only a small number of distinctions is made. This reflects the relatively low peak-bearer rates available to the user, when compared with the traffic needs of most applications. The result is that most transactions across the air interface take some time to complete, giving rise to unpredictable delays for other users whilst they queue for TBFs. However GPRS is very successfully used for many applications, such as internet access (using the *wireless application protocol* (WAP)) and email where near real-time performance is perfectly acceptable.

2.1.3 GSM network operations costs

In this chapter, various approaches have been described to support more users on the same amount of infrastructure and spectrum. In GSM, new voice *codec*s have been devised to support *half-rate* speech, almost doubling the number of users, whilst in GPRS, the packet network will allow somewhere between three and ten times the number of users to be supported for applications, such as web browsing and interactive gaming. Both sets of new functionality address the revenue side of the profit and loss account. It is, therefore, appropriate to understand the make-up of the total cost base for an established operator, as this can be used to identify areas to target in future cellular systems. Figure 2.10 shows the various types of costs that an established operator might expect.

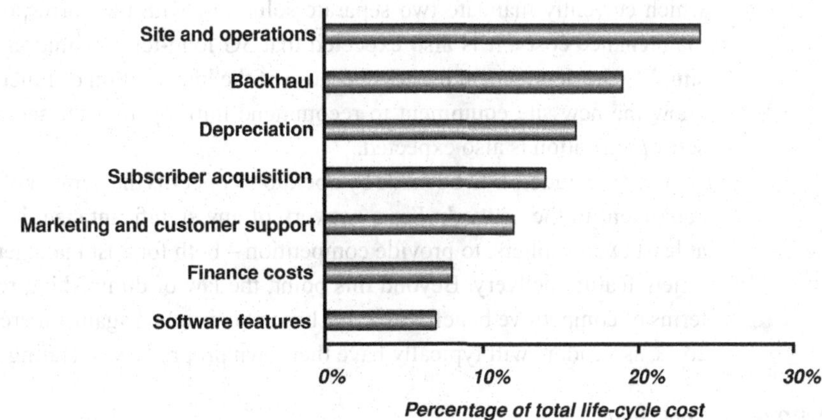

Notes
1) Source Motorola, for a typical GSM network, capital amortised over 7 years.
2) Site and operations comprises: site rental, power, maintenance.
3) Absolute costs vary from country to country but are consistently about 70% of revenue.

Figure 2.10 Total life-cycle costs for an established operator

2.1.3.1 Site and operations

The single largest component of costs in most networks is that associated with site rental, power and maintenance. With increasing regulation restricting the locations that may be used for cellular masts, new and existing site owners can (and do) charge an increasing rental premium. The most effective way to manage these costs is simply to have fewer sites in the network.

In developed countries, where the number of sites in most regions is driven by the need to support a given subscriber density, moving to technologies that can support higher capacity in a single cell can address this issue. Conversely, in developing markets, which account for a large percentage of new deployments, technologies that can support very large radius cells are important in establishing coverage over the required area for minimum investment, whilst network traffic is initially low. Constraining the number of cells has a further benefit in that it also reduces that element of maintenance that requires site visits. The number of such visits can also be reduced by the careful choice of base station equipment. As RF technology continues to improve, the power efficiency of BTS installations will soon reach a stage where they can be self-cooling in most configurations, eliminating air-conditioning equipment, a source of cost and regular maintenance visits.

A further goal is to minimise the number of operations staff necessary to maintain the network, over and above those required to perform site visits. Later chapters will review the activities necessary to plan, evolve and introduce new functionality into networks. Suffice it to say that at present staff with specialist skills are needed for each network technology. At the time of writing, for an operator with GSM and UMTS networks, this requires staff with radio skills in two different air interface solutions (TDMA and CDMA) and core network skills in circuit and packet engineering. It may soon be possible to employ one standard, 3G LTE, to meet the coverage and capacity extremes, which currently mandate two separate solutions, with the corresponding reduction in maintenance costs. It is also expected that 3G long-term evolution (LTE) will greatly simplify the deployment process because of the introduction of functionality which will allow the new site equipment to recommend initial parameter settings. Some level of self-optimisation is also expected.

A further trade-off to be made by operators concerns the number of vendors supplying equipment to the network. For a network of any significant size, it is essential to have at least two suppliers, to provide competition – both for cost management and to ensure timely feature delivery. Beyond this point, the law of diminishing returns can apply in terms of competitive benefits and this has to be weighed against increasing maintenance costs, as vendors will typically have their own proprietary operating procedures.

2.1.3.2 Back-haul

It is not immediately obvious that back-haul should represent such a large proportion of life-cycle operating costs. However, from the architecture discussion earlier in this chapter it is clear that point-to-point links will be required to each BTS site, and that the BTS is far and away the most frequently occurring network element. This situation is further aggravated by the requirement for network connectivity actually at the site – nearby is not good enough! This is clearly not an issue in urban areas but frequently requires that custom cabling is introduced in suburban locations and that dedicated microwave links are deployed for rural sites. The most obvious way of reducing this cost – ensuring that the minimum number of sites is deployed – has already been addressed for other reasons; so what more can be done? If it is expected that the cell sites will be fully loaded on deployment, there is not a great deal that can be done. However, for new deployments in developing markets, cells are unlikely to be fully loaded for some time and there are methods that reduce transport bandwidth requirements. In these markets, often the only practical method of linking widely dispersed cell sites is via satellite, and such transport is priced on a required bandwidth basis, so minimising the transmission bandwidth is all important.

In a new deployment with low subscriber penetration, typical sites will not, in fact, be omnidirectional but tri-sectored with one transceiver per sector. The selection of this sectored configuration is driven by the increased cell range possible with the additional antenna gain, rather than capacity. The disadvantage of a single transceiver for each cell is the low capacity if acceptable blocking rates are to be delivered. Recall that, once allowance is made for the BCCH, a single GSM transceiver supports seven time slots for traffic. Reference to Erlang tables reveals that for the seven physical channels in each cell and for 2% blocking probability, only 3 Erlangs of traffic can be supported, giving rise to an expected demand of 90 Erlangs for a coverage region of ten such sites. Yet connectivity for *21* physical channels per site needs to be provisioned, some 210 16 kbit/s links. In practice statistical multiplexers can be introduced, prior to back-haul, to reduce this requirement to about 120 links.

An additional opportunity arises from the GSM standard for the *transcoder and rate adaptor unit* [33]. It provides for voice and in-band signalling to be conveyed in 16 kbit/s channels, four in each DS0 slot of the E1 link. However, it also recognises that in a normal conversation, there are usually periods when one person is speaking and the other listening. During such periods of silence, the speech *codec* will not output data and *idle frames* are transported instead. In practice these frames can be removed prior to transport and reinserted after transport with savings of up to 30%. There is, in fact, a thriving market for products [34, 35], which are placed at each end of the A-bis link and that carry out both the statistical multiplexing and redundant frame removal. These offer capacity savings of up to 50%, depending upon the extent of traffic aggregation.

As discussed in Chapter 1, data traffic arising from applications is much less well behaved, statistically, than voice and special planning techniques are needed to crystallise transport savings while guaranteeing correct operation of the application. This methodology will be discussed in Chapter 3.

2.1.3.3 Depreciation and finance costs

Typically, the whole network deployment activity, including planning, site preparation and equipment costs together with deployment and initial optimisation, will be treated as a 'project' and capitalised. The project costs will also include the cost of the spectrum licence, if this is significant. Amortisation of these capital costs will take place over the expected lifetime of the project to give a yearly depreciation charge, which, when added to finance costs (representing the interest payable on the capital outstanding in any one year), will be charged as an operating expense. Given the previous discussion regarding the number of BTSs compared with other network elements, it will come as no surprise to find that BTS equipment typically represents some 70% of the capital expenditure for GSM operators. Every factor that influences the number of cell sites required, therefore, needs to be considered. The remaining area of flexibility is the decision regarding the services that will be supported.

In developing markets, the initial requirement is typically to maximise voice and low data rate coverage for the minimum cost; this ensures the earliest project payback during the period when traffic on the network is low, immediately following initial deployment. A specific configuration of GSM, known as *extended cell*, allows the possibility of sites with ranges of up to 70 km, depending on the propagation conditions. Such cells might typically increase the cell area by between three and four times in practical deployments. So that the link budget still works with standard mobiles, a number of additional features may be deployed to deliver improvements of up to 10 dB:

- Two-antenna transmit diversity (downlink),
- Four-antenna receive diversity (uplink),
- Use of the adaptive multi-rate (AMR) *codec*.

The diversity antenna systems provide 5 dB improvement or more. Under low signal-to-noise (S/N) conditions, the AMR codec can reduce the source rate to allow heavier coding to be used, enabling satisfactory operation in S/N ratios 5 dB lower than for the standard GSM full rate. This improvement in link budget through the use of AMR comes at the expense of some deterioration in voice quality.

2.1.3.4 Marketing, customer support and subscriber acquisition

Expenditure in this area is largely a matter of operator choice and will vary depending upon the maturity of the market and the perceived competitive environment. However, there is an increasing tendency to segment the market beyond traditional prepaid and postpaid classes. For instance, operators have established websites, which enable the computer literate to undertake the effort of completing new subscriptions on line and, in return, secure more attractive tariffs. This may mark the demise of 'phone shops' over some period of time. Another trend originated by the *mobile virtual network operators* (MVNOs) – those operators who choose to offer a service without owning their own network – is subscriptions with much reduced tariffs for those prepared to supply their own mobile phones.

2.2 IEEE 802.11 networks

The 802.11 series of wireless LAN standards was originated by the IEEE 802 LAN/MAN Committee, which develops local and metropolitan area network standards [36]. The original 802 project was set up in 1980 to address this area. It was intended that there should be one LAN configuration with speeds from 1 to 20 MHz. It was divided into media or physical layer (PHY), media access control (MAC) and higher level interface (HILI). More than 25 years on, this committee has generated many standards, which are found on copper, cable and fibre bearers throughout the world, perhaps most notably *Ethernet* and *Token Ring*. It was not until 1997 that the 802 project introduced its first wireless commercial standard – 802.11b. In line with its original project remit, this only defines layers 1 and 2 – although it does introduce a layer 2 sub-layer – the LLC or link-layer control.

In contrast to GSM, which was originally devised to support national and international voice coverage with one mobile device, 802.11 was conceived to provide a wireless alternative to Ethernet. In the office, cells having ranges of 10 to 20 metres can be expected, depending upon user rates and when attenuation from walls, etc., is considered. By deploying a number of overlapping sites, a 'hot-spot' region of coverage can be created. Good solutions are quickly adopted by others for new applications and the world is now familiar with Wi-Fi® hotspots in places where latter day nomads (aka businessmen) need sustenance in the form of connectivity to their home network.

The standard was developed to utilise the *Industrial, Scientific and Medical* (ISM) bands at 2.4 GHz and 5.8 GHz, which are usually *licence exempt*; this means, in practice, that the spectrum will be shared (without frequency planning) by many users. For this reason, and to allow system deployment by unskilled users, the *PHY* layer is designed to select the RF channel with the least interference and the *MAC* employs a 'collision avoidance' philosophy. Because there is a finite number of radio channels, and no 'admission control', the available network capacity is shared amongst all users, leading to lower data rates in times of congestion. No attempt is made to define mobility management between two or more access points.

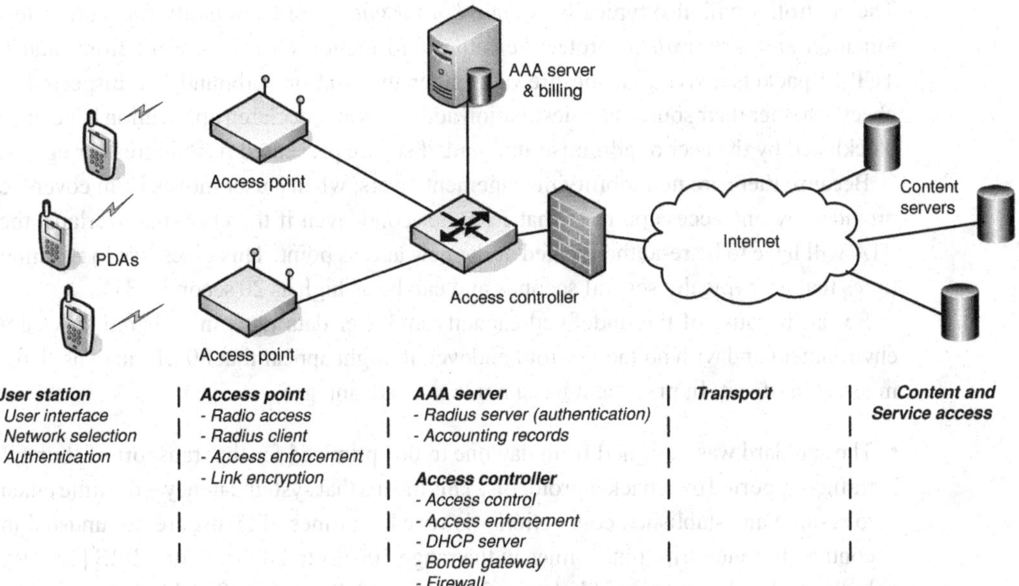

Figure 2.11 Example of public wireless access point architecture

Figure 2.11 outlines a representative network configuration that may be used to support public 802.11 hot spots. Reviewing the key steps in accessing such a network will illustrate the role of the network elements.

- The wireless station (STA) discovers an IEEE 802.11 access point (AP) and initiates a connection request.
- The authentication server responds with a request for the STA identity.
- The AP forwards the STA identity as an authentication request message to the local authentication server.
- The AAA server authenticates the user via an EAP-based challenge-response method running between the authentication server and the STA. A user database is consulted by AAA to verify the username and credential provided by the STA. The result of the authentication and session key material are communicated to the AP and STA.
- The AP configures link-layer session keys and signals that the STA has been successfully authenticated. The DHCP server in the access controller leases an IP address to the STA, which will be used for the duration of the hot-spot session. Prior to this time, the controller blocks any attempt by the STA to obtain an address or access the Internet.
- The AAA keeps accounting records and, if this hot-spot is not part of a larger regional network, may generate billing information as well. When the STA disconnects, an accounting stop message is sent as the last message for that session. The billing and settlement components process the accounting data and generate charging and billing records.

The controller will also typically contain *border gateway* functionality for route determination and a *firewall* to protect sensitive information on the network from attack. TCP-IP packets arriving at the firewall, either inbound or outbound, are inspected to check whether their source and destination addresses and declared application have been blacklisted by the user or administrator, and, if so, are prevented from ingress or egress.

Because there are no mobility management layers, when a user moves from coverage provided by one access point to that from a second, even if the coverage overlaps, the STA will have to be re-authenticated at the new access point. This gives rise to roaming times that are typically several seconds and can be as high as 20 seconds [37].

So far, because of the undefined capacity and user data rates in a shared spectrum environment and with no facility for handover, it might appear that 802.11 may not have much going for it. In practice it has a number of advantages.

- The standard was designed from day one to be optimised for the transport of 'bursty' traffic supported over packet protocols. This means that system latency – the time taken to resume an established connection – is *very* low; times of 50 ms are not unusual in contrast to round trip 'ping' times in the range 700 ms to 2000 ms for GPRS [38, 39].
- Individual cell capacities of between 5 Mbits/s and 20 Mbits/s (802.11 version dependent) and the 'bursty' nature of packet-based applications mean that, in practice, it is unusual for users to experience low data rates.
- An industry group, the Wi-Fi Alliance® [9] was established in 1999 to establish a product certification programme and work to establish other enabling infrastructure, such as improved security protocols, etc. This has guaranteed interchangeability between different vendors' equipment and facilitated its widespread adoption.

The Wi-Fi Alliance® has been so successful that its work is largely responsible for ensuring the availability of product at close to commodity prices. This, in turn, has spurred the IEEE 802.11 committees to develop an evolution path from the early standards to offer ever-increasing capacity and range. 802.11g offers data rates up to 54 Mbits/s and 802.11n provides ranges up to 100 metres through MIMO techniques.

The low latency of the 802.11 family is very attractive for bursty traffic; what makes this possible? Figure 2.12 contrasts the protocol stacks for GSM and 802.11. It is immediately apparent that GSM contains *four* more layer 2 protocols than 802.11. These all have their origins in requirements for GPRS:

- Packet data support and wide area mobility (MM context),
- Support of multi-protocol networks (PDP context),
- Enable network-originated traffic (BSSGP),
- GPRS had to be an upgrade to GSM (RLC/MAC).

The latency impact of these additional layers is made worse by two further factors:

- Two serial events that need to have taken place before the final PDP context can be activated (mobile–RAN connection and MM context),
- Each state change has a minimum of four message flows, and sometimes as many as 15 to 20.

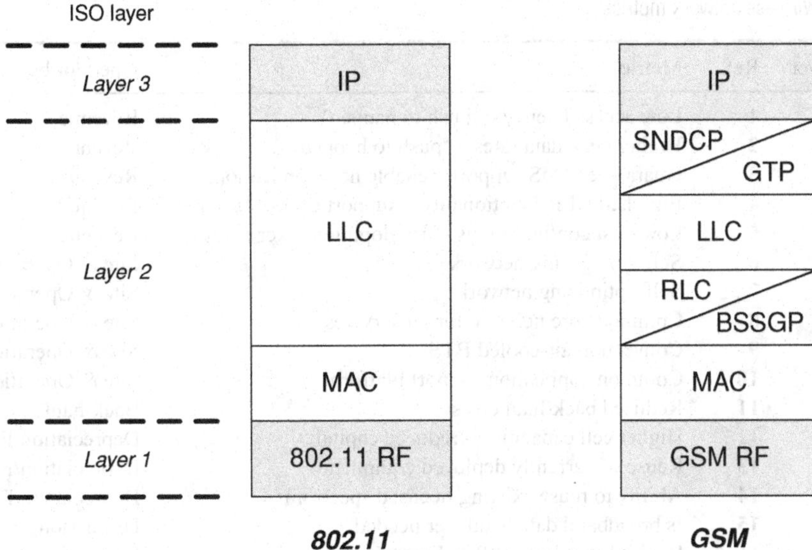

Figure 2.12 A comparison of GSM and 802.11 protocol stacks

As is often the case with two very successful standards operating in two distinct but adjacent product spaces, the standards organisations concerned aspire to add the good performance attributes of the competing standard to their own capability – without eroding their current strengths. The 3GPP product family responded first with UMTS, and now has 3G LTE in prospect. The IEEE organisation has or is delivering significant evolutions of 802.11g (802.11n and 802.11s) and is upgrading a broadband point-to-point wireless standard (802.16) to provide wide area mobility. The success or otherwise of some of these ambitions will be discussed in later chapters.

2.3 Operator and user issues with GSM

This brief concluding section is intended to focus the preceding discussion. A candidate heading for this section could have been 'What's wrong with GSM?' but for those of us who remember the pre-GSM days of suits with especially large pockets to accommodate the 'one phone per region' whilst travelling worldwide, that would not be fair. GSM has delivered so much that is good, but objectively it has its limitations. So, in identifying the areas in Table 2.4, the intention is to use these as a set of 'metrics' that can be used to assess the benefits of newer technologies. Many of these metrics apply equally to cellular and non-cellular technologies.

To make these results more digestible, the metrics have been split into 'user driven', 'operator driven' and 'other'. Almost all the metrics result in either cost or revenue benefits to the operator, so the criterion for classifying functionality as 'user' is that improvements will provide direct benefits to the user experience or enable new applications.

Table 2.4. Wireless network metrics

Metric driver	Ref.	Metric	Operator benefit
User	**1**	Low access latency – 'push to happen'	Revenue
User	**2**	Higher peak data rates – 'push to happen'	Revenue
User	**3**	Guaranteed QoS support – enable new applications	Revenue
Operator	**4**	Enriched HLR functionality – support of new features	Revenue
Operator	**5**	Low-cost configurations – developing market access	Revenue
Operator	**6**	Self-configuring network	Site & Operations
Operator	**7**	Self-optimising network	Site & Operations
Operator	**8**	Common core network for all services	Site & Operations
Operator	**9**	Convection air-cooled BTS	Site & Operations
Operator	**10**	Common application support platform	Site & Operations
Operator	**11**	Reduced back-haul costs	Back-haul
Operator	**12**	Higher cell capacities – reduced capital	Depreciation/Finance
Operator	**13**	Reuse of currently deployed equipment	Depreciation/Finance
Operator	**14**	Ability to reuse existing licensed spectrum	Depreciation/Finance
Other	**15**	Is broadband data handover needed?	Discussion
Other	**16**	Revised standards IPR policy	Discussion

The limitations in Table 2.4 have already been discussed or alluded to, with the exception of rows 15 and 16. Historically, a given assumption has been that handover was essential in cellular systems. This was appropriate when voice was the only revenue-generating traffic but it is already possible to envisage a time when the majority of revenue will be generated by other traffic. There is, inevitably, an infrastructure overhead required to support handover with its associated cost. Thus, the degree to which handover is supported is a debate the operator needs to have, bearing in mind the target market and the fact that the VoIP experience indicates that there is a quality–price trade-off that many subscribers are already making.

The consumer has of course been largely unaware of the issue of IPR policy in standards and until recently it has not been a matter of major concern to operators. However, as work has started in earnest on 3G long-term evolution, a number of operators are voicing their perception that the royalties payable in current systems are not compatible with the industry ambition to develop a thriving market for the new product. At the time of writing, there is a fundamental review of the policy, which has been in force for some years, and the operator goal is, '*to seek a new regime with greater transparency and predictability of future IPR costs, enabling selection of optimal technology in a commercial environment* [40].' No clear consensus has currently been reached, but one measure that is seen to be attractive by many is the proposal to execute a survey which identifies patents that might be potentially relevant to any proposed standard. This landscape, which would not be exhaustive or binding, would enable ETSI members to get a feel for the ownership of IPR that might be important to technologies under consideration for the standard and whether or not those companies had committed to licence under fair and reasonable terms. Individual ETSI members could then consider this information when making decisions on the selection of technical solutions for the standard.

References

1 European Telecommunications Standards Institute (ETSI), www.etsi.org/.

2 ETSI, *Digital Cellular Telecommunications System (Phase 2+); Network Architecture*, GSM 03.02 version 7.1.0, Release 1998.

3 ETSI, *Digital Cellular Telecommunications System (Phase 2+); Mobile Stations (MS) Features*, GSM 02.07 version 7.1.0, Release 1998.

4 ETSI, *Digital Cellular Telecommunications System (Phase 2+); Specification of the Subscriber Identity Module, Mobile Equipment (SIM-ME) Interface*, GSM 11.11 version 7.4.0, Release 1998.

5 3GPP, www.3gpp.org/.

6 3rd Generation Partnership Project, *Technical Specification Group Terminals, Specification of the Subscriber Identity Module – Mobile Equipment (SIM–ME) Interface*, 3GPP TS 11.11 version 8.13.0 (2005–06), Release 1999.

7 ETSI, *Digital Cellular Telecommunications System (Phase 2+); Subscriber Identity Modules (SIM); Functional characteristics*, GSM 02.17 version 8.0.0, Release 1999.

8 The Institute of Electrical and Electronics Engineers, www.ieee.org/portal/site.

9 Wi-Fi Alliance, www.wi-fi.org/.

10 ETSI, *Digital Cellular Telecommunications System (Phase 2+), International Mobile Station Equipment Identities (IMEI)*, GSM 02.16 version 7.0.0, Release 1998.

11 ETSI, *Digital Cellular Telecommunications System (Phase 2+); Mobile Station–Base Station System (MS–BSS) Interface; Channel Structures and Access Capabilities*, GSM 04.03 version 7, Release 1998.

12 ETSI, *Digital Cellular Telecommunications System (Phase 2+); Multiplexing and Multiple Access on the Radio Path*, GSM 05.02 version 7.1.0, Release 1998.

13 ETSI, *Digital Cellular Telecommunications System (Phase 2+); Functions Related to Mobile Station (MS) in Idle Mode and Group Receive Mode*, GSM 03.22 version 7.1.0, Release 1998.

14 ETSI, *Digital Cellular Telecommunications System (Phase 2+); Mobile Radio Interface Layer 3 Specification*, GSM 04.08 version 7, Release 1998.

15 ETSI, *Digital Cellular Telecommunications System (Phase 2+); Radio Subsystem Link Control*, GSM 05.08 version 7.1.0, Release 1998.

16 ETSI, *Digital Cellular Telecommunications System (Phase 2+); Basic Call Handling; Technical Realization*, 3GPP TS 03.18 version 7.5.0, Release 1998.

17 ETSI, *Digital Cellular Telecommunications System (Phase 2+); Signalling Procedures and the Mobile Application Part (MAP)*, GSM 09.10 version 7.1.0, Release 1998.

18 ETSI, *Digital Cellular Telecommunications System (Phase 2+); Location Registration Procedures*, GSM 03.12 version 7.0.0, Release 1998.

19 ETSI, *Digital Cellular Telecommunications System (Phase 2+); Security Related Network Functions*, GSM 03.20 version 7.1.0, Release 1998.

20 ETSI, *Digital Cellular Telecommunications System (Phase 2+); Security Aspects*, GSM 02.09 version 7, Release 1998.

21 ETSI, *Digital Cellular Telecommunications System (Phase 2+); Physical Layer on the Radio Path, General description*, GSM 05.01 version 7.1.0, Release 1998.

22 ETSI, *Digital Cellular Telecommunications System (Phase 2+); Modulation*, GSM 05.04 version 7.0.0, Release 1998.

23 ETSI, *Digital Cellular Telecommunications System (Phase 2); Network Management (NM); Part 1: Objectives and Structure of Network Management*, GSM 12.00, Release 1997.

24 ETSI, *Digital Cellular Telecommunications System (Phase 2+); Event and Call Data*, GSM 12.05 version 7.0.1, Release 1998.

25 ETSI, *Digital Cellular Telecommunications System (Phase 2+); General Packet Radio Service (GPRS); Service description; Stage 2*, GSM 03.60 version 7.0.0, Release 1998.

26 ETSI, *Digital Cellular Telecommunications System (Phase 2+); General Packet Radio Service (GPRS); GPRS Tunnelling Protocol (GTP) Across the Gn and Gp Interface*, GSM 09.60 version 7.1.0, Release 1998.

27 ITU-T, *Interface Between Data Terminal Equipment (DTE) and Data Circuit-terminating Equipment (DCE) for Public Data Networks Providing Frame Relay Data Transmission Service by Dedicated Circuit*, ITU-T Recommendation X.36 (2003).

28 ETSI, *Digital Cellular Telecommunications System (Phase 2+); General Packet Radio Service (GPRS); Mobile Station (MS) – Base Station System (BSS) Interface; Radio Link Control/ Medium Access Control (RLC/MAC) Protocol*, GSM 04.60 version 7.0.0, Release 1998.

29 ETSI, *Digital Cellular Telecommunications System (Phase 2+); General Packet Radio Service (GPRS); Mobile Station (MS) – Serving GPRS Support Node (SGSN); Sub-network Dependent Convergence Protocol (SNDCP)*, GSM 04.65 version 7.0.0, Release 1998.

30 ETSI, *Digital Cellular Telecommunications System (Phase 2+); General Packet Radio Service (GPRS); Mobile Station – Serving GPRS Support Node (MS-SGSN) Logical Link Control (LLC) Layer Specification*, GSM 04.64 version 7.0.0, Release 1998.

31 ETSI, *Digital Cellular Telecommunications System (Phase 2+); General Packet Radio Service (GPRS); Base Station System (BSS) – Serving GPRS Support Node (SGSN); BSS GPRS Protocol (BSSGP)*, GSM 08.18 version 7.0.0, Release 1998.

32 ETSI, *Digital Cellular Telecommunications System (Phase 2+); General Packet Radio Service (GPRS); Overall Description of the GPRS Radio Interface; Stage 2*, GSM 03.64 version 7.0.0, Release 1997.

33 ETSI, *Digital Cellular Telecommunications System (Phase 2+); In-Band Control of Remote Transcoders and Rate Adaptors for Enhanced Full Rate (EFR) and Full Rate Traffic Channels*, GSM 08.60 version 7.0.0, Release 1998.

34 M. Di Paolo, *GSM Over VSAT: Choosing the Right Backhaul Solution*, Application note, Comtech EF Data (2006).

35 NMS Communications AccessGate™: *The Performance Leader in RAN Optimization Solutions*, www.nmscommunications.com/NR/ rdonlyres/F22606DA-62E3-4235-9835-2EBDE85038D1/0/ AccessGate.pdf (2006).

36 IEEE 802 LAN/MAN Standards Committee, www.ieee802.org/.

37 F. Mlinarsky and G. Celine, Testing braces itself for voice over Wi-Fi, *Electronic Design*, July/August (2004).

38 D. Michel and V. Ramasarma, The application of trial methodologies for measuring and verifying KPIs in the railway environment, *Bechtel Telecommunications Technical Journal*, **2** 2 (2004).

39 Citrix Systems Inc., *Optimizing Citrix Technology for Operation over GPRS Networks – Recommended Best Practices* (2001).

40 A. Cox, *Next Generation Mobile Networks Beyond HSPA and EVDO – Mission, Vision and the Way Ahead*, Presented to ETSI Board 57, June 2006.

3 Principles of access network planning

In Chapters 1 and 2, the drivers for the development of the cellular system architecture were discussed and an overview of the key network elements and principles of operation for a GSM cellular solution was provided. The remainder of the book will address the activities necessary to design and deploy profitable wireless networks. Figure 3.1 summarises where these key processes are to be found by chapter.

In this chapter, the principles and processes that are used to plan wireless access networks will be developed. The major focus will be on cellular networks, as these usually represent the most complex planning cases, but an overview of the corresponding processes for *802.11* is also provided. Circuit voice networks will be examined initially; the treatment will then be extended to understand the additional considerations that come into play as first circuit-data and subsequently packet-data-based applications are introduced.

With the planning sequence understood, the way in which information from such processes can be used to explore the potential profitability of networks well in advance of deployment will be addressed. Choices regarding which applications are to be supported in the network and the quality of service offered will be shown to have a major impact on the profitability of network projects.

3.1 Circuit voice networks

In most forms of retailing, the introduction of new products follows the 'S curve' sequence first recognised by Rogers [1]. Rogers stated that adopters of any innovation or idea could be categorised as innovators (2.5%), early adopters (13.5%), early majority (34%), late majority (34%) and laggards (16%), based on a bell curve. Adopting this model, initial sales would come from the 'early adopters', with volume sales from the 'early and late majorities'. In conventional retail products, this means that all of the development and initial production costs have to be incurred before any revenue is earned; but it is likely that the costs of volume production can be deferred until there is some feedback on whether the product is likely to be a success. For cellular network operators, the investment profile is different; the service offered is 'connectivity' over an entire country or region. Thus the major part of the network coverage needs to be in place before the service becomes attractive and significant revenue starts to flow. Indeed, this was a major issue for operators in Europe in transitioning customers from a good analogue network to what was then the new GSM network, which, at that time, had less

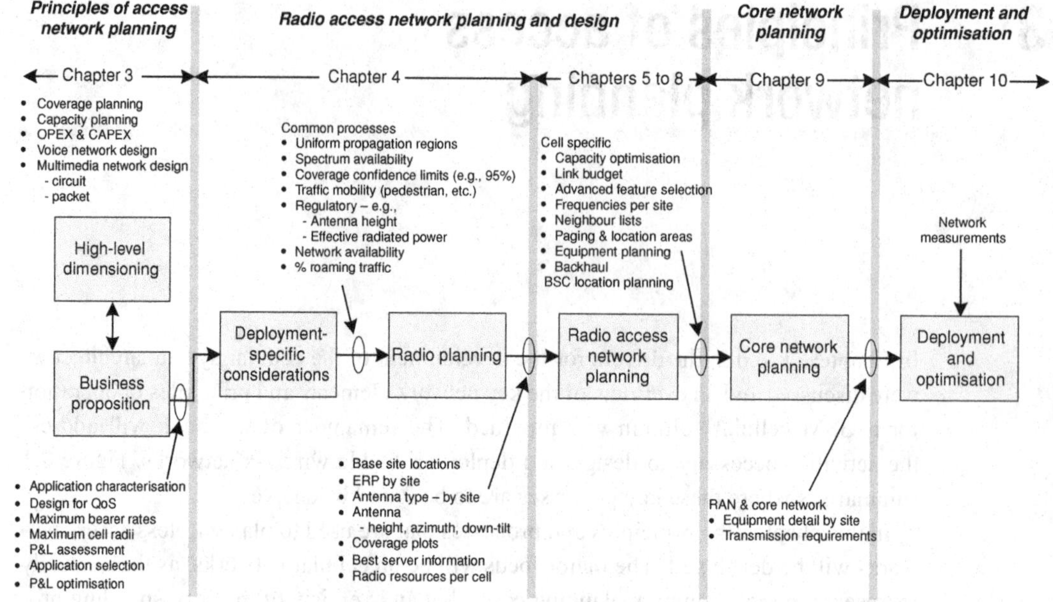

Figure 3.1 Overview of the cellular network planning process

comprehensive coverage. Considerable attention must, therefore, be paid to the way in which coverage is rolled out if the project 'payback' period – the period that must elapse before the initial investment has been recovered through annual cash flows – is to be kept short. This 'payback' period, or the more comprehensive Project *Net Present Value (NPV)*, which reflects the current value of a series of projected future cash flows at a specified discount rate, are key metrics in assessing whether a particular project is viable. Periods as short as 5 years can be used as go/no-go investment criteria for new networks, depending upon the maturity of the technology and the region to be covered [2].

There are at least four dimensions that can be exploited by the operator to minimise investment in the early phases of a project:

(i) Network coverage,
(ii) Network capacity,
(iii) Network features,
(iv) Handset subsidies.

3.1.1 Coverage

The sequence and extent of network rollout can have a major impact on project profitability. Addressing regions of high population density first (such as city and urban areas) can maximise potential revenue for the minimum infrastructure investment. This attractive result arises both because the area to be covered (and hence the number of BTS sites) is minimised and also because this same area is likely to contain a high percentage of businesses, which are typically early adopters. This initial phase is usually followed by coverage of regions where large numbers of people live and work, and the routes used to

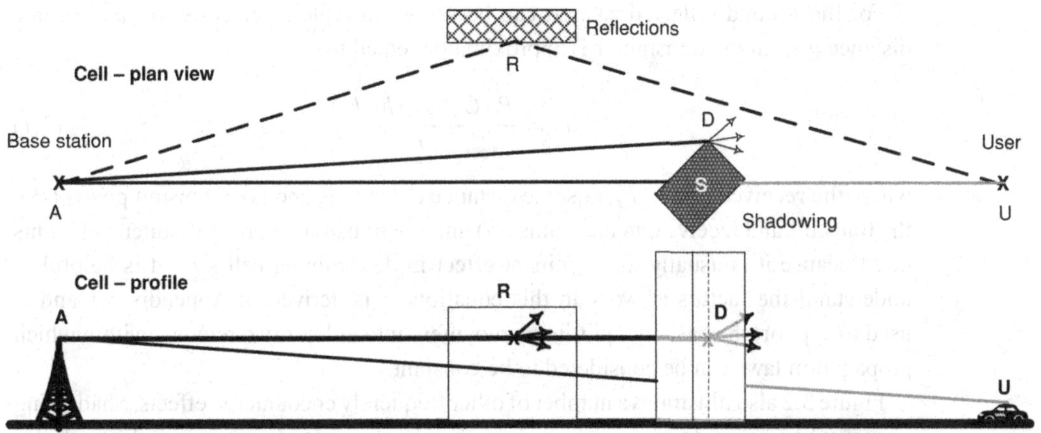

Figure 3.2 Macrocell propagation

travel between these locations. Coverage of rural areas with small numbers of potential subscribers typically occurs last and may only happen because of legal obligations to provide such coverage, arising as part of the spectrum licence conditions. Initially, coverage in rural areas will be far from comprehensive because not all cells will overlap – perhaps because of local topography (e.g., hills attenuating RF signals). At a later stage, when revenues can justify it, 'in-fill' sites may be introduced to address these limitations.

It is clearly desirable to maximise the cell radius in order to provide coverage in sparsely populated regions, or during the early stages of network rollout. So what, in practice, limits the maximum achievable range? For a given air interface, there will be a minimum signal-to-noise ratio (S/N) that is needed to decode the received signal at a specified bit error rate (BER). Consistent with Shannon's Law, the required S/N will increase as the number of bits/symbol increases. For GSM systems utilising only 1 bit/symbol, this is in the region of 9 to 11 dB. The received power at the input to a perfect mobile thus needs to be at least 9 dB above the noise power. Knowledge of the losses that will be encountered as signals propagate from the transmitting base station antenna to the mobile will enable the maximum practical cell size to be estimated.

One estimate of the path loss from the base station to the mobile, at some distance r, can be obtained by projecting the power transmitted by the BTS antenna onto the surface of a sphere of radius r (assuming that the antenna radiates power equally in all directions). This 'loss' is due to the intrinsic reduction in power density with range and is known as the 'free space' path loss. However, except at short ranges from the BTS and with a few other exceptions, this form of simple propagation loss is rarely encountered in cellular systems because of a number of additional effects.

The most frequently encountered effect is that caused by the presence of at least one path in addition to the direct line-of-sight ray. This can arise from lateral reflections from large objects close to the line-of-sight path, as shown in Figure 3.2; represented by the path ARU in addition to the direct path ASU. However, the most frequently encountered effect is caused by reflections from the ubiquitous ground surface.

For the ground-reflected case, it can be shown that the received power P_r at some distance d from the transmitter is approximately equal to

$$P_r(d) = \frac{P_t \cdot G_t \cdot G_r \cdot (h_t \cdot h_r)^2}{d^4}, \tag{3.1}$$

where the received power, P_r, at some distance d depends upon the transmit power (P_t), the transmit and receive antenna gains (G) and the transmit and receive antenna heights (h). Because it is usually the dominant effect in determining cell size, it is helpful to understand the factors at work in this equation. It is derived in Appendix 3.1 and is used to support the case for splitting deployment into at least two regions, within which propagation laws can be considered to be constant.

Figure 3.2 also illustrates a number of other frequently encountered effects. Shadowing occurs when a building or other feature that attenuates radio signals is located at a distance from the base station or mobile such that it significantly obscures the line-of-sight propagation path. Propagation does occur, despite the shadowing object, and is the sum of the attenuated direct ray *ASU* and diffraction around the object *ADU*. A further effect needs to be considered when the user is moving. For locations where there are several reflecting surfaces (most deployments) it is apparent that movement of the user will alter the phase of signals from the various paths so that when the signals combine at the receiver, they can give rise to fading. The fading may exhibit Rician or Rayleigh characteristics [3] depending on the relative strength of the direct and multi-path rays. There will also be a shift in the frequency of the received signal, from Doppler effects caused by user movement.

A real-world deployment will usually include impairments from all of the above sources, with large numbers of reflecting and scattering surfaces together with shadowing from objects of arbitrary size and location. The definition of models that may be used in planning wide-area networks is, therefore, carried out on the basis of curve fitting to extensive empirical data. This approach was first proposed by Hata [4], in his landmark paper 'Empirical formula for propagation loss in land mobile radio services' but has since been developed by many contributors. In calculating the maximum cell range possible, all of the above effects need to be considered in the *link budget*, a convenient expression representing the losses that may be incurred between the base station and mobile. So for a system with a BTS equivalent isotropic radiated power (EIRP) of P_t dBW – the radiated power once transmit antenna gain is considered – and mobile sensitivity P_r dBW (mobile antenna gain = 1) the downlink budget may be expressed as

$$\text{link budget} = (P_r - P_t)\ \text{dB}.$$

This link budget can be equated to the appropriate improved Hata models [5, 6, 7, 8] to develop an equation that can be solved for distance. The resulting equation takes the general form:

$$(P_r - P_t) = \underbrace{A + 10b\log_{10}\left(\frac{d}{d_0}\right)}_{\text{propagation loss}} \underbrace{+S}_{\text{shadowing loss}} \underbrace{+R}_{\text{fade margin}}, \tag{3.2}$$

Table 3.1. Propagation parameters for a representative cellular system

Parameter	Definition	Typical values	Units
d	The distance at which propagation loss is to be evaluated	–	m
d_0[a]	The distance at which propagation becomes close to free space; this will depend on operating frequency and BTS antenna height.	100–300	m
A[a]	Propagation loss at the distance d_0	80	dB
b[a]	Path loss exponent – this varies and is heavily influenced by antenna height and the terrain over which propagation occurs – e.g., is the region flat or hilly, dense foliage or free of trees, etc.	2–5.5	–
S[a]	Shadow fading loss; it is log–normally distributed, is BTS antenna height dependent and can also alter with the deployment region. With 95% confidence, shadowing will be:	<15	dB
R	Fade margin (terrain dependent)	~10	dB

[a] Transmit antenna height = 25 m; frequency = 1.9 GHz

where the equation is in dB and the terms are defined in Table 3.1 along with representative values for a cellular system.

It is perhaps worth highlighting at this point that the reader will note that many different values will be used in this text for the exponent of range b used in Equation 3.2 to determine propagation loss. There is no universal 'correct' value, but the exponent usually lies in the range 2 to 5.5, depending on the particular propagation environment.

In concluding this section on coverage it is appropriate to recognise that comprehensive software packages, which address 'network planning', are available [9, 10]. However, such systems still need the system planner to select appropriate options that represent good choices for the particular region to be planned. Some of the key decisions to be made are summarised below.

Classification of propagation in the planned deployment area

From the discussion above and in Appendix 3.1, it is very clear that this is a critical parameter to get right. Table 3.1 shows that the variation of loss with distance is a power law whose exponent can lie between 2 and 5.5, driven by this classification. If a network is to be rolled out in a country new to the network operator, it may be advisable to carry out a limited measurement programme to characterise propagation behaviour. A publication from the *WINNER* programme [11] provides a good insight into the classification of propagation regions. For GSM, COST 259 identifies six distinct propagation regions characterised by differences in path loss, delay spread, shadowing and user speeds. ETSI specifications [12] reference a sub-set of the COST 259 models to be used in conformance testing of equipment.

Antenna considerations

The decision regarding the height of the base station antenna is one of the few choices that the operator may be able to make that *can* influence propagation loss. When the antenna height is significant compared with the mobile range, the angle of the ground-reflected ray to the ground is no longer small and the amplitude of the reflection becomes much lower . . . and in the limit propagation loss changes from $1/d^4$ to $1/d^2$ (free-space propagation). Additionally, under these conditions, further benefits accrue because the antenna height means that in such regions shadowing is much less significant. However, in many countries there are restrictions on the mast heights that may be used, particularly in urban and suburban areas. This can limit the scope for range extension, although in Germany some operators [13, 14] have been able to introduce exceptionally high masts for UMTS (known as 'high sites') even in these regions, with great benefits.

Another possible choice is to deploy *sectored* antennas. Although *sectorisation* is normally introduced as a capacity measure, tri-sectored antennas provide a further 5 dB gain over the equivalent omnidirectional configuration, albeit at the cost of additional transceivers (a minimum of three vs. one for the omni case). However, the tri-sectored configuration can typically provide a range extension of between 30% and 50% compared with the omnidirectional deployment.

System fade margin

In the heavy multi-path environment of most deployment regions, it is essential to include a fade margin if acceptable call quality is to be maintained everywhere in the cell. However, this margin can be considerably reduced if *antenna diversity* is provided [15, 16, 17], with fade margin reductions of close to 10 dB if the diversity antenna spacing allows full decorrelation. This benefit again needs to be weighed against some additional equipment costs but is usually worthwhile in cells with significant multi-path.

The discussion so far on cell coverage has been from a downlink perspective. In practice, it is obviously necessary to maintain a link budget that works both for the downlink *and* for the uplink at the cell edge. Some of the factors discussed, such as antenna height, benefit the uplink and downlink equally. Others, such as transmitted power, can only benefit the downlink, as mobiles will conform to a standardised maximum power class.

3.1.2 Capacity

As discussed in Chapter 1, the central tenet of cellular systems is that they use the spectrum available many times over the planned coverage area, with a key figure of merit being the cellular reuse factor. If every tenth cell uses the same frequency as the first cell in a deployment, the network is said to have a reuse factor of nine. This figure is important as, for a given amount of spectrum, it determines the maximum number of frequencies per cell and, hence, cell capacity. This, in turn, determines the number of sites for a given subscriber density. For instance, for a given amount of spectrum, moving from a reuse pattern of nine to three would enable three times as many carriers to be deployed in each cell. Factors determining the reuse pattern will be discussed further in Chapters 5, 6 and 7, but for macrocells it is usually simple geometry on the downlink – the ratio of the serving cell radius to the distance to the first cell which reuses that frequency – that determines the reuse factor for a given air interface.

It is useful at this point to recall the way in which circuit networks, historically used to carry voice or video traffic, are dimensioned. Regardless of whether the network is supporting fixed or mobile subscribers, a relationship needs to be established between

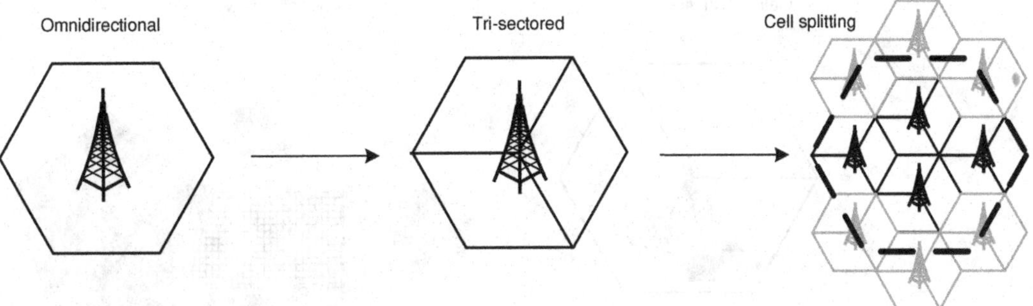

Figure 3.3 Cell capacity evolution

the transmission capacity n (the number of circuits available), the grade of service $B\%$ (the acceptable blocking level) and the offered busy hour load E, in Erlangs. Then for a GSM single carrier, and recognising that one time slot is needed to support the BCCH channel, the offered load in Erlangs can be found, by reference to Erlang 'B' tables [18] for seven circuits and 2% blocking probability, to be:

$$\text{offered load} = 2.93 \text{ Erlang}.$$

In a green-field network, the key requirement is to provide coverage over the planned region as quickly as possible and for the minimum financial outlay. GSM cells with radii up to 20 km are easily practical, depending upon the propagation characteristics of the deployment area and so, for a single carrier, omnidirectional cell and 25 mE demand, subscriber densities supported in such cells can be calculated as:

$$\text{subscriber density} = \frac{2.93\,\text{E}}{\pi.20^2.25 \times 10^{-3}} = 0.09 \text{ subscribers/km}^2.$$

These cells are often 'noise limited' in early deployments – i.e., the BTS signal falls close to thermal noise and can only just be detected at the cell edge – rather than being limited by interference from overlapping coverage of adjacent cells, which is more typical of mature networks.

Figure 3.3 shows a capacity evolution path, which it is often convenient to follow. As more subscribers join the network, capacity can be increased by converting the omnidirectional configuration to a tri-sectored site, with one or more GSM carriers in each sector. More detailed discussions in Chapter 5 will show that frequency allocations found in typical GSM licences will allow up to two carriers per sector and, exceptionally, four carriers per sector. In the latter case, capacities of up to 2.2 subscribers/km^2 are possible (2% blocking, 20 km cell radius and 25 mE demand), once the Erlang benefits of larger numbers of channels, known as 'trunking efficiency', are considered.

The move to the final configuration of Figure 3.3, by splitting each tri-sectored site into a number of smaller radius cells, will only be made once all of the frequencies available to the operator have been utilised and traffic has risen to the point where unacceptable blocking – the failure of subscriber calls because of insufficient system capacity – will soon occur. The concept of cell splitting is straightforward. If the cell radius, and thus

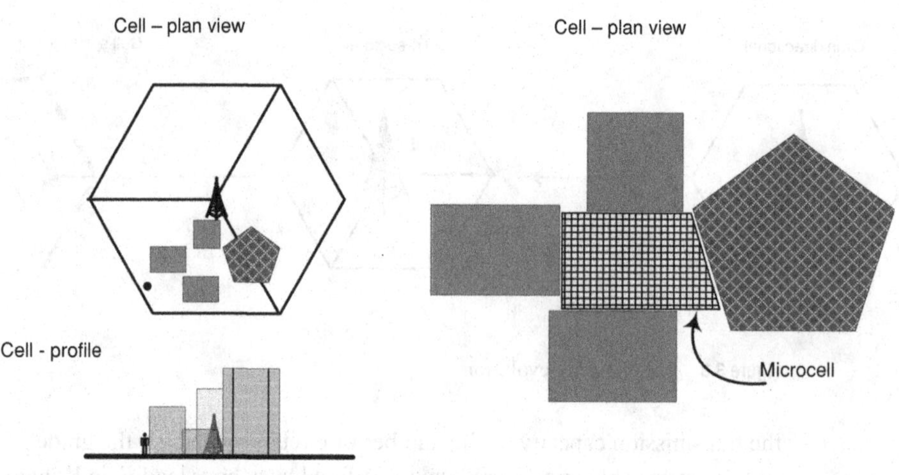

Figure 3.4 Limits on macro-cell size

cell area, is reduced, and the cell capacity (number of GSM carriers) is left unchanged, the subscriber density that can be supported must increase. However, cell splitting brings with it the need for new sites and new transmission to each of the sites. Discussion in Chapter 2 showed the number of sites to be one of the major elements in determining the operating expense (OPEX) of the network. Nevertheless, cell splitting is a necessary part of network evolution to address the subscriber densities found in urban and dense urban areas. By reducing cell radii to 100 m and with other assumptions as before, it is possible to achieve subscriber densities of up to 87 000/km^2. Cell splitting may occur a number of times as network traffic increases, depending on the population densities in regions to be covered and the ultimate penetration (the percentage of the population adopting the service) achieved. Cell radii as low as 100 metres are found in central city areas but at ranges much smaller than this other problems are encountered, as depicted in Figure 3.4.

The working assumptions for macrocell planning are that the antenna at the centre of the cell site is higher than most of the surrounding obstacles, and that buildings and other obstructions are 'small' compared with the cell area. These assumptions mean that there is usually at least partial line of sight to the subscriber, and when this is not the case reflection or diffraction from one or two buildings is sufficient to regain close to line-of-sight communication. Under these conditions there is still a dominant path to the user, with perhaps only 5 dB to 10 dB excess loss over the direct ray. For the situation shown in the left-hand diagram of Figure 3.4, the cell radius has now been reduced to the point where the buildings are large compared with the cell size and thus it is quite likely that one or more buildings will be directly between the user and the base station. This can easily result in losses of 40 dB or more relative to line-of-sight communication and gives rise to unpredictable coverage. In city centres containing large numbers of tall buildings an additional approach known as *microcellular* is therefore adopted.

Microcells are characterised by deployments that deliberately site the cell antenna *well below* the prevailing building height. The concrete and steel construction of city centre buildings means that there is no significant signal propagation *through* buildings; the buildings themselves thus define the cell boundaries in these deployments. These cells can achieve very high subscriber densities, depending upon the particular street geometry, and propagation is typically close to line of sight. Microcells are never the only type of deployment in a coverage region; macrocells must also be present to provide coverage, as users move between microcell canyons. Configuring algorithms to manage traffic between macrocells, microcells and picocells (even smaller cells *within* buildings) is a key aspect in delivering high-quality networks.

3.1.3 Network features and handset subsidies

So far, the discussion has concentrated on how to minimise the cost of rolling out the network infrastructure to support voice and basic telephony services, such as voicemail, call forwarding and other functionality hosted by the MSC that typically forms part of the basic feature bundle. However, rather like the early experience of fixed telephony operators with video calls, service revenues only take off if people find that the terminal cost is acceptable and are thus prepared to buy them. When rolling out services in developed countries, the practice has been to subsidise the cost of the handset and recover this over a fixed term contract. At some later time, generally as the technology matures to the point where handset costs drop dramatically, pay-as-you-go tariffs become available with the end user bearing all or most of the device cost but no fixed monthly tariff.

With most of the green-field deployments now taking place in countries where the annual income of the population is much lower than in the developed world, particular attention is being paid to this market. They are attractive from a revenue perspective because the governments concerned have often decided not to roll out copper to the home and rely on cellular coverage instead. However, operators need to be able to make a profit when the *average revenue per user (ARPU)* is in the region of \$10 per month. Planning for such networks is quite distinct from the corresponding activity in developed countries. Figure 3.5 shows the make-up of OPEX for a typical *low revenue market (LRM)*. The cost of handsets (within the subscriber acquisition category) assumes key importance in these markets and has given rise to a much publicised initiative from *Motorola, Nokia* and the *GSM Association (GSMA)* to develop lower cost handsets targeted at these markets. Handsets that sell for <\$30 are forecast for 2010 as a result of this work.

From an infrastructure perspective the challenges encountered are also different, as Figure 3.5 highlights. Whilst the same elements of cost are usually present as for the developed markets, the relative size can vary quite dramatically from one requirement to the next. The cost of back-haul usually features high on the list of OPEX components in most bids – yet on specific bids this can fall to almost zero – perhaps as a result of the way the sunk costs of a government optical-fibre network initiative are treated. Similarly, operators are familiar with the high costs of site rental. In general, these costs are much lower in developing countries except in rural areas, where bids sometimes have to cover the costs of road and power infrastructure to service the site, as well as site acquisition. Fundamentally different technical and business approaches are needed here, which address each situation creatively. The use of very large antenna heights

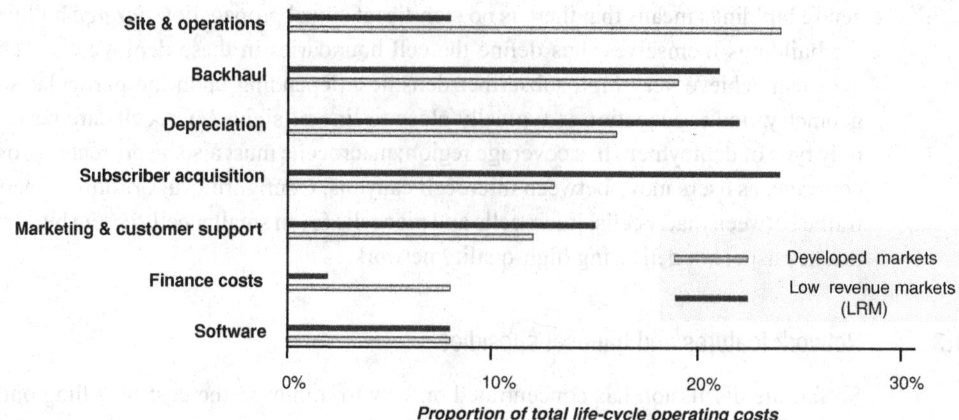

Figure 3.5 Life-cycle costs for low-revenue markets

with extended range cells of up to 70 km and beyond is fairly well established in such cases but creativity is often present in winning bids. For instance, it may require that the successful operator or vendor offsets the cost of site infrastructure in isolated locations by providing additional (non-telecoms) services, such as power, water, shops, etc. to the local community.

3.2 Planning for circuit multimedia services

A short discussion of circuit data services is included at this point for completeness. In practical terms, such services are becoming less relevant with time, as effective QoS management for packet services becomes a reality. Such bearers, with guaranteed low latency and assured bit rate, can efficiently deliver most media.

In GSM, the initial approach for data bearers was to retain the standard GSM GMSK modulation, and to optimise the channel coding and interleaving to offer a variety of circuit data bearers with rates from 2.4 kbits/s through to 14.4 kbits/s [19], all within the framework of the link budget for speech. This has clear advantages, in terms of planning simplicity. However, to provide a higher rate circuit bearer, it was decided to take advantage of the new 8PSK (phase shift keyed) modulation proposed for EDGE [20] and offer circuit EDGE bearers at rates up to 43.5 kbits/s alongside the packet services [21]. Because EDGE delivers this performance by transmitting up to 3 bits/symbol (rather than 1 bit for GSM), the increased C/I needed for satisfactory BER means that the highest bearer rates are only available to subscribers away from the cell edge. Thus, for all GSM circuit data solutions, the support of multimedia applications, albeit at fairly low bearer rates, does not alter the cell size from that planned for voice.

UMTS was conceived from the outset to be a multimedia-capable system. It offers a variety of bearers with rates between 15 kbits/s and 1920 kbits/s on the downlink and 7.5 kbits/s and 960 kbits/s on the uplink. However, unlike GSM, the maximum cell radius is directly impacted by the choice of bearer rate. For practical systems with finite mobile power and equal uplink and downlink data rates, it is usually the uplink that limits the maximum cell size. The signal to noise plus interference power ratio (C/I) required at the base site for satisfactory operation is dictated by

$$C/I = E_b/N_0 \cdot D,$$

where E_b/N_0 is the energy per bit to noise spectral density ratio needed for BPSK at the required BER in the particular multi-path environment and D is the data rate of the required bearer. Since the available transmit power from the mobile is finite, one can expect to see the maximum size of cells supporting applications requiring high-rate bearers to be significantly smaller than the corresponding cell where voice is the only service deployed. This result has significant consequences for operators; more cells will be required to support specified multimedia services over a given deployment area than a voice-only network. Worse yet, subscribers are only willing to pay a small premium over voice for applications such as video calls. The profit and loss benefits of deploying new services are, thus, far from transparent and will be discussed further in the next section on packet services.

3.3 Planning for packet multimedia services

It is expected that the current traffic mix, dominated by voice and other codec-driven traffic, such as video, will change to include an increasing proportion of traffic generated by applications such as VoIP, web browsing, gaming and, in the future, machine-to-machine communications. This traffic is characterised by aperiodic bursts of data separated by relatively long periods of inactivity, as shown in Figure 3.6. In a cellular system, dedicating one circuit bearer per user, matched to the peak data rate, to carry such traffic would be grossly inefficient. A far more attractive approach would be to convey such traffic over a packet bearer whose average throughput is well below the peak application rate, provided the delay and packet loss requirements are still met. The lower transmit

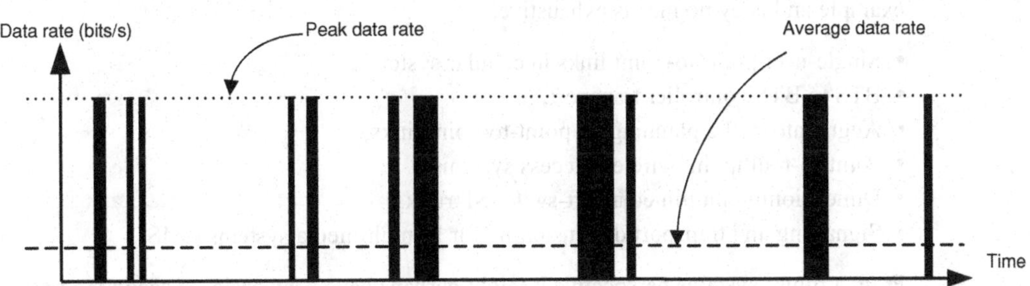

Figure 3.6 Traffic with arbitrary burst size and inter-burst intervals

power (associated with the reduced bearer rate) and corresponding reduced interference releases capacity in the cell for other users.

More generally (for example in the case of cellular back-haul or point-to-multipoint access to the home) it is possible to multiplex bursts of traffic from many users on to one bearer. This is conceptually easy to envisage by thinking of superimposed time-shifted versions of the traffic shown in Figure 3.6. However, because the inter-burst-intervals are aperiodic, there will inevitably be times when two or more users are contending for the same circuit. Another consideration for such a burst-multiplexing system is the choice of data rate for the bearer. If the rate is chosen to equal the peak data rate from the application source and two users are available to share the bearer, the system will still be very inefficient. Although it would be desirable to reduce the bearer rate to closer to twice the long-term average rate for the application, the bearer would then be unable to deal with the instantaneous data rate from the application, giving rise to traffic loss. Because of this and the contention issue, packet systems are invariably configured as a 'buffer–pipe' arrangement. This is, in essence, a means of smoothing the incoming data demand before it is carried by the bearer – the penalty for this benefit is the introduction of a queuing delay. The data buffer typically comprises memory address space and traffic is usually read in and out on a 'first-in–first-out' basis, clocked at a rate that fully loads the shared bearer or data 'pipe'. The trade-off between bearer rate reduction on the one hand and increased delay and packet loss on the other occupies most of the rest of this chapter.

It should be stressed that the packet modelling process does not replace the coverage and capacity planning discussed in Sections 3.1 and 3.2; rather it is an *addition* to these processes and precedes them. It is clear that the maximum cell radius is often dictated by the highest bearer rate in use and that the number of users that can simultaneously be supported falls as the bearer rate increases. Thus a high-level view of the planning process for packet multimedia systems is a complex trade-off and will comprise:

- Selection of the services to be deployed,
- Packet modelling of each application, to determine the highest bearer rate needed,
- Link-budget planning for the highest-rate bearer (needed by any of the applications),
- Cell-capacity planning,
- Deployment strategy.

Before addressing the detail of the packet-planning process, it is useful to consider where such planning will be required in future systems. The list below is included by way of example and is by no means exhaustive:

- Single-user point-to-point links in cellular systems,
- BTS to BTS controller transport,
- Aggregate traffic planning on point-to-point links,
- Point-to-multipoint wireless access systems,
- Dimensioning multimedia soft-switched trunks,
- Signalling and transport dimensioning for IP multimedia systems (IMS).

From a future systems perspective it might perhaps have been easier to scope the applications by defining them to be any information transport problem where QoS and transmission costs are important, as almost all media will be transported over packet networks in the future. But in some locations (e.g., in towns), over-provisioning of transport poses no problem and is cost-effective; this can be the case when bulk transport solutions such as multicarrier optical fibre networks are used.

In cellular systems for developed countries, it has been seen that costs related to the number of sites are the single largest OPEX component. The ability to reduce the number of sites by constraining the highest rate bearer used, consistent with delivering the required QoS, and, hence, maximising the cell radius is therefore a key consideration. Dramatic reductions in back-haul transport, the second-largest cost component, are also possible through statistical multiplexing of users within a traffic class. In other fields, such as point-to-point links or point-to-multipoint access, the ability to reduce the overall transport bandwidth required manifests itself as wider spacing between link sites (hence, less equipment) or the ability to share multipoint cell capacity amongst more users whilst still ensuring the correct operation of all user applications. However, these system cost reductions are only valuable if the savings can be delivered whilst offering *guaranteed* QoS. As ever, there is no such thing as an absolute guarantee and it is typical to plan against such outage criteria as:

- Queuing delay <500 ms (for 95% of delivered packets),
- Packet loss <2% (of total offered traffic).

The remainder of Section 3.3 will focus on the activities and processes necessary to deliver guaranteed QoS over packet bearers whilst offering significant cost reductions over current planning techniques.

3.3.1 Potential planning approaches

3.3.1.1 Measurement test beds

For some problems it is often quicker to build a test bed to measure performance rather than spend time developing potentially complex simulation models. Indeed, this was one of the initial approaches used by the authors. The test bed employed two computers interconnected by an Ethernet link of defined bandwidth. The source computer selected bursts from the relevant statistical population and converted these to sequences of time-stamped packets of constant size, which were passed to a buffer of predetermined capacity. The second computer was connected to the first by the Ethernet link and contained software to time-stamp packets on arrival. This enabled a distribution of delay times and packet loss rates to be established. However, it soon became obvious that this was not going to be a useful approach. Consideration of the characteristics for a representative Pareto distribution, together with the requirement to obtain accurate performance statistics for QoS confidence limits up to 99%, means that some 200 million instances of bursts drawn from the distribution would have to be captured. If the data originated from a web-browsing application, as might well be the case, an average burst arrival rate of 10 bursts/second would be appropriate and thus the test bed would need to run for:

$$= \frac{200 \times 10^{+6}}{10 \times 86\,400} \sim 200 \text{ days.}$$

This is clearly not a practical solution for applications with such heavy-tailed statistical distributions. When the test bed is run for shorter periods, of a few days, limitations in the results become apparent as very significant errors in the results for the probability of large delays. This is because insufficient samples have been captured for these long-delay cases, which occur only very infrequently.

3.3.1.2 Simulation

Simulation can sometimes be an acceptable solution, at least in some engineering investigations, and has again been used by the authors. To ensure the fastest possible run times, excellent general modelling frameworks, such as MATLAB®, were discarded in favour of specifically optimised 'C' code. Depending upon the tail parameters, this too can sometimes take a day or two to run but more typically one to two hours. A bigger disadvantage is the level of engineering expertise necessary to configure and run the tool and ensure that the results are 'sensible'. Nevertheless, run times of several hours are not amenable to the sort of 'what if' analysis proposed earlier.

3.3.1.3 Modelling

The limitations of the two conventional approaches discussed above in addressing the planning requirements for the new generation of applications mean that modelling is the preferred way forward. Unfortunately, classical queuing theory relies on a number of assumptions, which will not be valid in the majority of future planning scenarios [22]. Most notably, the convenient assumptions of an exponential burst size distribution and infinite buffer size are unlikely to be met. The authors have developed a proprietary tool, called *ARC*, specifically for this purpose and used within Motorola. It relies upon transformations to map the problem to a space that is amenable to analytic approaches. The *ARC* tool can address the applications that are currently foreseen in future networks. It allows the exploration of required bearer rates and buffer sizes necessary to deliver packet traffic within specified QoS parameters and confidence limits in the course of a few minutes. This process will be discussed further in Section 3.3.4.

3.3.2 The buffer-pipe model

The buffer-pipe model of Figure 3.7 is central to the management of packet traffic and it is important to optimise buffer size and bearer rates in order to minimise the cost of transport in future systems. So what drives these choices? A given application will only provide an acceptable user experience under certain conditions:

- A maximum delay time – the time taken for a burst leaving the user's device to arrive at the application server some time t_D later.
- An acceptable packet loss ratio l_R – the percentage of packets transmitted by the user that do not arrive at the server.

If the reader is in any doubt about this, recall the VoIP telephony experience when the internet is busy, preventing either or both of the packet loss and delay criteria from being met!

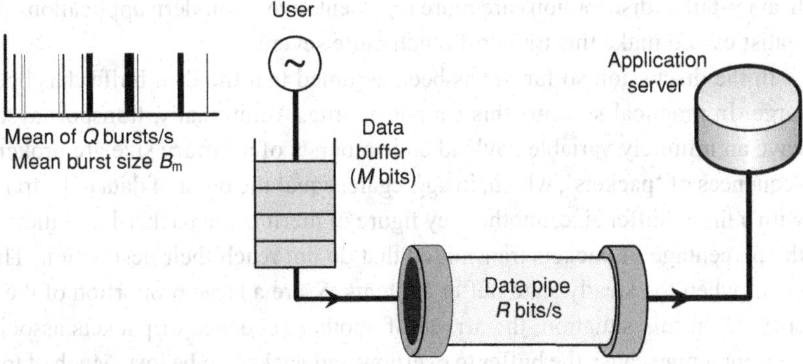

Figure 3.7 Buffer-pipe model used for queuing analysis

The delay time, t_D, comprises two parts. The first component, t_t, reflects the time taken for a burst of data to travel across the bearer, as typically the whole burst must be received at the server before it can be processed – i.e., it cannot be used on a byte-by-byte basis. An indication of the expected transport delay can be calculated from:

$$t_t = \frac{B_m}{R},$$

where B_m is the mean burst size in bits and R is the bearer rate in bits/s. This component of the overall delay time can be managed through the choice of bearer rate relative to the mean burst size.

The second part of the total end-to-end delay, t_b, recognises that, at any point in time in a system with significant load, the buffer is likely to contain some bursts waiting to be transmitted. If it can be assumed that the burst distribution and burst arrival rate exhibit Poisson behaviour, the steady state buffer contents, S bits, can be estimated using Markov chain techniques [22] and t_b can be evaluated.

Consider first a single application being handled by the buffer. If an application has a mean burst size B_m and a mean burst arrival rate of Q bursts each second, then the long-term average data rate generated by the application is $B_m \times Q$ bits/second. The most efficient use of the available transport would be to choose the bearer rate R, to equal $B_m \times Q$. However, by definition, this relationship only makes sense provided it is considered over the whole population of traffic events – i.e., is averaged over an infinitely long time period with a correspondingly large accumulation of bursts in the (infinite) buffer. In practical terms, this means that this ideal transport bandwidth, $B_m \times Q$ bits/second, can only be approached if the possibility of significant delays is acceptable (any new burst of data may have to queue behind a sequence of previously received bursts that have been delayed whilst above average bursts have been served). This is clearly not a practical solution and a trade-off has to be made between acceptable delay and a transport bandwidth larger than the long-term average data rate. For traffic conforming to Poisson arrival rate and burst distributions, and depending upon the delay and packet loss ratio that may be acceptable, the application traffic load may typically not exceed more than 70% to 80% of the available bearer rate. It will be shown later that heavier-tailed distributions are more representative of modern applications than Poisson statistics and make this trade-off much more severe.

In the discussion so far, it has been assumed that the data buffer has been infinitely large. In practical systems this cannot be true. Additionally, transport systems do not have an infinitely variable payload and so bursts of arbitrary size are broken down into sequences of 'packets', which, in aggregate, equal the burst of data to be transported. So with a finite buffer size, another key figure of merit is the packet-loss ratio, representing the percentage of packets transmitted that do not reach their destination. This loss may occur when the steady state buffer contents, S, are a large proportion of the total buffer size, M. In this situation, the arrival of another sequence of packets associated with a new burst may cause the buffer to overflow and packets to be lost. Much of today's traffic is routed using the *Internet Protocol* (IP) where, depending upon the physical bearer, packets of up to 1500 bytes can be carried. For a specific application (if it is highly interactive and thus requires low delay) the buffer size may deliberately be kept low and will require careful modelling if unacceptable packet loss ratios are to be avoided.

Modelling of the expected packet loss ratio is conceptually straightforward. Assuming that the steady state buffer contents, S, are known, the probability of packet loss occurring is simply the probability that B is greater than $(M - S)$, where B is the size of the current burst being transported and is a direct function of the application statistics. Depending upon the choice of packet size and the extent to which the burst size exceeds the available space in the buffer, $(M - S)$, the number of lost packets from a given burst can be calculated. By evaluating this figure over an appropriate sample of the traffic population, a good estimate of the packet-loss ratio can be obtained. It is clear that, once again, the packet-loss ratio can be managed to a desired level by the *designer's* choice of the mean traffic level $(B_m \times Q)$ relative to the bearer rate R and the choice of buffer size.

This short introduction to the concepts and parameters that influence the performance for services delivered over packet media has served to highlight the key issues. The devil, as is often the case, is in the detail. Central to the evaluation of both delay and packet loss is a knowledge of the steady-state buffer contents, S. As was discussed, this can only be analysed conveniently for Poisson processes, and in the next section it will be seen that these are atypical of most popular applications and so different modelling approaches need to be developed.

3.3.3 Characterisation of applications

In the introduction to Section 3.3, where the diversity of applications was discussed, it will have become clear that an accurate characterisation of the application is central to the delivery of guaranteed QoS. How is this characterisation best undertaken?

Figure 3.8 shows a structure that is convenient to adopt and is sufficiently general to address arbitrary applications. It is perhaps easiest to understand by using a real application, web browsing, as an example. The session level is driven by user behaviour, thus, a research worker might have several sessions a day, each lasting for 60 minutes or more, whereas the average user might only have one or two sessions a week, to check Hotmail, with each session lasting less than 10 minutes.

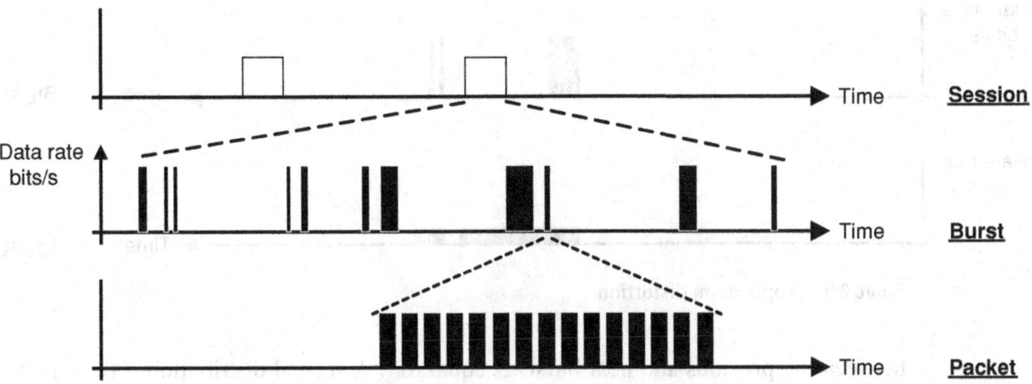

Figure 3.8 Application characterisation

Burst behaviour is heavily influenced by the particular application. In the case of web browsing, one might expect to see clusters of bursts with each representing one of a number of 'objects' on the selected page. The interval between the sub-bursts representing each object is unpredictable and will often reflect the fact that they are stored on different servers. Finally, the inter-burst interval is again very application dependent. For web browsing this might reflect the 'read time' that the user spends on the page; which, in turn, is influenced by whether the page is relevant and, if so, how much detail is on the page, and may be as much as 10 seconds. By way of contrast, machine-to-machine communications typically have much shorter inter-burst intervals, reflecting processing and transport delays.

The final layer is the packet layer. The number of packets per burst and the number of packets per second reflect the transport network parameters and in principle are of less interest. However, in practice this layer is important because it is at this level of granularity that the limitations of packet-based applications become apparent in terms of packet-loss ratios and packet delays.

3.3.3.1 Data collection approach

Planning for guaranteed QoS over packet bearers is a relatively new process and data describing all the applications of interest may not be available. As has been seen, the required data fall into two types, the session data and the burst data. The session data are of interest as they describe the number, duration and timing of sessions, enabling the number of sessions per user to be calculated both on average and in the 'busy hour' if this is relevant. The session data represent a load multiplier once the resources needed to support one instance of the application are known. Session data are best obtained by market surveys and cannot be characterised in the laboratory.

In contrast to session data, burst data are key to the definition of resources needed to support a single instance of a given application with a specified QoS. The parameters of interest are the burst-arrival rate and the burst size behaviour. These can be defined in a completely general way by two independent statistical distributions. A distribution, $P(t)$, can be defined to represent the probability that the inter-arrival interval (the time

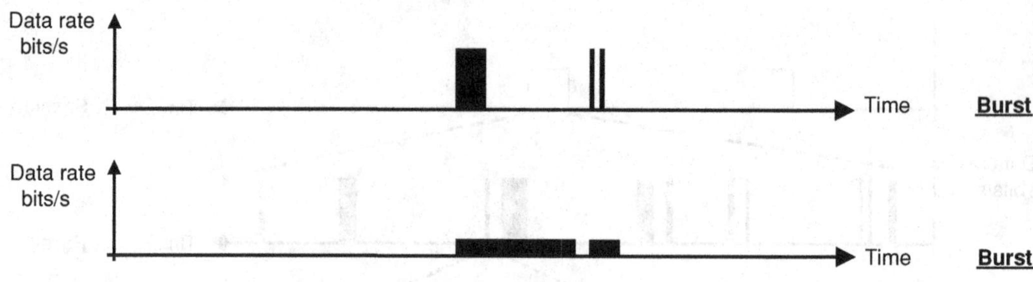

Figure 3.9 Application distortion

between the previous and next burst) is equal to t. A second distribution, $P(B)$ can be established to represent the probability that the next burst size equals B.

Returning to the example of web browsing, $P(t)$ will exhibit a range of values, which are influenced by:

- The driver data-rate capability on the master server,
- Communication times between the master and subsidiary servers,
- Processing and location times on the servers,
- System load (in the case of shared servers).

It is important to measure burst characteristics accurately in the absence of effects that can typically be introduced by the transport higher layers, such as TCP. This will allow the network operator complete freedom to choose QoS design goals and determine the allowable degradation from the ideal (infinite bandwidth) performance client–server transactions. These goals will be specified in terms of additional delay Δt and packet loss Δl, which is exceeded no more than x% of the time (e.g., 5%). It will be seen that making Δt and Δl arbitrarily small is very expensive in terms of transport bandwidth.

Data should therefore be collected over a network that has a much higher bandwidth than the driver capability of the application server, if distortion of the sort shown in Figure 3.9 is to be avoided. A narrow-band transmission link not only alters the inter-burst arrival interval but also delays the point at which objects can be displayed at the client device. This might be critical in ensuring the correct operation of some applications, such as interactive gaming. It is also important to ensure that the system is configured to prevent distortion of the basic application statistics by higher layer protocols. TCP, in particular, can introduce significant effects unless considerable care is taken.

3.3.3.2 Application statistics

It is appropriate to provide some insight into the statistical behaviour that can be expected from popular 'data' applications. Research that began to gather critical mass in the early 1990s identified that local-area network (LAN) traffic exhibited behaviour that differed from the Poisson arrival rate and Poisson service rates familiar to circuit telephony planners [23, 24]. A plot of packet rate vs. time for traffic exhibiting Poisson behaviour

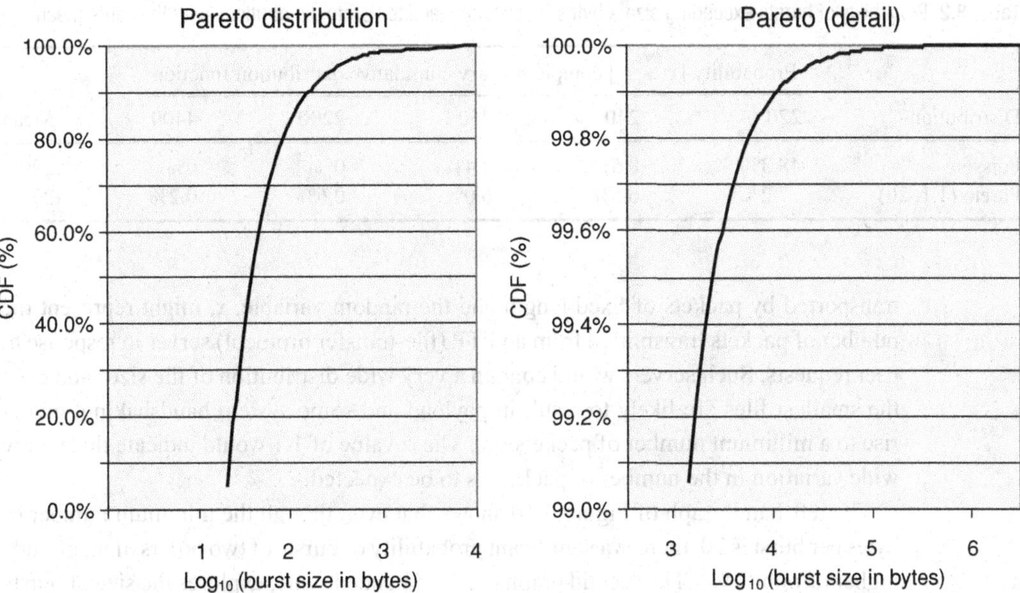

Figure 3.10 Burst sizes for traffic conforming to a Pareto distribution

may show bursty variations when viewed over short periods but when viewed over a longer period the plot will converge towards the average value. Such traffic also has the convenient characteristic that the aggregate traffic becomes smoother (less bursty) as the number of independent traffic sources increases. Packet LAN traffic, by contrast, demonstrates behaviour that is known as statistically '*self-similar*' and exhibits above-average packet rate bursts even when viewed over many orders of magnitude of time. Additionally, when the number of independent self-similar sources increases, it is found that the 'burstiness' of the aggregate traffic can *increase*. The significance of self-similar traffic, which can be characterised as having 'fractal' burst shapes that recur *regardless of the time period over which traffic is viewed*, should be understood. Over very short periods, these peaks could represent individual transactions. However, as the time window in which the fractal recurs gets longer (and self-similarity, in principle, occurs over an infinite period) the implication is that bursts are occurring above the long-term average packet rate for longer and longer periods. To look at the behaviour of LAN traffic another way, if this traffic were to originate from a single application, it would imply the possibility of almost infinitely large bursts of data being generated with a small but finite probability. This burst behaviour can be captured by 'heavy-tailed' distributions, such as those exhibiting Pareto or Weibull statistics.

Figure 3.10 shows the cumulative distribution function (CDF) for a *Pareto* burst distribution with a *location* parameter $x_m = 20$ and *shape* factor $\alpha = 1.1$ (for a discussion of Pareto and Poisson statistics and parameters, see, for instance, [25]). Suffice it to say that x_m is the minimum burst size expected and α is a measure of the shape, with small α corresponding to distributions with long tails – i.e., significant probabilities of very large burst sizes in this case. To link this to the real world, bursts will, in practice, be

Table 3.2. Probability of bursts exceeding size *X* bytes for exponential and Pareto distributions with the same mean

	Probability $[x > X]$ complementary cumulative distribution function					
Distribution	220	240	260	2200	4400	Mean
Poisson	48.3%	8.5%	0.4%	0%	0%	220
Pareto (1.1, 20)	7.2%	6.5%	6.0%	0.6%	0.2%	220

transported by packets of fixed length and the random variable, x, might represent the number of packets transmitted from an FTP (file-transfer protocol) server in response to user requests. Such servers would contain a very wide distribution of file sizes and even the smallest files are likely to result in payload and some system handshaking, giving rise to a minimum number of packets, x_m. The α value of 1.1 would indicate that a very wide variation in the number of packets is to be expected.

The left-hand graph of Figure 3.10 shows that even though the minimum number of bytes per burst is 20, there is a significant probability of bursts of two orders of magnitude higher (around 1%). The second graph, on an expanded scale, explores the size of bursts likely to be encountered in the last 1% of the CDF. Note that the shape of the curve is similar to the whole distribution and that there is still a 0.2% probability of bursts *exceeding* 4400 bytes. This is 20 times the mean of the distribution and 220 times the minimum value! This behaviour is perhaps more clear in Table 3.2, where it can also be contrasted with the normally used Poisson statistics that essentially predict no bursts larger than twice the mean value. Traffic with Pareto statistics makes life really difficult for packet network planners trying to design networks with guaranteed QoS!

Much effort has been devoted to investigating the origin of self-similarity. A key paper [26] argues that 'The self-similarity observed in packet traffic arises from aggregating individual sources, which behave in an on–off manner with heavy-tailed sojourn times in one or both of the states.' More detailed analysis in support of the same argument is presented in [27]. The self-similar behaviour of aggregate LAN traffic has also been observed in individual applications, such as web browsing [28, 29], modern variable-bit-rate video codecs [30, 31] and, most recently, in VoIP traffic, with voice traffic streams produced by codecs using silence compression and generating active (on) and inactive (off) periods alternately (e.g., G.723.1, G.729 B, GSM-FR) [32, 33].

3.3.4 Practical modelling methodologies

The discussion at the beginning of Section 3.3 made it clear that, to ensure the correct operation of real applications and services, it would be necessary to guarantee the network packet transport performance with the critical metrics being delay and packet-loss ratio. Armed with statistical descriptions of the applications to be deployed, how can this information be used to optimise a highly cost-sensitive parameter, such as the transport bearer bandwidth? Consider a single subscriber using a cellular system for web browsing. What bearer rate should be assigned, to ensure a satisfactory user experience?

The total delay t_D for any traffic utilising a buffer-pipe structure of the form shown in Figure 3.6 can be written:

$$t_D = t_t + t_b,$$

where t_t is the transport time and t_b is the buffer queuing delay. The transport time is the minimum delay that the user will experience before the first object (burst) is available for display from the server and, therefore, the bearer rate needs to be chosen to make this acceptably short. Often a useful starting point in the planning process is to allocate equal budgets to transport and queuing delays. Knowledge of the application statistics can then be used to choose a starting point for the modelling process. Thus, if it is desired to deliver 99% of the burst traffic within $t_D = 0.2$ seconds, and using the example statistics from Table 3.2:

$$\text{initial bearer rate} = \frac{2200 \times 8}{0.1} = 176\,\text{kbits/s}.$$

Similarly, a starting point for the buffer size would be for it to be set equal to the burst size used in the bearer-rate calculation, which will only be exceeded 1% of the time. The actual choices of bearer rate and buffer size are likely to be significantly different from these, depending upon the average rate of bursts per second. At one extreme, if bursts are infrequent compared with the time taken to empty the buffer, the bearer rate can be reduced, although it is unlikely to go much below 100 kbits/s. Conversely, a high intensity of bursts per second can cause the steady-state contents of the buffer to increase, with the corresponding requirements for larger buffer size and much increased bearer rates if the packet-loss requirements are to be met and delay budget is to be maintained. Therefore, bearer rates in the range from 100 kbits/s to 1 Mbit/s might be allocated to address an application having a mean burst size of less than 2 kbits.

Additionally, in practice, the delay of interest to the user is the delay between their initial action (e.g., a mouse click) and 'gratification' in the form of a refreshed web page. Therefore, as well as the downlink delay, the budget needs to include time for uplink transport from the user to the server (although this traffic is often much lighter), server-latency and processing delay, and possibly typical server-queuing delays. Because of the large implications for network costs, in terms of both the number of cells required to cover a given area (driven by smaller radius cells for higher bearer rates) and the number of users that can be supported in a cell (which reduces as the bearer rate is increased), one can expect to spend a considerable period optimising the trade-off between:

- The services deployed (which service limits the cell size?),
- Underlying cost per service instance,
- The default QoS offered for each application (implications on cost per instance),
- Profit and loss impact of application mix (cost vs. revenue per service instance).

Methods of quickly estimating the bearer-rate and buffer-size requirements for applications as a function of offered QoS are, therefore, needed to evaluate the impact on profit and loss as a function of service mix and QoS choices.

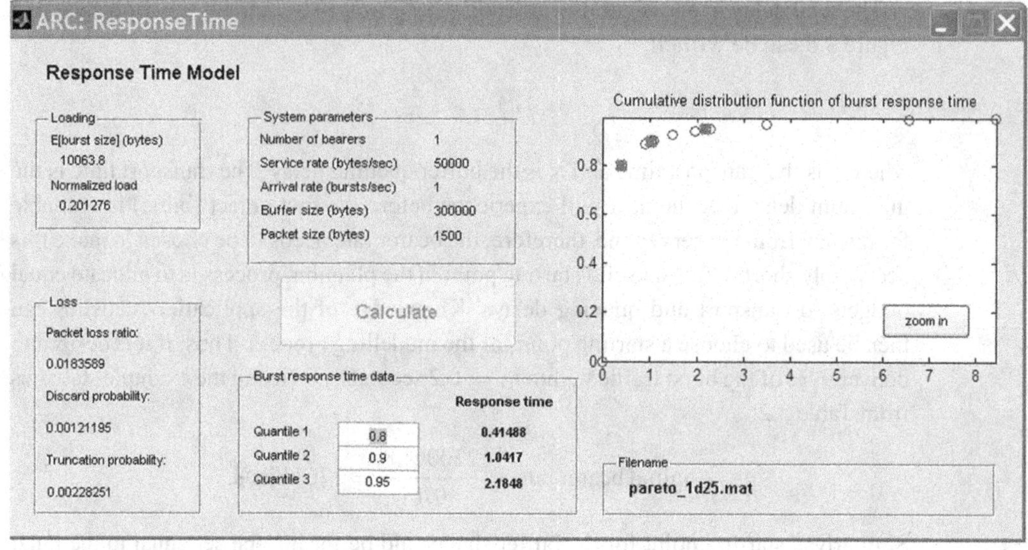

Figure 3.11 Packet modelling tools – Motorola ARC

3.3.4.1 The ARC tool

In Section 3.3.1 it was argued that a modelling-based methodology was the only practical approach to examine the trade-offs in the buffer-pipe model in anything approaching real time. In developing the *ARC* tool for this purpose, the target audience of business development staff, perhaps with a broad engineering background, was kept very much in mind. No specialist knowledge of the analytical techniques employed in the tool is needed. This approach should be contrasted with simulations that invariably have finite ranges over which the results are valid, and, therefore, require engineers with very specific (and expensive) backgrounds to generate the results.

Again, to cater for the target audience, the tool runs on a PC. It can either be operated via the graphical user interface (GUI) shown in Figure 3.11, or as an Excel 'plug-in' for automated optimisation routines. The GUI, with its display of delay versus the percentile of the samples lying within this limit, is particularly useful in exploring the sensitivity of the application to the proposed bearer rate and buffer combination. Given the application statistics, the *ARC* tool will typically enable the trade-off between capacity (network cost per application instance) and offered QoS to be fully explored in less than 30 minutes, with some variation depending upon the particular distribution parameters. The tool can thus be used by system vendors or operators to ensure that networks support profitable traffic.

Key input parameters to the tool include:

- Distribution type,
- Distribution parameters,
- Burst arrival rate,
- Bearer or service rate,

- Buffer size,
- Packet size,
- Number of bearers,
- Delay percentiles of interest.

Key outputs include:

- Response time delay (by percentile),
- Packet-loss ratio,
- Normalised load,
- Expected burst size.

The tool also has the facility to store distribution parameters as they are used, so that they may be quickly recalled for further analysis at a later date.

3.3.4.2 Modelling – the two-step process

It is useful to consider how modelling using the buffer-pipe paradigm fits into the total planning solution. To do this, consider again our ubiquitous mobile web browsing user. This user is one of many sharing the resource provided by a particular cell in a cellular network. The discussion in this section has established that, provided a user is given ownership of a buffer, and a wireless bearer of a guaranteed minimum bit rate and maximum delay, applications with known burst-arrival rates and arbitrary burst-distribution sizes can be optimally delivered whilst guaranteeing delivery within specified delay and packet-loss windows. Is this the only planning required? Clearly not – our user is one of many contending to access the finite resources of this cell and requires ownership of the specified bearer for the duration of the application session. If it is supposed for the moment that web browsing is the only application offered by the operator, this problem can be solved straightforwardly by:

- Evaluating the number of wireless bearers that can be supported by the cell,
- Calculating the probability that a bearer can be made available when requested.

If this contention issue sounds familiar, it is because it is exactly the problem that fixed-telephony engineers have been addressing for years using Erlang 'B' theory. So if the operator decides to offer the web-browsing service with only a 2% probability of being blocked (not available), the corresponding Erlang capacity for the available number of web-browsing bearers can be evaluated, which in turn will dictate the number of users that can be supported, given the average browsing session duration.

So, to deliver applications with a guaranteed quality of service, it is proposed first to model, using the application statistics, and then complete the calculation, using conventional Erlang 'B' methods! Is this reasonable? The approach becomes clearer when it is recognised as addressing two separate problems. The first step, addressed using the buffer-pipe model, transforms the application behaviour from one of arbitrary and varying burst bit rates with potentially long periods of inactivity to one of guaranteed maximum bit rates with shorter periods of inactivity. This is shown in Figure 3.12 and operates on a timescale of hundreds of milliseconds. If there was only ever one user

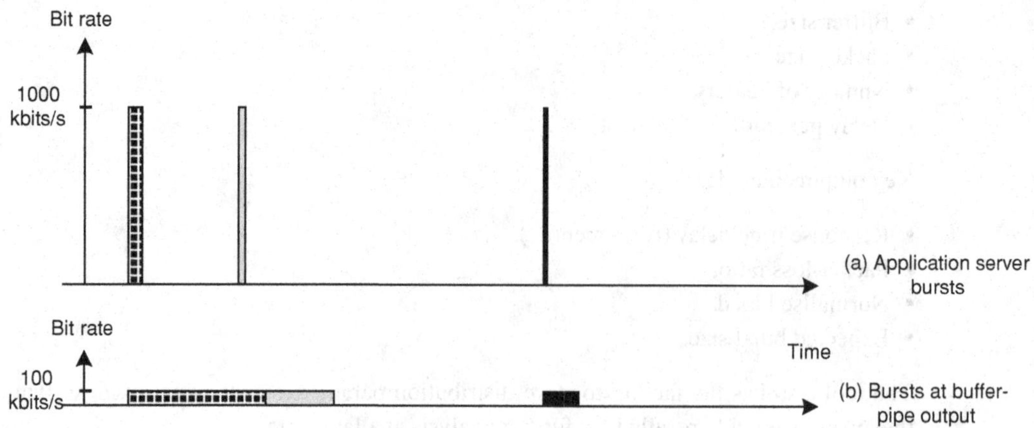

Figure 3.12 Application behaviour after buffer-pipe transformation

and the cell capacity was adequate, that would be the only calculation required, as the bearer would always be available. The second step will operate on a different timescale (perhaps tens of minutes) and can be treated separately. It addresses the likelihood of a suitable wireless bearer being available when requested, given the cell capacity and web-browsing session durations.

In practice, the operator will offer a number of different application classes and the total cell resource will have to be partitioned in a way that reflects the anticipated use of applications provided by market research in the planning phase (e.g., 10% gaming, 20% web access, 30% multimedia and 40% voice). Because of the 'non-linearity' introduced by the bearer capacity to Erlang translation, this partitioning is non-trivial and a computer-based iterative technique is needed.

Depending upon the nature of the dimensioning problem, it is not always necessary to consider the probability of blocking. For instance, if the transport requirements between the serving gateway and the PDN gateway of the evolved packet core are being addressed, the traffic will reflect the aggregated data from a number of mobiles. To ensure that the QoS needs of the different application classes are met, the total transport capacity will be evaluated as a series of parallel pipes. Thus, if the capacity of a number of cells served by one gateway is 3000 users, and using the application probabilities in the paragraph above, the pipe transporting gaming traffic would need to support 300 times the average burst rate of one gaming instance. Thus, even though individual gaming users come and go (as for users sharing a single cell) the transport pipes need to be dimensioned for the substantially invariant aggregate burst rate, which can be equated to one user who is always present but has a higher burst rate (although an unchanged burst size distribution). Example calculations of the number of users that can be supported in a cell and of dimensioning back-haul requirements are provided in Chapters 7 and 9, respectively.

A final consideration is the bearer class that should be selected. For applications that require the minimum air interface latency or the transmission of a fixed amount of data at regular intervals, the equivalent of a circuit bearer would be appropriate. However, if traffic of the sort shown in Figure 3.12(b) formed the payload, a bearer service that provided *guaranteed* access at a *minimum* specified rate on every frame *if* required *but* made it available to best-efforts users when it was not required would be more efficient. A hierarchy of bearer classes is typically provided by modern systems: the extended real-time polling service class of WiMAX is well suited to the web-traffic application, for example.

3.3.5 Multi-user packet transport configurations

Before concluding this chapter it is perhaps useful to look at practical deployment configurations, which can support multiple applications. To date, the single user model of Figure 3.7 has been assumed. This simple model is, indeed, appropriate for a large variety of real planning cases including:

- Single-application dimensioning for cellular mobile users,
- Dimensioning of individual point-to-point signalling links.

However, what would be the planning approach and configurations for use in the design of cellular back-haul or dimensioning of multimedia soft switch transport? Figure 3.13 shows what might be an appropriate configuration for multimedia cellular packet back-haul.

The underlying concept here is to recognise that although there will be an almost infinite number of application types, it will be possible to group them into a small number of classes that have similar QoS needs. They will, therefore, be able to share a common buffer and bearer without adversely impacting their behaviour. Note that this approach does not require all users in a specific class to use the same application or even application server. The *N* class 1 users in Figure 3.13, in this instance, are using one of two servers with packets associated with a particular user being addressed to the

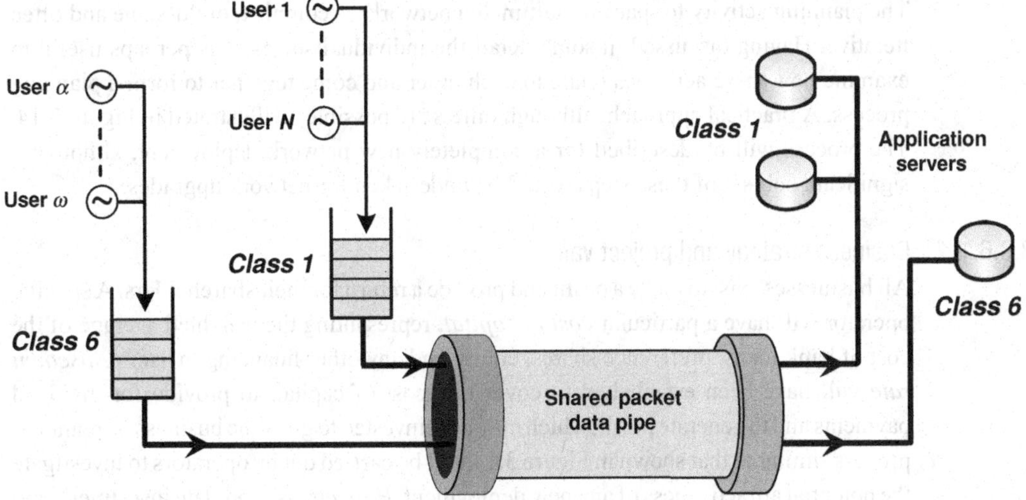

Figure 3.13 Example configuration for cellular packet back-haul

appropriate server (or servers) by means of the IP address and port number ('socket') in the packet header.

Dimensioning of the bandwidth required to guarantee QoS for the aggregate traffic in each service class is carried out using the techniques discussed in Section 3.3.4. This process is extended to all services classes and, once complete, defines the total data rate required in the shared pipe. Although omitted for clarity, in practice a scheduler is required to multiplex traffic from the individual application-class buffers on to the common bearer in a way that recognises their relative priority. De-multiplexing at the far end will occur using the socket information in the usual way.

In practice, the QoS specification applies to the round-trip performance between the user device, perhaps a mobile phone or PC, to the application server and back to the device. It is not only necessary to dimension the uplink and downlink bandwidths separately but also to recognise that each of these links may comprise several separate links in series. For instance, in a cellular system, the stages might comprise:

- Mobile to base station,
- Base station to base-station controller,
- Controller to core network,
- Core to IMS,
- Transport to application server.

In principle, a common QoS budget might be allocated to each stage and these stage budgets could be aggregated statistically to reflect the unlikelihood of each stage encountering its worst-case delay simultaneously. In practice, the bulk of the QoS budget will be allocated to the most expensive sections of the end-to-end transport. For a cellular system, this usually turns out to be cell and back-haul capacity, and the QoS budget will normally represent the sum of these effects without much statistical averaging benefit.

3.3.6 Outline of the overall packet-planning process

The planning activity for packet multimedia networks is complex, multi-stage and often iterative. Having discussed in some detail the individual stages, it is perhaps useful to examine how these activities relate to each other and come together to form a planning process. A practical approach, although others are possible, is illustrated in Figure 3.14. The process will be described for a completely new network deployment, although a significant sub-set of these steps would be undertaken for network upgrades.

3.3.6.1 Business strategy and project value

All businesses exist to make a profit and provide a return for their shareholders. A specific operator will have a particular *cost of capital*, representing the weighted average of the cost of bank loans, preference shares, equity and any other financing in use. A *discount rate* will have been established to cover this cost of capital, to provide for dividend payments and to generate profit, which may be reinvested to grow the business. A planning process similar to that shown in Figure 3.14 will be carried out by operators to investigate the potential attractiveness of any new deployment. *Projects*, as candidate investments are

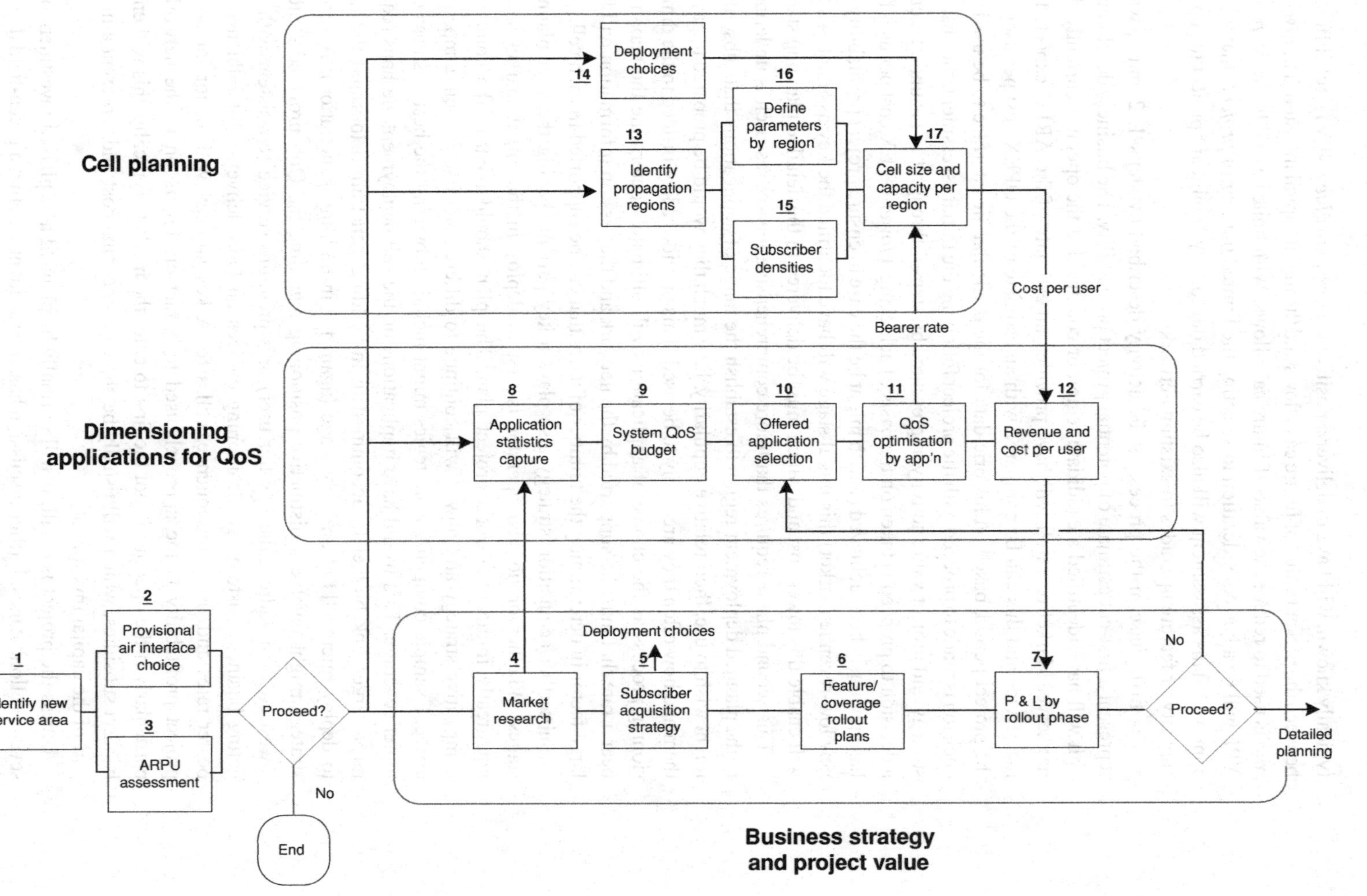

Figure 3.14 Network planning process

typically known, will have to deliver a positive *net present value* (NPV) – the difference between the present value of future cash flows and the initial capital invested. The *discount rate* is used to reduce the value of future cash flows. With finite capital to invest, *project NPV* will be a key consideration in selecting the limited number of *projects* that will be approved. This assessment will also be carried out for any significant project to support capacity or feature upgrades to existing networks.

The first phase in the process is the activity described by steps 1, 2 and 3, which represent a first-cut estimate of potential project profitability. The business development unit will have identified a candidate region or country that the operator currently does not serve. Market research will have provided an estimate of the ARPU expected for the service and this can be compared with an estimate of the OPEX cost per user over the project period, based on historic data for the particular air interface to be used. If a decision is made to proceed with the project following the initial assessment, the business strategy and project value activity, represented by steps 4 through to 7, is undertaken. If a decision to proceed is taken on the basis of achieving a *lower* OPEX cost per user than has previously been achieved (which might be the case to ensure project profitability in a new low-revenue market) this initial stage will have identified the key assumptions that will enable this lower cost and which must be delivered in the detailed planning stage.

This second phase requires that more comprehensive market research is undertaken in the planned deployment region to establish the range of applications that subscribers might wish to use, their relative popularity, how much they would be prepared to pay for them and how often they are likely to be used. This provides a key in-feed to the dimensioning process as the choice of applications will ultimately determine the maximum bearer rate that must be supported by the network and the session information will be a large factor in determining the number of users that can be supported in each cell.

Subscriber-acquisition strategy also plays a key part in shaping the later planning stages. Whether or not the operator is first to deploy in the target market should not greatly influence a good project plan. The plan should identify the company's competitive strategy and how it will continue to differentiate itself when competitors arrive. Example competitive strategies might include providing the best business customer coverage. This would have implications on the cell deployment, perhaps requiring a larger link budget to accommodate in-building penetration loss and the need to deploy microcells at an early stage (again for in-building penetration). A second strategy might involve a distinctive positioning regarding the QoS provided – either lowest cost (to address the low-tier user) or a premium service (consistent with targeting business customers). Positioning choices on QoS delivery will influence the bearer rates and, as a consequence, cell size. A key strategy with major impact on project profitability is the approach used to attract customers on to the network – particularly if the operator is not the first to enter the market. Typically, this will entail handset subsidies, which will usually be shown as a capital cost and represent a major part of the capital investment.

Unless the project is small, it will normally be planned as a phased investment over several rollout stages. Earlier discussion has already identified that a green-field deployment will initially comprise a rollout focused on coverage, with subsequent phases to provide further capacity in the regions that are ultimately expected to require this. The project profit and loss analysis will, therefore, be carried out for each rollout phase, although some costs, such as the investment to acquire planned sites for all stages will normally have to be included in the initial phase. This enables decisions to be made during the network deployment on whether to proceed to later phases based on *actual* profit and loss vs. plan for the early phases. One imagines that, at the time of writing, a number of operators worldwide are considering their expansion options for UMTS whilst assessing the likely impact of WiMAX and alternate strategies involving LTE.

3.3.6.2 Dimensioning applications for QoS

This process, comprising activities 8 through 12, is key to the success or failure of the operator's business in the marketplace. It determines the applications that will be rolled out and heavily influences the *actual* QoS that will be offered in response to the subscriber acquisition strategy. Steps 8 and 9 will establish end-to-end system budgets (in terms of delay and packet loss) for both the uplink and downlink to ensure that the required QoS will be delivered. A practical budget will be specified for each stage of the end-to-end transport with the major part of the budget being allocated to those stages that dominate the rollout costs (typically the mobile-to-BTS and BTS-to-BSC links or their equivalent). If necessary, characterisation of any new applications will take place and tools such as *ARC* will be used extensively to determine the bearer rates and buffer sizes for each stage of the link.

Stage 10, the offered application selection, compares the resources that would be needed in the most expensive segments of the link with the initial market research, indicating the popularity of such applications and the revenue they might generate. It may be apparent at this stage that one or two applications for which the subscriber has relatively low interest consume most resources. With the agreement of the strategy team, these could be removed from the feature rollout plan.

Step 11 examines whether a small change in the offered QoS could make a significant reduction in the use of resources in the most expensive stages of the end-to-end link.

3.3.6.3 Cell planning

Step 13 defines the characteristics for the project rollout area, or each region within the rollout area. If the project covers a significant area, it may be necessary to establish a series of planning 'regions' within which the propagation characteristics, subscriber densities, feature rollout, etc. can be considered uniform. Step 16 will determine the propagation exponent, together with appropriate values for shadowing and fade margin parameters for the region. The corresponding subscriber densities to be supported will be defined in Step 15 and will determine whether cells will be range or capacity limited. Step 14 is one of the few activities where the planner can make some choices; these will include decisions on antenna height, building-penetration loss, whether to deploy BTS diversity, etc., although in most cases only the last decision is typically not already constrained. All of these decisions then come together in Step 17 along with the key input from Step 11 regarding application-bearer rates.

The highest bearer rate from Step 11 along with the other cell-planning parameters will determine the maximum cell radius in coverage limited regions. Cells below this radius may be chosen if necessary to meet the subscriber density requirements. With the cell radius now defined, the numbers of BSCs and other network elements can be determined to quantify the capital costs. These costs will normally also include the civil works to build each cell site.

The application mix and the number of subscribers per cell can be combined with the bearer rates necessary to support each application instance (evaluated in the dimensioning activity) to determine a total back-haul requirement for each cell and for the other links in the network. These, and other recurring costs such as site rental and maintenance, can be combined with capital depreciation and finance costs over the project life to give a project cost per user for the region. Activities 14 through 17 will then be repeated for each of these homogeneous regions. The cost per subscriber for the project will then be established by forming a weighted average reflecting the relative areas of each region. It may be that the project NPV goals are not met with the initial strategy. It will then be necessary to revise some parameters in the project until this goal is met. Steps 5, 10 and 14 will probably be the focus of activities to improve project NPV.

Steps 4 through 17 will be repeated for a project with a phased rollout.

Although it is quickest and most convenient to carry out this planning process using analytic tools such as *ARC*, it should be stressed that exactly the same process would be followed if simulators were used to carry out the buffer and bearer optimisation.

3.4 Planning for 802.11x deployment

In the introduction to Section 2.1, the driver for the development of cellular networks was identified as '... to enable *people* to communicate whilst they were away from their 'normal' fixed-line home or office phones,' with the implicit requirement for connectivity *wherever* people were. The corresponding terms of reference for IEEE 802.11 [34] state:

> The 802.11 WG's charter is to develop physical layer and MAC sub-layer specifications for wireless local area networks (WLANs) carried out under project authorization requests (PAR) approved by the IEEE Standards Board and assigned to 802.11 WG.

The expectation at the inception of the standard in 1990 was that the 802.11 equipment would be deployed to provide an alternative to *local* connectivity provided by *wired* LANs, such as Ethernet. Only the physical (PHY) and medium access control (MAC) layers were to be specified [35], as user equipment (PCs, etc.) would be quasi-stationary; i.e., they would be used in one location, taken to another and then used in the new location. There was no expectation of use in transit and, hence, no need for the cellular concept of handover. In terms of the IEEE 802.11 standard, this is still true today, although various companies (outside of the IEEE standards body) are investing a lot of effort to reduce the dead period (the period of no connectivity) as a user moves between overlapping areas of LAN coverage. The driver for this is the prospect of supporting real-time applications, such as VoIP.

The 802.11 series of WLAN standards was designed to be deployed in the *industrial, scientific* and *medical* (ISM) bands. These bands exist in most countries and are usually *licence exempt*. This means, in practice, that the spectrum will be shared (without frequency planning) by many users. In Europe, equipment to be deployed in this band is governed under *The Radio Equipment and Telecommunications Terminal Equipment Directive 99/5/EC* (R&TTE Directive) [36]. A useful document specifying the spectrum allocated to the ISM band (2.4 GHz to 2.4835 GHz) and clarifying the typical allowed radiated powers (100 mW), together with other information, is published by UK Ofcom [37]. A second ISM band at 5 GHz is also defined [38] and there is a related UK Ofcom document [39]. Although the references here have focused entirely on the European market, there are corresponding documents published by the US *Federal Communications Commission* (FCC) and equivalent agencies throughout the world.

3.4.1 Coverage planning

During the discussion on cellular coverage planning, much importance was attached to the use of statistical models that could usefully describe wide-area propagation, provided the appropriate model was selected (urban, suburban, rural, etc.). Modelling of indoor coverage is much more complicated. A number of studies [40, 41, 42] have shown that it is possible to model indoor propagation accurately but that it is heavily dependent on:

- The location of doors, windows and furniture (and these need to be accounted for in the model),
- The electrical properties of the building materials,
- The number of walls in the region to be covered,
- The type of walls (plasterboard partition, brick, stone, etc.).

For practical network deployment, these models are not useful, as they would require a detailed information-gathering campaign about the specific locations in which 802.11 coverage was to be provided and skilled staff to interpret the results. Despite the absence of models equivalent to those used in macrocell planning, the situation is not as serious as it might seem, as Wi-Fi ‘base stations’ are physically very small and can easily be repositioned in the event of coverage holes. Instead, network planning is carried out on the basis of typical coverage, which will include some allowance for losses associated with internal partition walls. The 802.11 access-point vendors usually provide ranges to be expected under these conditions and these are often shown as a function of user data rate vs. range and the type of access point (802.11a, 802.11g, etc.) because of the impact of operating frequency on range. For user data rates greater than 10 Mbits/s, ranges in the region of 10 m to 30 m can be expected, although coverage up to 100 m can be possible at the lowest user rates [43]. In most practical deployments, stone or brick load-bearing walls, which typically exhibit losses in the region of 20 dB to 40 dB, usually define an 802.11 access point cell boundary.

3.4.2 Capacity planning

A fundamental difference between cellular systems and 802.11 access networks is that cellular systems are designed to guarantee a specific grade of service. For voice systems, the network will typically be planned to provide 2% blocking or less and 95% area coverage; this is delivered through the use of spectrum dedicated for this purpose (e.g., the GSM band) and licensed to specific operators, who will usually pay a significant amount of money for the privilege. In contrast to cellular networks, 802.11 systems operate in the ISM bands, as discussed above, and may share this band with many other users. What is the impact of using shared spectrum and what does it mean for 802.11 users?

Central to the effectiveness of the 802.11 series of standards is the design of the physical (PHY) and media access control (MAC) layers. These address two specific classes of interferers found in the ISM band. The first class of interferers are those who are using the ISM band for other purposes – a key user in this class is the microwave oven user – that is, almost all of us. The second class of interferers are other users of an 802.11 system.

The distinguishing characteristic of the first class of users is that they are independent; i.e., provided they conform to the general licence conditions for the ISM bands they can use this spectrum in any way they wish. A recent paper looked at the impact of microwave-oven use on Bluetooth communications [44]. It showed that whilst the nominal operating frequency of these devices was 2.45 GHz (in the middle of the ISM band) with a bandwidth of 2 MHz, during the period immediately after switch on, the spectrum could be as wide as 20 MHz and its centre frequency might move around from its nominal frequency by as much as 8 MHz. The 802.11 system has two mechanisms to reduce the impact of such interference. The first is known as 'clear channel assessment'. This looks at the noise power in each of the available channels (three 20-MHz channels at 2.4 GHz and up to 19 20-MHz channels at 5 GHz) and chooses the one with least interference power. The second mechanism is the ability to select from a number of modulation schemes at the physical layer. For instance, 802.11g offers modulation schemes from 1 bit/symbol (BPSK) through to 6 bits/symbol (64QAM) and from discussion in earlier chapters it is clear that this will enable the system to operate with acceptable bit-error rates over a signal range of almost 20 dB. This flexibility, combined with the attenuation afforded by load-bearing walls and free space, means that 802.11 systems can normally manage this class of interference, albeit at reduced user rates in some cases.

The second source of interference is self-interference. This might arise from other access points in the same area using the same channel (on the downlink) or two or more users trying to access the same access point at the same time (on the uplink). The downlink interference problem points to there not being much benefit in providing more access points than there are available ISM channels in a given coverage area (room, etc.), although propagation loss and load-bearing wall attenuation will often allow these same channels to be reused in an adjacent room. For the uplink, 802.11 incorporates a number of protocols to avoid and manage collisions when these do occur. These include:

- Carrier sense multiple access with collision avoidance (CSMA/CA),
- Request-to-send and clear-to-send (RTS/CTS),
- Back-off following collision.

The first of these is the same approach as is used on Ethernet. The RTS/CTS protocol causes a potential user to wait for authorisation and transmit only following receipt of CTS, which also serves to cause other users to wait. The last mechanism causes users to wait a random number of time slots (1 to 31) after their last (collision) transmission before retrying.

Hopefully, the foregoing discussion has made it clear that the PHY and MAC layers of 802.11 are sufficiently intelligent to manage interference that is likely to occur in the ISM bands efficiently. The only downside of shared spectrum use is that there can be no guarantees of quality of service, particularly under heavy load conditions – commercial Wi-Fi hot spots offer access to the Internet with no QoS guarantees in terms of user data rate or delay. Having said that, most applications are bursty and very acceptable user access is provided, albeit coverage, even in the intended coverage area, can often be patchy. Published data from Cisco [43] indicates 'cell' capacities for 802.11b, 802.11g and 802.11a as 6 Mbits/s, 22 Mbits/s and 25 Mbits/s respectively for each available 20-MHz channel.

The planning procedure for indoor systems is, therefore, fairly straightforward, with cell radii being determined by the maximum data rate required and 'cell' boundaries often defined by load-bearing walls in enterprise buildings. Capacity can be increased by adding access points up to the number of channels available. For outdoor Wi-Fi mesh systems the situation is more complicated with wireless back-haul typically also needed from each access point to the nearest convenient fixed broadband access. Some of these considerations are discussed in [45, 46] and will be addressed in Chapter 8, together with a detailed worked example.

References

1 E. M. Rogers, *Diffusion of Innovation*, 4th edition (Free Press, 1995).

2 *The Business Case for Fixed Broadband Wireless Access based on WiMAX Technology and the IEEE 802.16 Standard* (WiMAX Forum™, 2004).

3 IEEE Vehicular Technology Society Committee on Radio Propagation, Coverage prediction for mobile radio systems operating in the 800/900 MHz frequency range, *IEEE Transactions on Vehicular Technology*, **31** 1 (1988) 57–60.

4 M. Hata, Empirical formula for propagation loss in land mobile radio services, *IEEE Transactions on Vehicular Technology*, **29** 3 (1980) 317–325.

5 V. Erceg, L. J. Greenstein, S. Tjandra *et al.*, An empirically-based path loss model for wireless channels in suburban environment, *IEEE Journal on Selected Areas in Communications*, **17** 7 (1999) 1205–1211.

6 V. S. Abhayawardhana, I. J. Wassell, D. Crosby, M. P. Sellars and M. G. Brown, *Comparison of Empirical Propagation Path Loss Models for Fixed Wireless Access Systems*, UK Ofcom Study Ref: AY4463 (2004).

7 D Har, A. M. Watson and A. G. Chadney, Comment on diffraction loss of rooftop-to-street in COST 231-Walfisch–Ikegami model, *IEEE Transactions on Vehicular Technology*, **48** 5 (1999) 1451–1452.

8 J. M. G. Linnartz, *Wireless Communication – The Interactive CD-ROM*, 2nd edition (Baltzer Science Publishers, 1997).

9 Andrew, *Optum™ Automated Cell Planner Tool*, www.andrew.com/products/measurement_sys/optum/.

10 Aircom International, *ENTERPRISE Suite*, www.aircom.co.uk/.
11 J. Meinilä, T. Jämsä *et al.*, A set of channel and propagation models for early link and system level simulations, *IST-2003–507581 WINNER*, www.ist-winner.org.
12 ETSI, *Digital Cellular Telecommunications System (Phase 2+); Radio Transmission and Reception*, GSM 05.05 version 7.1.0, Release 1998.
13 A. Hecker, M. Neuland and T. Kuerner, Propagation models for high sites in urban areas, *Advances in Radio Science*, **4** (2006) 345–349.
14 P. Schneider, F. Lambrecht and A. Baier, *Enhancement of the Okumura-Hata Propagation Model Using Detailed Morphological and Building Data*, IEEE Document 0-7803-3692-5/96 (1996).
15 E. P. Mogensen, *GSM Base-Station Antenna Diversity Using Soft Decision Combining on Up-link and Delayed-Signal Transmission on Down-link*, IEEE Document 0-7803-1266-x/93 (1993).
16 T. Chen, M. P. Fitz, W.-Y. Kuo, M. D. Zoltowski and J. H. Grimm, A space-time model for frequency non-selective Rayleigh fading channels with applications to space-time modems, *IEEE Journal on Selected Areas in Communications*, **18** 7 (2000) 1175–1190.
17 A. I. El-Saigh and J. W. Burnett, *Evaluation of diversity performance in GSM* (IEE, 1996).
18 H. Leijon, *Extract from the Table of Erlang Loss, ITU*, www.itu.int/home/index.html.
19 ETSI, *Digital Cellular Telecommunications System (Phase 2+); Physical Layer on the Radio Path, General description*, GSM 05.01 version 7.1.0, Release 1998.
20 ETSI, *Digital Cellular Telecommunications System (Phase 2+); Modulation*, 3GPP TS 05.04 version 8.4.0 (2001–11), Release 1999.
21 ETSI, *Digital Cellular Telecommunications System (Phase 2+); Physical Layer on the Radio Path*, 3GPP TS 05.01 version 8.6.0 (2001–11).
22 T. Janevski, *Traffic Analysis and Design of Wireless IP Networks* (Artech House, 2003) pp. 91–112.
23 W. E. Leland and D. V. Wilson, High time-resolution measurement and analysis of LAN traffic: implications for LAN interconnection, *Proceedings of the 10th Annual Joint Conference of the IEEE Computer and Communications Societies (INFOCOM 1991)* pp. 1360–1366.
24 W. Leland, M. S. Taqqu, W. Willinger and D. V. Wilson, On the self-similar nature of Ethernet traffic, *IEEE/ACM Transactions on Networking*, **2** 1 (1994) 1–15.
25 R.-D. Reiss and M. Thomas, *Statistical Analysis of Extreme Values from Insurance, Finance, Hydrology and Other Fields* (Birkhäuser, 2001).
26 P. Pruthi and A. Erramilli, Heavy-tailed on/off source behavior and self-similar traffic, *IEEE International Conference on Communications, ICC 1995*, Seattle.
27 W. Willinger, M. S. Taqqu, R. Sherman and D. V. Wilson, Self-similarity through high-variability: statistical analysis of Ethernet LAN traffic at the source level, *IEEE/ACM Transactions on Networking*, **5** 1 (1997) 71–86.
28 M. E. Crovella and A. Bestavros, Self-similarity in World Wide Web traffic: evidence and possible causes, *IEEE/ACM Transactions on Networking*, **5** 6 (1997) 835–846.
29 H. Choi and J. O. Limb, A behavioural model of Web traffic, *Proceedings of the Seventh International Conference on Network Protocols (ICNP'99)*, (1999).
30 H. Hassan, J. Garcia and O. Brun, Generic modelling of multimedia traffic sources, *HET-NETs 2005 Third International Working Conference, Performance Modelling and Evaluation of Heterogenous Networks* (2005).

31 S. Ledesma and L. Derong, Synthesis of fractional Gaussian noise using linear approximation for generating self-similar network traffic, *Computer Communication Review*, **30** (2000) 4–17.
32 T. D. Dang, B. Sonkoly and S. Molnar, Fractal analysis and modelling of VoIP traffic, *NETWORKS 2004*, Vienna, Austria (2004).
33 L. Ding and R. A. Goubran, Speech quality prediction in VoIP using the extended E-Model, *IEEE Globecom* (2003) 3974–3978.
34 S. J. Kerry, Operating rules of IEEE Project 802 – Working Group 802.11, wireless LANs, *IEEE 802.11–00/331r4*, www.ieee802.org/11.
35 IEEE, Standard for local and metropolitan area networks – part 11: wireless LAN medium access control (MAC) and physical layer (PHY) specifications, *IEEE Standard 802.11g* (2003).
36 ETSI, Wideband transmission systems; technical characteristics and test conditions for data transmission equipment operating in the 2.4 GHz ISM band and using spread spectrum modulation techniques, *ETS 300 328*.
37 UK Radio Communications Agency, UK interface requirement 2005 wideband transmission systems operating in the 2.4 GHz ISM band and using spread spectrum modulation techniques (version 1.0), *98/34/EC Notification Number 2000/274/UK* (2000).
38 ETSI, Broadband radio access networks (BRAN); 5 GHz high performance RLAN; part 2 harmonised EN covering essential requirements of article 3.2 of the R&TTE Directive, *Draft ETSI EN 301 893*.
39 UK Radio Communications Agency, UK interface requirement 2006 short range, broadband, data services (high performance RLAN) operating in the frequency range 5150–5725 MHz (version 1.02), *98/34/EC Notification Number: 2002/0254/UK* (2002).
40 B. S. Lee, A. R. Nix and J. P. McGeehan, Indoor space-time propagation modelling using a ray-launching technique, *11th International Conference on Antennas and Propagation* (2001).
41 B. Sujak, D. K. Ghodgaonkar *et al.*, Indoor propagation channel models for WLAN 802.11b at 2.4 GHz ISM Band, *2005 Asia-Pacific Conference on Applied Electromagnetic Proceedings* (2005).
42 D. Dobkin, Indoor propagation and wavelength, *WJ Communications*, **1** 4 (2002).
43 Cisco Systems, *Capacity, Coverage and Deployment Considerations for IEEE 802.11g* (Cisco Systems, Inc., 2005).
44 T. W. Rondeau, M. F. D'Souza and D. G. Sweeney, Residential microwave oven interference on Bluetooth data performance, *IEEE Transactions on Consumer Electronics*, **50** 3 (2004) 856–863.
45 Sputnik Inc, *RF Propagation Basics* (Sputnik Inc, April 2004).
46 M. Hope and N. Linge, Determining the propagation range of IEEE 802.11 radio LANs for outdoor applications, *LCN 1999* (1999).

Appendix 3.1 Propagation with significant ground-reflected rays

In most environments, line-of-sight rays associated with free-space propagation are accompanied by a reflected ray of some magnitude. This causes the fall off of power with distance d to increase from $1/d^2$ to $1/d^4$ or higher. The point at which propagation transitions between these two behaviours is greatly influenced by the base-station antenna

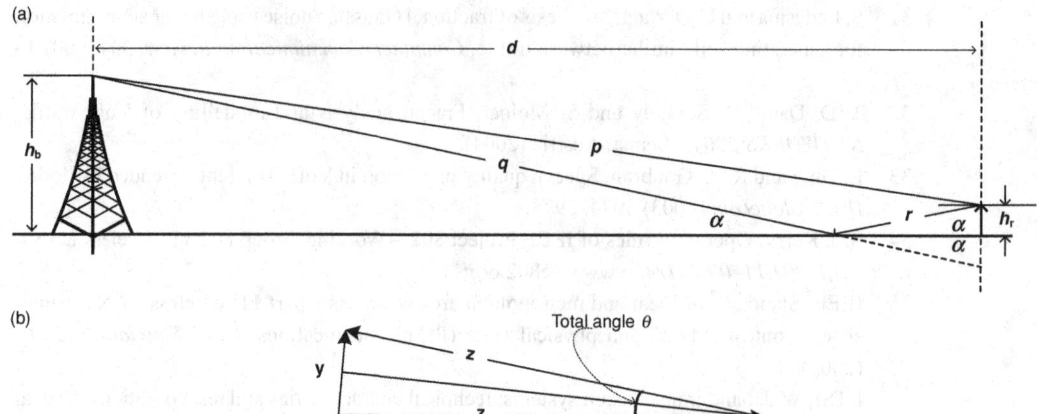

Figure 3.15 The two-ray model

height is known as the 'turnover point'. This Appendix develops a simple two-ray model and uses some elementary maths to explain the origin of this behaviour.

Figure 3.15(a) shows a base station of height h_b, which transmits a signal at a power P_T watts through an antenna of gain G_T (relative to isotropic) to a user at height h_r some d metres away. The user's antenna gain is G_R and the wavelength of the propagating signal is λ. In this model, there are two rays; a direct ray p and a ground-reflected ray qr. The distance d is large compared with the transmitter antenna height h_b, so grazing incidence can be assumed at the point of reflection (α is small). Thus almost perfect reflection occurs, where the signal amplitude is unaltered and the phase is reversed (i.e., changed by π radians). Additionally, since α is small, $p \cong (q + r)$; as a result, the propagation loss of the direct and reflected rays is essentially the same.

Overall, therefore, the strengths of the two received rays can be considered equal. In terms of power, each is received at a level given by:

$$P_r = \frac{P_T G_T}{4\pi d^2} \cdot \frac{G_R \lambda^2}{4\pi} = P_T G_T G_R \cdot \left(\frac{\lambda}{4\pi d}\right)^2.$$

The corresponding amplitude (z) of each component is given by:

$$z = (P_r \cdot \eta_0)^{1/2},$$

where η_0 is the impedance of free space and equals 120π.

z may then be written as:

$$z = k \cdot \left(\frac{\lambda}{4\pi d}\right), \qquad \text{where } k = (P_T G_T G_R \cdot 120\pi)^{1/2}.$$

Because p and $(q + r)$ are very nearly but not quite equal, a phase difference θ arises between the two signals (see Figure 3.15(b)) given by:

$$\theta = \frac{2\pi}{\lambda} \cdot ((q + r) - p),$$

where $(q + r) = \left(d^2 + (h_b + h_r)^2\right)^{1/2}$ and $p = \left(d^2 + (h_b - h_r)^2\right)^{1/2}$.

Then

$$\theta = \frac{2\pi d}{\lambda}\cdot\left\{\left(1+\frac{(h_b+h_r)^2}{d^2}\right)^{1/2} - \left(1+\frac{(h_b-h_r)^2}{d^2}\right)^{1/2}\right\}.$$

For $d > 10h_b$, the binomial expansions of the form $(1+\delta)^{1/2}$ can be simplified to $(1+\delta/2)$ with little error and:

$$\theta = \frac{2\pi d}{\lambda}\cdot\left\{\frac{2h_b h_r}{d^2}\right\} = \frac{4\pi h_b h_r}{\lambda d}$$

Because of the phase inversion on reflection, the rays add destructively as shown in Figure 3.15(b). The resulting vector, y, can thus be expressed as:

$$y = 2z\sin(\theta/2) = 2k\cdot\left(\frac{\lambda}{4\pi d}\right)\cdot\sin\left(\frac{2\pi h_b h_r}{\lambda d}\right).$$

The net received power $P_{\text{net}}(= y^2/120\pi)$ can then be expressed as:

$$P_{\text{net}} = 4P_T G_T G_R\cdot\left(\frac{\lambda}{4\pi d}\right)^2\cdot\sin^2\left(\frac{2\pi h_b h_r}{\lambda d}\right).$$

It is instructive to understand the behaviour of the resulting power P_{net}. For $d = 12h_b h_r/\lambda$, known as the 'turnover' point, the sine^2 term $= 1/4$ and the equation for P_{net} simplifies to:

$$P_{\text{net}} = P_T G_T G_R\cdot\left(\frac{\lambda}{4\pi d}\right)^2.$$

At this point, the path loss is equal to the free-space path loss.

Either side of the turning point, the propagation law alters. As d increases beyond this distance, θ becomes smaller, and for $\theta/2 < 0.3$, $\sin(\theta/2)\approx\theta/2$, and under these conditions the net received power may be approximated by:

$$P_{\text{net}} = P_T G_T G_R\cdot\frac{h_b^2 h_r^2}{d^4}.$$

For d less than the 'turning point' distance, the path difference quickly increases and the characteristic interference pattern of sharp nulls and peaks of constructive interference can be anticipated. This change in range also increases the angle of incidence of the ground-reflected ray with the result that the magnitude of the reflected ray reduces, gradually reducing the power law exponent.

It is thus clear that there are two distinct regions of propagation. For GSM macrocells with antenna heights in the range 5–15 m and receivers 1 to 2 m above ground, the onset of monotonic power fall-off in line with the $1/d^4$ law is predicted for ranges in the region of 300 m to 2 km. In practice, the fourth-power fall-off can be modified by additional effects, such as shadowing, to give rise to power laws in excess of this.

At ranges closer than the turnover point, coverage is much less predictable, because of the interference patterns and the effects of shadowing from buildings. For these reasons, and also to improve spectral efficiency, microcells characterised by antennas mounted below the prevailing roof-top height are almost exclusively used if 'cell' sizes much less than 500 m are planned.

4 Introduction to RAN planning and design

In Chapter 3, the generic principles, processes and deployment configurations applicable to cellular network planning were developed. In this chapter and the following four, a *detailed* design process will be developed to address, step by step, the practical deployment process for four different wireless networks.

This chapter describes the steps in the planning process that are essentially common, regardless of the specific air interface under consideration. The high-level planning of Chapter 3 will have estimated the total number of cell sites and the maximum cell size, and made decisions on the applications to be deployed. The detailed plan will define the *actual* locations of cell sites, antenna types, mast heights, etc., using topographical data for the specific regions. The plan will also assure guaranteed levels of coverage, capacity and availability for the applications to be supported.

The changing relative implementation cost and impact on battery life of particular technologies at points in time over the last 25 years has given rise to three distinct RAN standards:

- TDMA (as employed in GSM, GPRS, EDGE),
- CDMA (as employed in UMTS releases 99, 4, 5, 6, 7 and CDMA 2000),
- OFDMA (as employed in 802.11, 802.16e (WiMAX) and as planned for 3G LTE).

These technologies are expected to dominate wireless deployments over the next 20 years and it is a comprehensive understanding of major factors, such as coverage, capacity and latency, that will enable system designers to exploit their potential fully. For instance, the discussion on planning for capacity in Chapter 3 highlighted the desirability of low-frequency reuse patterns because, if there are Q frequencies available, by making the reuse pattern, N, small, the number of frequencies per cell, Q/N, and hence capacity, can be increased. Many readers will already have wondered what determines the reuse factor – can it be arbitrarily small? As with most things in life, the answer is yes . . . but only under certain conditions and these vary from one air interface to another.

The system analysis approach used to determine downlink and uplink capacity will be outlined for each air interface, and the way in which settings of particular parameters within each standard can be chosen to maximise capacity or meet other goals will be discussed. In each case, this will be followed by brief comments on the control-plane message structure required to manage cellular traffic efficiently. Such structures can sometimes adversely impact system capacity or introduce undesirable behaviour as far as the user is concerned.

Each air interface-specific chapter will conclude with a detailed design example to illustrate the points made during the discussion. These examples will address how the specific standard can best meet the requirements set out in the *application scenarios*. Such scenarios describe the mix of services to be provided, the total service demand and the key attributes of the deployment region (physical area, subscriber density and propagation characteristics). The scenarios to be used in the examples will be discussed in Section 4.1.1.

4.1 Data collection

The first step in planning any network is to ensure that there is a clear definition of the required network performance. This should include the area to be covered, the applications to be deployed and the subscriber densities to be supported in each part of the network. The specification will also normally include coverage requirement confidence limits, which may vary by application, and QoS requirements for each application (again with confidence limits); all of these parameters will also typically vary with time! Whilst this list is not exhaustive, it does give an idea of the amount of data to be collected and approved before planning can start. The system planner is unlikely to have all these data to hand at the outset, and thus a variety of different organisational units will have to be consulted to construct a complete set of design goals. Although the data used in developing the business case discussed earlier will be a good starting point, there will usually have been some changes since the project was approved and, in any event, more comprehensive information will be needed.

The network designer should ensure that all relevant groups are contacted and canvassed for their views of the design goals, and that consensus is reached. Small changes in some parameters can make a major difference to the cost, complexity and feasibility of the final design; consequently it may be advisable to perform several iterations of the initial system dimensioning process, for example, to enable cost vs. quality trade-offs to be explored. The sources of the design inputs should be documented and recorded to ensure full traceability in case problems arise later. Two activities, the definition of user applications and the development of a comprehensive topographical description of the intended network coverage area, typically represent the majority of effort during this data-gathering phase and these activities will be discussed next.

4.1.1 User applications

The *application scenarios*, shown in Table 4.1 and Table 4.2, will be used in the worked examples for all standards and will serve to highlight differences in their capabilities. As suggested by the use of these tables, there are at least two dimensions necessary to describe user behaviour:

- A description of the applications (Table 4.1),
- A description of application usage (Table 4.2).

Table 4.1. Typical application characteristics

Application	Burst statistics	Mean rate (kbits/s)	Latency (s)	Jitter (s)	BER	Asymmetry (UL/DL)	Session duration (s)	Burst arrival rate (/s)	Burst size (kbytes)
Circuit voice	Poisson	4–64	0.1–0.15	N/A	10^{-2}–10^{-4}	1	60–180	50	N/A
WAP	Truncated Pareto	2–5	2–6	1–5	0.03% (BLER)	0.01	30–120	–	1–32
Interactive gaming	Game dependent	10–1000	0.05–0.1	0.02	10^{-5}–10^{-6}	0.1–0.5	1800	Game dependent	1–100
Video clip download	Pareto/log–normal	64–384	1–5	N/A	10^{-4}–10^{-6}	0.0001	–	–	40–1000
Streaming video	Pareto/log–normal	64–384	1–5	N/A	10^{-4}–10^{-6}	0.0001	20–60	2	40–1000
VoIP	Pareto (aggregated traffic)	4–25	<0.1	0.01	10^{-3}–10^{-4}	1	60–180	50	1–12
Web browsing	Pareto	64–2000	2–6	1–2	10^{-4}–10^{-6}	0.0001	30–1800	0.025–0.1	30–2000
Messaging	Exponential	8	1–30	N/A	10^{-3}	1	60–300	0.01–0.02	0.1–10

Table 4.2. The common application planning scenarios

Chapter	System demand	Busy hour call attempts	Application usage (% 'calls' during the busy hour)							System characteristics			
			1	2	3	4	5	6	7	8	9	10	11
			Voice	Mobile web	Interactive gaming	Video	VoIP	Web browsing	Messaging	Population density ($/km^2$)	Penetration (%)	Propagation environment	Mobility (km/hour)
5	–	**1**	90	3	0	1.5	0	2.5	3	1000	70	Suburban	0–240
6	512 kbps (20:1)	–	–	–	–	–	–	–	–	1000	100	Suburban	0–50
7	–	**1**	0	0	0	0	90	10	0	5000	100	Urban	0–120
8	512 kbps (20:1)	–	–	–	–	–	–	–	–	1000	20	Urban, campus	0
9	–	**0.5**	35	0	5	10	30	15	5	5000	100	Urban	–

4.1.1.1 Application description

Historically, the application description has been simple:

e.g., 'Voice using the G711 codec' [1].

The application planning information was similarly straightforward:

'50 calls/hour (in the busy hour) with a Poisson arrival rate, and 25 mE demand.'

Today, and increasingly in the future, users expect a much richer portfolio of applications, as most communication and business activities previously carried out only in the home or office are increasingly executed 'on the move'. As discussed in Chapter 3, most of the new applications are most efficiently transported using packet protocols and are likely to be carried over a common data pipe for at least part of the end-to-end transport. It is, therefore, necessary to have a full 'description' of the application so that appropriate capacity can be provided to guarantee the QoS chosen for the application. A European *IST* project, SEACORN [2], carried out an activity to characterise applications that are expected to be important in the future and some of the results from this study are presented in Table 4.1.

The first two columns in this table identify the name of the application and, where known, the statistical distribution that usually best describes the population of burst sizes. Unless the application is standardised, it will usually be necessary to collect data to establish the parameters of the distribution. In the early stages of planning, it is often convenient to get a feel for the total volumes and rates of data transfer involved; this may enable some transport options – e.g., existing 2 Mbit/s microwave links – to be discarded. The third column recognises this need and an average data rate will usually be established, either from the intrinsic nature of the application, e.g., a circuit voice codec, such as G711, or by estimating this using the burst distribution parameters. Note that, with the exception of simple media, such as non-VAD codec-driven traffic over circuit bearers, transport at this average rate is almost always insufficient to ensure correct operation of the application.

The next three columns address requirements for latency, jitter and bit-error rate (BER). Although, in many cases, there are industry 'norms' for these parameters (e.g., to ensure the correct operation of international communication), it is always the network operator who ultimately sets the target for these parameters and these are often used to differentiate services in the marketplace. A perverse example of this at the present time is that some operators clearly dedicate so little capacity to supporting SMS, that messages arrive half a day or later after transmission, and in some cases not at all!

Asymmetry is an issue of increasing importance in network planning as the proportion of non-voice traffic becomes ever larger. Even though earlier 'wisdom' that non-voice traffic would be downlink dominated is now suspect, with the rise of user-generated content, managing asymmetry will still be a key issue – but how should asymmetry be defined? It can exist in voice traffic, when one person dominates the 'conversation' (most people can probably relate to this but let's not go there) although it is more usual in data-driven applications, for instance web browsing. Vignali, Malavasi, *et al.* [3] propose a measure called the *traffic asymmetry factor*, which is defined as:

$$\text{TAF}\,(S, T, L) = \frac{\text{DL bits}\,(S, T, L)}{\text{UL bits}\,(S, T, L)}$$

– illustrating that TAF is a function of S, the service under discussion, T, the time period over which the asymmetry is observed, which might typically be a *session*, and L, the level of aggregation (single or multiple users, etc.). Whilst this is useful in the context of total volumes and rates of data transfer involved, in practice it integrates two factors that need to be considered separately: the burst-size distribution and the burst arrival rate. For instance, *DL bits* could be high because of a high arrival rate of bursts with a low mean and small variance or because of a low arrival rate of large bursts with large variance. The discussion in Chapter 3 has highlighted that the first scenario could be addressed with a low-rate bearer whilst the second would require a bearer many times this size. The implications for wireless network design are quite different.

The final three columns give information to complement the application burst statistics description. The *session duration* and *burst arrival rate* are self-explanatory, whilst the *burst size* column gives an indication of the range of burst sizes that might be encountered with each application. These sizes very much reflect the current situation and will typically increase as time goes by. The parameter is of interest because it can be useful in the QoS modelling process. For instance, if it is known that most mobile devices in the plan period will not be able to handle individual WAP pages whose size is greater than 2 kilobytes successfully, WAP servers will certainly not support pages any larger than this. This puts an upper limit on burst sizes for WAP, at a particular point in time, which can be used to truncate the Pareto distribution, often the best fit for WAP traffic. The distribution would otherwise indicate a small but finite possibility of much larger pages . . . with the corresponding adverse impact on bearer rates that would be needed.

Table 4.2 sets out, in separate rows, five hypothetical planning scenarios that will be used to illustrate planning techniques and the capabilities of air interfaces and transmission solutions in the chapters referenced in the first column. As discussed earlier, a key trend over the period of the *scenarios* will be the increasing use of services other than mobile voice. The nomenclature used in the table retains the concept of 'calls' (these could equally well have been called sessions) but to recognise that each call will, in fact, be one of a number of applications shown in the first seven columns. The probability that a particular 'call' will be a specific application is reflected in the probability percentages for each service, which add to 100%. The *total* call-arrival rate is reflected in the busy-hour call-attempt column.

Traditionally, traffic over circuit networks is expressed in Erlangs, making no distinction between a few short calls and one longer call in a busy hour. However, for multimedia systems, the traffic description must embrace not only circuit but also data for which Erlang figures are inappropriate. Accordingly, the concept of 'busy hour call attempts' (BHCA) per subscriber is used and then apportioned between the various applications.

The summed figure for BHCA will arise from market research and represents the aggregate of 'all' call attempts whether due to voice or other applications. Note that, in the data for Chapter 5 worked example, the BHCA figure for voice of 0.9 (1 × 90%) equates to an Erlang demand of 0.025E if 100 second call duration is assumed.

Note that the scenarios for the worked examples of Chapter 6 (HSDPA) and Chapter 8 (mesh networking) do not contain data on busy hour call attempts. In these cases, no attempt is being made to support the QoS for specific applications. Instead, a broadband data pipe is shared with many users to provide 'best-effort' connectivity. In moderately loaded networks, and for tolerant applications, like downloading email or web browsing, shared 'data pipes' can give a very useful service, as the proliferation of Wi-Fi hot spots attests. In these cases, it is appropriate to define a coverage area over which a specified bearer rate will be delivered and to indicate a maximum 'contention' ratio. The contention ratio is defined to be the maximum number of subscribers that will be contending for or sharing the common bearer – thus if the bearer is 1 Mbit/s and the contention ratio is 20, the average data rate would be 50 kbits/s, even though the peak rate would be much higher.

Columns 8 to 11 relate to the type of deployment region and are substantially independent of the applications specified in the *scenarios*; indeed it would have been perfectly possible to have the same deployment region in each of the three cases. In practice, it is convenient to make each region different to illustrate the distinct nature of challenges that will be encountered.

Column 8 (the population density) and column 9 (percentage penetration of the mobile service) are combined in the planning stage to give rise to the 'subscriber density' that must be supported by the network. In practice, a 'penetration' figure versus time is often provided to allow for plan evolution as subscriber densities increase. Columns 10 and 11 together define critical parameters in the link budget such as average propagation loss, shadowing losses, multi-path fading levels, etc., and together with other inputs define the cell size.

4.1.2 GIS data

A GIS is a *geographic information system* and is software that displays digital map data and allows users to query and analyse the data. GIS data can be purchased for almost any location in the world, and can typically provide all of the geographic data needed for network planning. The GIS has evolved from a combination of two well-established types of software. Map features are handled using graphics and computer-aided-design (CAD) technology; location-attribute information is supported by extensions of conventional spreadsheet and database technology.

There are many different formats and sources of such data but, for RF planning purposes, the main categories are as follows:

- Digital terrain models (DTM) or digital elevation models (DEM): an array of measurements of the elevation of the earth's surface above mean sea level, for a given area,

- Clutter models: similar to a DEM, the clutter model contains details of the land cover at different locations: forest, water, high-rise urban, suburban development, and so on,
- Demographic and morphology models of land use and population behaviour: like clutter models, these models indicate the activities of the potential subscribers at different locations, for example light industry, residential, retail, agriculture,
- Vector data – roads, railways, rivers, coasts, canals, national and regional boundaries,
- Detailed 3-D models: these indicate the dimensions and construction of buildings, accurate to less than 1m – used for in-building propagation and line-of-sight studies,
- Photographic imaging data: pictures of a region of the earth's surface, usually taken by satellites.

Terrain, clutter and morphology data are usually provided as a raster or matrix of equally spaced observations covering a 'tile' of the area of interest. Usually, these measurements are given as values at locations equally spaced in x and y, relative to some offset coordinate. For wider areas, greater than a few tens of kilometres, it may be necessary to use coordinate systems that take into account the curvature of the earth, possibly requiring conversions and transformations between different coordinate systems before the mapping data can be used in a planning tool. It may also be necessary to have different resolution data sets for different regions – for example, sparsely populated rural areas may be modelled with measurements 50 metres apart, whereas dense urban areas are modelled at higher resolution, with measurements at 2-metre intervals or closer. Most commercial planning tools are designed to integrate such variations in resolution with no difficulties, and some provide the tools to convert between different projections and coordinate systems.

Terrain and elevation models are usually derived from satellite photographs and survey measurements. These models are used to estimate the degree to which RF propagation is attenuated by obstructions and reflections along the path. The accuracy and precision of available data, and indeed the cost, will vary greatly from country to country. In rapidly developing markets, the available mapping data may significantly lag the pace of construction, and this can have a major impact on the network design process – the network designer must ensure that recent data are used wherever possible. In some circumstances it may be necessary to commission surveys to obtain data of sufficient accuracy but this is a costly and time-consuming process.

Clutter models are an attempt to simplify the complexity of the wide variety of structures and artefacts present on the earth's surface for planning purposes. Propagation models use these data to estimate how much, if at all, the radio signal is attenuated or reflected. Simple clutter models differentiate only a few clutter types, but more sophisticated models may classify dozens of different land-use types.

Vector data are useful for verifying coverage along important highways, or for estimating levels of on-street coverage in a given area; however, they are not normally directly used in the empirical models discussed below. Similarly, photographic data can be useful to confirm coverage (or lack of coverage) at important landmarks, or to confirm that the clutter model is correctly representing what is actually present on the earth's surface. Coverage plots are often overlaid onto photographs as an aid to interpretation.

Three-dimensional models of buildings and obstacles are very important when line-of-sight studies are needed. Most planning tools include features that allow the designer to see which parts of a region are in direct line-of-sight from a particular location, such as a tall building or tower. Ray-tracing techniques require such models when simulating scattering of rays off building walls and roofs. Some planning tools can also use information on building construction when simulating the degree to which rays are reflected or absorbed. Municipal authorities and utility companies often compile digital maps containing the whereabouts of items such as power lines, water pipes, traffic lights and other street furniture. This can be very useful for the selection of candidate site locations in urban environments.

In addition to the mapping data sets described above, it is sometimes necessary to get more detailed information about subscriber behaviour. Examples include counts of numbers of people entering or leaving shops or train stations at different times of day, or volumes of traffic flowing along thoroughfares. These data can be used to improve estimates of traffic likely to be offered to the radio network. Such information may be available from national mapping agencies, but more usually it is commissioned to be gathered for specific areas of interest.

4.2 Propagation modelling

A critical element of the radio-planning process is the estimation of path loss between the cell-site transmitter and the receiving subscriber unit. Section 3.1.1 introduced the subject of path-loss estimation using radio-propagation models. Such models give the radio network designer an efficient and practical method for simulating the radio network. Propagation models can be considered to fall into two types: *empirical* and *deterministic*.

Empirical models, such as Okumura–Hata [4] and Walfisch–Ikegami [5] are derived from observations of how the free-space propagation of radio signals is modified by different factors, such as terrain and land use. The free-space propagation loss is modified with various attenuation factors, which represent the sum of the effects to which radio signals are prone (scattering, reflection, fading, and so on). For each environment (e.g., urban, suburban, rural) a set of coefficients is provided in the empirical model, to allow the received signal strength at each location to be estimated quickly.

Deterministic models use a ray-tracing approach to follow the paths of individual radio 'rays' transmitted from the antenna, simulating their journey as they are reflected, scattered and absorbed by the obstacles in their path; these obstacles are modelled in significant detail in terms of their physical size and shape, down to a resolution of a few metres. The signal received at different locations is the sum of the contributions from the various rays simulated in the planning tool. A hybrid category of *semi-deterministic* models uses a more fine-grained classification of terrain and building structure to estimate path loss, without incurring the extra computational burden associated with deterministic models.

An overview of the most common models is now presented. The choice of which models should be adopted will usually be driven by the characteristics of the 'substantially constant propagation region', listed as a key step in classifying *uniform planning areas* in Section 4.3. If the UPA is characterised by a small, densely populated inner-city area, it is more likely that deterministic models will be most useful. On the other hand, a UPA characterised by large, rural areas can often be adequately modelled using empirical formulae, although there are no hard and fast rules.

4.2.1 Empirical models

Empirical models are fitted to match the behaviour of propagation under particular conditions and so it turns out that there are in fact several models in this category; these are usually presented in a dB format.

4.2.1.1 Okumura–Hata

The Okumura–Hata model is derived from the work of Okumura *et al.* [4] and Hata [6]. Extensive measurements of field strength at various distances from an elevated antenna were taken in an urban environment (Tokyo), and a set of empirical curves developed to fit the rate of change of field strength with distance. These curves were parameterised by Hata to produce a convenient set of formulae for calculating the field strength at different locations in urban and suburban environments. The widely used model given by Hata is of the general form:

$$\text{path loss} = A + B \log_{10} R,$$

where A and B represent functions of frequency and antenna height, and R represents the distance between the transmitter and receiver. More specifically, the propagation loss L_p is estimated to be:

$$L_p = 69.55 + 26.16 \log_{10}(f_c) - 13.82 \log_{10}(h_b) - a(h_m) + (44.9 - 6.55 \log_{10}(h_b)) \log_{10} R,$$

where:

f_c = frequency (150–1500 MHz),
h_b = base station antenna height (30–200 m),
h_m = vehicular station antenna height (1–10 m),
R = distance (1–20 km),
a = correction factors for small or medium cities.

Additional correction factors are applied to the above formula for suburban and open areas. As can be seen from the model, it is highly empirical in nature in that it attempts to model the observed path loss without explaining how it occurs. In practical applications, the correction factors associated with different terrain and clutter types need to be carefully tuned (often on a city-by-city basis) if the propagation predictions from the planning tool are to be useful. For frequencies above 1.5 GHz up to 2 GHz, an extended Hata model is sometimes used which has a similar form.

4.2.1.2 Walfisch–Ikegami

The Walfisch–Ikegami model is often used for frequencies between 800 and 2000 MHz, and distances from a few tens of metres up to about 5 km. This model distinguishes between line-of-sight (LOS) and non-line-of-sight (NLOS) situations, and is a mixture of empirical and semi-deterministic models, in that it uses more detailed information about average building heights and street widths, compared with a generic 'urban' or 'suburban' classification. For the LOS situation, where there is no obstruction to the direct path between the transmitter and receiver, the following formula for path loss is used:

$$L_{p(LOS)} = 42.64 + 26\log_{10}(d) + 20\log_{10}(f),$$

where:

d = distance between transmitter and receiver (for $d > 20$ m),
f = frequency (between 800 MHz and 2000 MHz).

The model assumes that there is little impairment to the first Fresnel zone and hence that the typical antenna height is of the order of 30 metres. For the NLOS case the Walfisch–Ikegami model uses a set of formulae that takes into account the building separation and street widths, which will need to be approximated for each city being designed. For more details on this model, consult [5].

4.2.2 Deterministic models

For situations where a more accurate estimate of signal strength is required at specific locations (and not just an estimate of overall coverage levels) it may be desirable to use a ray-tracing model. These use a detailed 3D terrain and building model instead of generalised building types, and are much more computationally intensive. The ITU has formally adopted a set of deterministic models for use in radio planning, such as P. 525 [7] and P. 526 [8], which estimate the free space attenuation and propagation by diffraction.

4.2.2.1 Model tuning

As mentioned earlier, if reasonably accurate results are required it will be necessary to tune any propagation models present in radio-planning tools to reflect the prevailing geographic and topographical structure of the region in which the network is being deployed. This is usually done by setting up a reference or test base site at a known location, and then performing a survey of the received signal strength at various distances from the transmitter, by either walking or driving around the region with a test receiver equipped with a GPS. Most modern radio-planning tools include the facility to import such measured data back into the planning tool, allowing propagation predictions to be compared with observed values. If enough data are gathered and fed back to the planning tool, the selected propagation model can be 'tuned' to reflect the environment – usually this involves modifying the correction terms in the empirical model.

Even after a propagation model has been carefully tuned, it is important to realise that propagation predictions are at best an approximation to the results that are likely to be found in practical deployment. Several factors that will influence the actual propagation cannot be easily simulated or modelled – examples include reflections and shadowing due to moving traffic, new building construction, and seasonal changes in foliage or vegetation. Consequently, it is wise to use planning tools to estimate overall system coverage and performance levels, rather than as a method for predicting performance at specific locations in the network.

4.3 Uniform planning areas

In the early stages of network planning it is desirable, whenever possible, to identify *uniform planning areas* (UPA). A UPA is defined to be an area where the parameters that have a major impact on cell site design are substantially constant. When it *is* possible to identify such areas, certain parts of the planning process become 'step and repeat'. Uniform planning areas are characterised by having similar parameters in at least the following areas:

- Propagation environment,
- Subscriber densities,
- Market penetration,
- Applications,
- User mobility attributes,
- Regulatory environment,
- Antenna height,
- Effective radiated power (ERP),
- Coverage confidence limits (e.g., 95%),
- Spectrum availability,
- Percentage of roaming traffic,
- Network availability,
- Deployment timescales.

Mobility requirements need to be carefully understood during the planning phase. Is it likely that the users will be physically moving around whilst making voice or data calls? Is it required that users should be able to maintain a session seamlessly when moving from one cell to another? If so, the cell density estimates are likely to increase further as a degree of cell coverage overlap must be provided in the radio design to allow handovers to take place. Nomadicity and roaming requirements will largely have an influence on the core network design for registration, authentication and accounting purposes. However, it is possible that the feature set offered to roaming users may also be different, which could influence the detailed cell design. Additional considerations include the provisioning of cells at gateway locations, such as airports and train stations, to make the capture of roaming subscribers (and the associated revenue streams) more likely.

Availability requirements impact the design process as the resilience and failure rate of the network components need to be considered. In some cases, for example where system availability figures of 99.99% or above are required, it may be necessary to include redundancy in the system design, and possibly to plan the coverage such that users may still be able to access the radio network even if the nearest available base site is out of service.

Quality-of-service (QoS) requirements will place considerable constraints on design choices, and these should be explicitly stated and agreed with the network operator early in the planning process. Examples of quality targets include voice quality *mean opinion score* (MOS), dropped-call rate, call set-up success-rate targets, worst-case end-to-end delay, packet data BER and packet loss ratio. Many others are possible. Detailed QoS requirements may call for the ability of the system to support different traffic classes such as background, best effort, expedited service or, increasingly in the future, to guarantee the correct operation of particular applications.

Regulatory requirements are normally expressed in terms of the permitted *effective radiated power* (ERP) – a combination of transmit power at the BTS antenna terminals and antenna gain in the chosen radio frequency band. The regulator will also directly or indirectly specify the allowable transmitted interference or noise. A limitation in the maximum ERP will inevitably reduce cell coverage, unless solutions such as repeaters are employed. There may also be international cross-border regulations and agreements in place to minimise inter-system interference with neighbouring networks. Such agreements may reduce the number of radio channels available for use in border areas. Furthermore, historically, some countries have placed restrictions on which radio access technologies and modulation schemes are permitted in a given frequency band. In many EU countries, at present the 900 MHz band may currently only be used for GSM under the terms of the licence granted to the operator. Other systems such as WiMAX or UMTS may not be deployed in this band, although this restriction is gradually being lifted. Further constraints on cell design can arise from local planning regulations. In areas judged to be of architectural or historic interest, regulations may only permit 'discreet' antennas up to a maximum height limit. There are ways round such regulations, such as siting antennas inside church spires . . . but even then, allowance must be made for the attenuation loss of the spire covering.

The deployment timescale is an important consideration, as it may dictate the rate and order in which base sites are to be deployed. If the initial need is to provide coverage for metropolitan areas, followed by increased capacity and subsequently coverage for less densely populated areas, then the designer may be able to begin with a smaller number of base sites, followed by more advanced techniques for capacity enhancement as the demand grows. Examples, discussed in Chapter 3, include cell splitting, sectorisation, and microcells.

4.4 Radio planning

The radio-planning process is intended to culminate in a baseline network layout satisfying the coverage and capacity requirements defined during the high-level planning detailed in Chapter 3. During this process, the area to be covered will have been identified, and a rough estimate of the number of base sites needed to provide sufficient coverage and capacity will have been established. The outputs from the radio-planning process will vary depending on the access technology, but typically include the following items:

- Candidate base-site geographical locations,
- Antenna heights, tilts and azimuth,
- Antenna type: manufacturer and model,
- Base-site transmit powers,
- Coverage plots showing estimated signal strength over the target area,
- 'Best-server' plots showing which cells are most likely to provide service,
- Location of possible coverage 'holes',
- Radio resources required at each cell site.

For some access technologies, additional design parameters, such as frequency allocations, pilot powers and neighbour lists are required. Further information is provided in Chapters 5 to 8, where the detailed operation of particular air interfaces is discussed.

Whatever the radio access technology, a primary concern is to ensure radio coverage for the area of interest with the specified confidence limits. The definition of coverage is now more complex, as subscribers expect to access multiple applications, each with its own unique QoS requirements. Nevertheless, the fundamental requirement is largely unchanged and means that at a given location, the radio signal transmitted by the network infrastructure, for the applications of interest, can be successfully received and decoded by the subscriber equipment. Similarly, the corresponding signals transmitted by the subscriber can also be received and decoded by the network infrastructure.

Although significant space has already been devoted to the prediction of cell coverage, the purpose here is to provide the reader with insights into the key parameters affecting cell radius and to indicate techniques for improving coverage that can be used, regardless of the specific air interface. In practice, proprietary radio-planning tools are usually employed to explore the coverage of candidate network configurations. There is a variety of commercially available RF planning tools, of widely varying complexity, power and cost; see for instance [9, 10]. Any tool will provide estimates of the received radio-signal strength across the target coverage areas. This is achieved by simulation of the propagation and attenuation of radio signals using the model of the geographic area under study. It is appropriate to note that the accuracy and usefulness of any planning tool is ultimately constrained by the quality of the GIS data and propagation model developed by the planner.

Once coverage has been provided, design effort will be focused on delivering the capacity required.

4.4.1 Antenna characteristics

A wide variety of antennas is available for use in wireless networks, each offering different characteristics in terms of gain, radiation profiles, tilting capability (mechanical or electrical), diversity, and so on. As part of the planning and design process it is

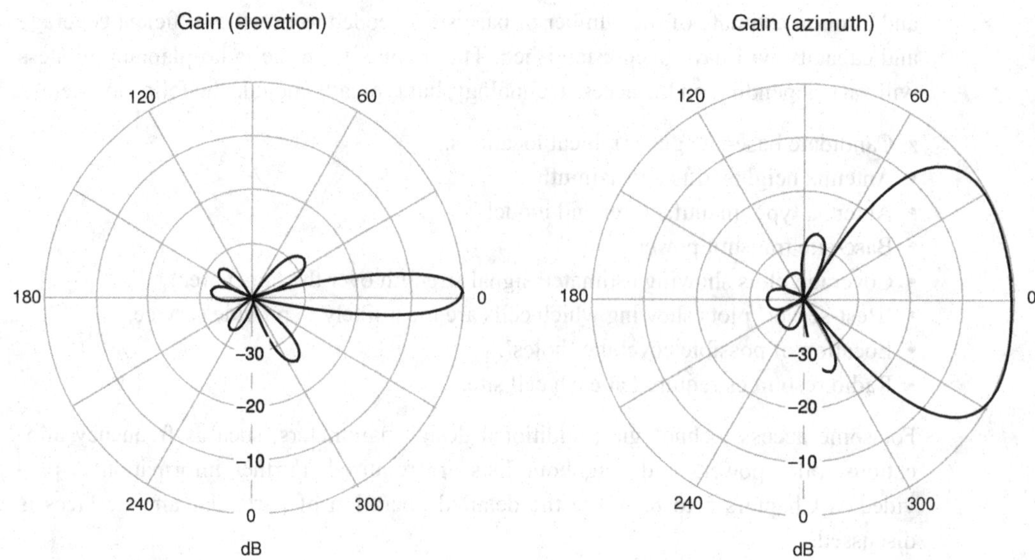

Figure 4.1 Example antenna radiation patterns

necessary to estimate the effects on the radio network of using different antennas so that their relative merits and cost versus benefit to the network can be evaluated. In the early stages of deployment it is likely that, for most networks, omnidirectional antennas will be used, as these can provide wide area coverage with minimum BTS equipment. As subscriber densities increase, techniques such as sectorisation and, ultimately, cell splitting (see Chapter 3) can be used to increase capacity and spectrum efficiency. Sectored sites will require the use of high gain antennas, which are designed to focus power in a particular direction, known as boresight (usually referenced to zero degrees on the antenna patterns). Unfortunately, limitations in design mean that radiation will also occur away from boresight. However, this unwanted radiation can be managed to occur in particular directions and, indeed, the main beam shape can also be adjusted, making some configurations more useful for particular deployments. To assist in the selection process, manufacturers provide *radiation patterns* for their antennas, and a carefully chosen antenna will play a key role in delivering network performance.

Antenna patterns show the relative power radiated as the receiver moves away from boresight. The results are usually presented as two polar plots, covering 360° in the horizontal and vertical planes. Example antenna patterns are shown in Figure 4.1. The plot is normalised to the power produced on the vertical and horizontal axis at 0 degrees (where the emitted power is highest) and the power emitted in other directions is seen to be less as one moves away from the axis. This is shown on the profile in terms of dB reduction as the angle from boresight increases. In the horizontal plane, the angular range over which the power is within 3dB of the maximum is said to be the 'beam width' of the antenna – in this case the beam width is about 90°, which makes the antenna suitable for use in a sectored configuration. In the vertical plane, the power is seen to be much more sharply focused and the roll-off with angle is much more pronounced;

this ensures that no power is wasted transmitting to areas high above the earth's surface or down into the ground. Note that the power does not roll off monotonically however; there are occasional rises and falls in the profile as the vertical angle increases. These can be experienced as 'near-field nulls' as a mobile moves nearer and nearer to a base site.

4.4.2 Coverage prediction

Radio planning tools use the radiation profiles of the selected antenna to attenuate the radiated (and received) power as a function of direction from the base site. A number of different antenna profiles can be evaluated on an existing radio plan without the need to re-run the full propagation model each time. This reduces the simulation costs and time taken to explore different planning options. Similarly, it is possible to change the relative orientation of antennas in the network model (for example by changing the azimuth or down tilt) to investigate the effect on coverage; this may be necessary to ensure that important areas are given adequate radio coverage.

Once the propagation has been modelled, the planning process continues by identifying the areas to be covered by each cell. The planning tool will usually provide a 'best-server' plot, essentially a coloured map indicating the cell best able to provide service at each geographic location. This is constructed by comparing the received signal strength figures from each cell at each location, and identifying the strongest. Cell coverage boundaries can be adjusted by altering the radiated power, adjusting tilts and azimuths, altering cell site locations, and so on. Note that the actual coverage boundaries will not exactly align with the best-server plot as there will be a blurring of boundaries due to handover margins and hysteresis values.

As well as identifying the best server at a specific location, coverage predictions can be used to estimate the extent to which neighbouring cells may interfere with each other, should they be allocated similar radio channels. The signal received at each location from the best server is compared with the estimated signal received from all other cells, and the carrier-to-interference (C/I) ratio is calculated. In areas where there is no clear best server, this value could be quite small, and, hence, frequency allocation schemes will need to manage the interference carefully if the network performance is to be acceptable. The C/I plots from the planning tool give a good visual indication of which areas are likely to receive a strong, high-quality radio signal offering good speech quality or high data-throughput rates. The radio designer can use the tool to adjust base-site locations, powers and antenna configurations to ensure that the best quality is delivered to regions with high subscriber density. However, it is inevitable that a proportion of total coverage will experience poor C/I, and it is desirable that these areas should be located where there is unlikely to be much traffic. The C/I plot (or matrix) is often used as input to an automated planning tool to assist with allocation of radio resources. The radio network designer is invariably working towards a coverage goal – for example, to provide a C/I value of 12 dB or greater across 95% of the coverage area. Planning tools will provide a cumulative C/I plot or CDF, which shows the proportion of the network covered as a function of C/I ratio. CDFs are thus a useful method for comparing potential radio coverage and signal quality in different network configurations.

4.5 RAN-specific planning

The planning described in Section 4.4 yields general data on signal coverage, C/I ratios and candidate base-station locations. The next stage, covered in Chapters 5 to 8, is to couple this with the characteristics of particular access technologies. Nevertheless, the principal steps can be summarised as follows:

- Determination of required traffic capacity per cell (voice and data),
- Allocation of radio resources to meet this capacity,
- Allocation of RF frequencies for each site,
- Determination of neighbour relationships,
- Radio link budget calculation,
- Determination of back-haul requirements,
- Selection of advanced features,
- Detailed equipment planning.

Once the cell designs have been completed, the higher-order network elements can be configured. This phase determines how each base site is to be controlled and configured, how its back-haul requirements are to be met, and how the cells will interface to the transmission and core network. Considerations for this phase include:

- Paging and location-area design,
- Back-haul configurations,
- Signalling-link design,
- Operation and maintenance interfaces,
- Location and capacity of base-station controllers and core network interfaces,
- Transmission requirements (copper, microwave links, fibre, etc.),
- Transcoding requirements.

In the space available it is not possible to address all these aspects. Instead, in Chapters 5 to 8, the text will focus on coverage and capacity planning at the cellular level for GSM, UMTS and OFDM systems. This is not to ignore the importance of matters such as CPU dimensioning on BTSs and BSCs or the significance, in terms of life-cycle cost, of minimising transmission costs. Rather, it recognises that many of these aspects will be vendor or transmission specific and that further advice will be available from the companies supplying the equipment or services.

References

1 ITU-T, *G 711 – Pulse Code Modulation (PCM) of Voice Frequencies*, www.itu.int/ITU-T/index.phtml (published November 1998).

2 SEACORN IST project, *Characterisation Parameters for Enhanced UMTS Services and Applications* (October 2002).

3 Vignali, Malavasi, *et al.*, *Asymmetric Traffic Management in UMTS FDD with Additional Downlink Spectrum*, IST project IST-2001–35125 OverDRiVE (Published 2003).

4 T. Okumura, E. Ohmori, T. Kawano and K. Fukuda, Field strength and its variability in VHF and UHF land-mobile radio service, *Review of the Electrical Communication Laboratory*, **16** Sep.–Oct. (1968) 825–873.

5 E. Damosso, editor, *COST Action 231: Digital Mobile Radio Towards Future Generation Systems, Final Report* (Luxembourg: Office for Official Publications of the European Communities, 1999).

6 M. Hata, Empirical formula for propagation loss in land mobile radio services, *IEEE Transactions on Vehicular Technology*, **VT-29** 3 (1980) 317–325.

7 ITU-R, *Calculation of Free-Space Attenuation, Recommendation* ITU-R PN.525–2 (1994).

8 ITU-R, *Propagation by Diffraction, Recommendation ITU-R P. 526–8* (2003).

9 Andrew, *Optum*™ *Automated Cell Planner Tool*, www.andrew.com/products/measurement_sys/optum/.

10 Aircom International, *ADVANTAGE Automatic Cell Planning Tool*, www.aircominternational.com/Automatic_Cell_Planning.html.

5 GSM RAN planning and design

5.1 System coverage and capacity

The development of a framework to assess system capacity for any air interface typically falls into two parts: firstly estimation of the S/N or C/I necessary to deliver the required bit error rate (BER) for the system in a point-to-point link and secondly understanding the limiting system conditions under which the most challenging S/N or C/I will be encountered. Once these two conditions are understood, calculations of maximum ranges and capacities for the system can be made.

5.1.1 C/I assessment for GSM

The base modulation scheme employed by GSM is Gaussian minimum shift keying (GMSK). This is a form of binary frequency shift keying with the input bit stream used to drive the frequency modulator filtered by a network with a Gaussian impulse response. This filter removes the high frequencies, which would otherwise be present because of the 'fast' edges in the modulating bit stream. If BT, the product of the filter − 3 dB bandwidth and the modulating bit period is chosen to be around 0.3, it enables most of the energy from the 270 kbits/s bit rate of GSM to be accommodated in a 200 kHz channel with low interference from the adjacent channels and negligible interference from those beyond.

Figure 5.1 illustrates the effect of the Gaussian filtering on the original rectangular pulse train and goes on to show the way *inter-symbol interference* arises, as multi-path introduces longer delays than that due to propagation of the direct 'ray'. An outline of GMSK, and a discussion of some of these trade-offs is included in [1].

Although Gaussian filtering can help reduce the effects of adjacent-channel interference, there is still the potential for inter-symbol interference, which can occur if the multi-path delay is significant compared with the bit period. It will be recalled that the underlying GSM bit period is about 3.7 μs and that the GSM system is required to operate in a wide variety of terrains where delay spreads of up to 16 μs can be expected. In practice, inter-symbol interference becomes a significant issue for channels with RMS delay spreads greater than 0.1 of the bit period and thus adequate performance for GSM in most practical deployments requires *equalised* receivers.

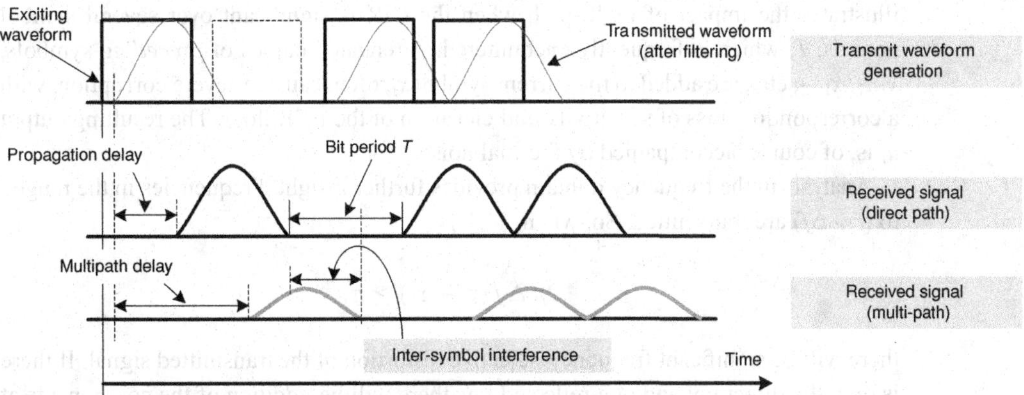

Figure 5.1 The benefits of GMSK

5.1.1.1 Principles of equalisation

A generic multi-path model is shown in Figure 5.2(a). The signal received from the base station and summed at the mobile is found to comprise a direct ray, subject to some attenuation factor α_1 and delay τ_1 and a number of other rays $(r-1)$ subject to some attenuation $\alpha_2 \ldots \alpha_r$ and delay $\tau_2 \ldots \tau_r$, which are in fact attenuated and delayed copies of the transmitted signal. The multi-path components α_r, τ_r in aggregate essentially represent the *channel impulse response* (CIR), $h(t)$.

Figure 5.2(b) recasts the continuous-time model of Figure 5.2(a) into a discrete-time model, since this corresponds to the way in which the GSM receiver operates. It also

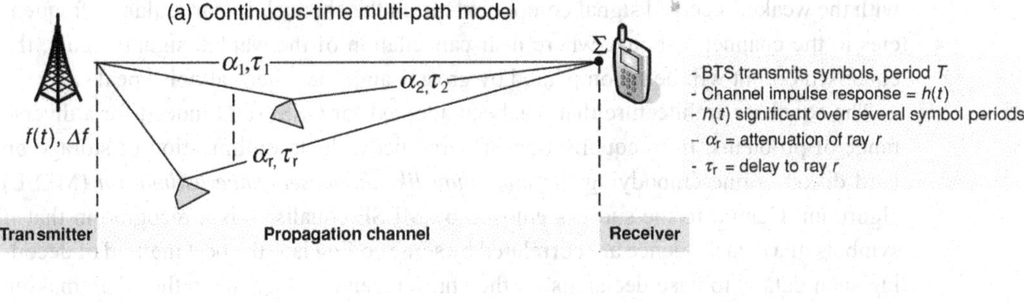

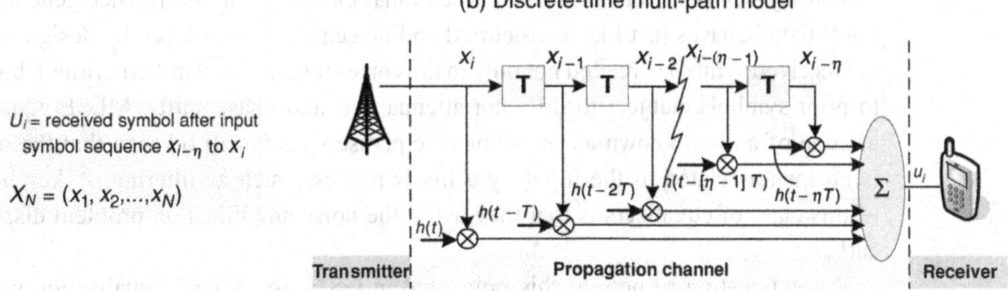

Figure 5.2 Channel multi-path

illustrates the impact of multi-path when the CIR is significant over several symbol periods, T, which is frequently encountered. Attenuated copies of *preceding* symbols, x_{i-1}, x_{i-2}, etc., are added to the current symbol x_i, often causing severe corruption, with a corresponding loss of sensitivity and elevation of the BER floor. The resulting output u_i is, of course, accompanied by thermal noise.

Analysis in the frequency domain provides further insight. Frequencies in the range f to $(f+\Delta f)$ are transmitted. So, when

$$2\pi\Delta f(\tau_r - \tau_1) \geq \pi,$$

there will be significant frequency-selective distortion of the transmitted signal. If there is only the direct ray and one reflected ray, there will be *addition* of the components at some points in the band and *subtraction* of the components at other points in the band. Indeed, many readers will recognise Figure 5.2(b) as having the general form of the finite impulse response (FIR) digital filter. Thus, another way of looking at multi-path is as a frequency-selective channel. Measurements, even within microcells, show that the *coherence bandwidth* (the information bandwidth in a narrow-band channel over which the correlation between amplitude envelopes remains above 0.9) lies between 30 kHz and 130 kHz [2].

Not surprisingly, the first attempts at equalisation, known as the *linear transversal equaliser* [LTE], introduced an FIR filter in series with the received signal, with the delay taps equal to the bit periods. The coefficients of the filter were chosen to provide the inverse of the channel frequency-selective profile and, in so doing, compensate for the effects of multi-path propagation. This seems a good solution until it is remembered that thermal noise present at the input to the filter will be selectively amplified along with the weaker received signal components from the channel. In particular, at frequencies in the channel response where near cancellation of the wanted signal occurs, the equalised output will be accompanied by grossly amplified thermal noise peaks.

The equaliser architecture that has been adopted for GSM (and indeed for a diverse range of problems, from equalisation of wired networks to optimisation of storage on hard disks) is one embodying the *maximum likelihood sequence estimation* (MLSE) algorithm. Central to the success enjoyed by MLSE equalisers is a recognition that if symbols of a data sequence are correlated by some coding law, the best method of decoding such data is to base decisions on the entire received sequence, rather than making symbol-by-symbol decisions. The unintentional inter-symbol interference generated by multi-path behaves just like a structured coding sequence introduced by design, in that the received symbol is related not only to the corresponding transmitted symbol, but also to prior symbols, subject to different attenuations and phase shifts. MLSE equalisers are one of a class known as 'non-linear' equalisers, as it will be seen that the output is no longer related to the input by a linear process, such as filtering. A key benefit of this class of equalisers is that they avoid the noise amplification problem discussed earlier.

It is interesting to note at this point that in GSM, the MLSE equalisation process can be elegantly incorporated alongside the same type of estimation function used to

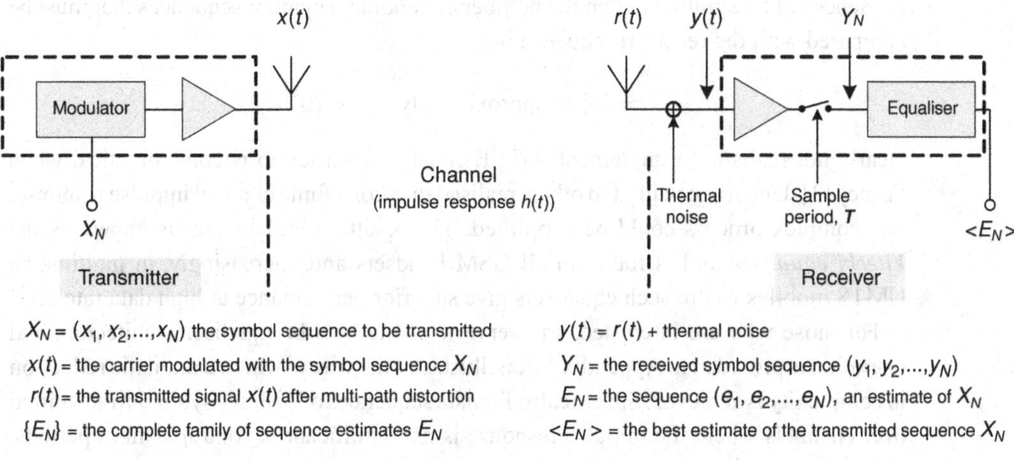

Figure 5.3 The role of the equaliser in GSM

decode the intentionally introduced structured convolutional coding carried out prior to transmission.

A system view of the equalisation process in GSM is shown in Figure 5.3. The symbol sequence $X_N = (x_1, x_2, \ldots, x_N)$ to be transmitted is modulated onto a carrier and the resulting signal $x(t)$ propagates via the radio channel to the receiver. The signal, $x(t)$, is subject to multi-path and thermal noise, giving rise to a received signal $y(t)$. Over a period of time a complete sequence of received samples $Y_N = (y_1, y_2, \ldots, y_N)$ is available to the equaliser.

In principle, because the number of bits per symbol and the number of symbols transmitted is known, the MLSE equaliser can generate all possible sequences that could have been transmitted, one of which *will be* X_N. Further, if it is assumed that the equaliser has knowledge of the channel impulse response, it can apply the effects of multi-path to each and every symbol and to each of the possible symbol sequences in turn to generate the complete family of possible sequence estimates, $E_{N,1}, E_{N,2}, \ldots, E_{N,\Omega} = \{E_N\}$. Once $\{E_N\}$ is available, each candidate sequence, $E_{N,q}$, is compared with the received sequence Y_N, which includes not only the *actual* channel multi-path effects (those that have been replicated by modelling using *estimates* of channel multi-path) but also thermal noise, which cannot be predicted or modelled. If the most similar sequence from $\{E_N\}$ turns out to be $E_{N,7}$, then $E_{N,7}$ is chosen as $<E_N>$ the best estimate of the transmitted sequence X_N. Although the MLSE equaliser relies on an *estimate* of the channel impulse response (since the *actual* CIR is never known), introduction of predetermined (and thus known) symbol sequences in the middle of the unknown symbol stream, known as training sequences, mean that very good estimates can be obtained in practice.

The process outlined above accurately represents the ambitions of the MLSE algorithm. In practice, although the benefits of decision making using complete sequences were well understood for some time, decision making on a symbol-by-symbol basis continued to be used until the early 1980s because of the complexity of implementing a full MLSE equaliser. Consider such an equaliser working on binary GMSK modulation with sequences of 58 symbols. Then the number of candidate symbol sequences that must be compared with the received sequence is:

$$= 2^{58} \text{ or approximately } 2.9 \times 10^{17},$$

clearly not realistic to implement. MLSE equalisers started to become practical when Forney [3], Ungerboeck [4] and others realised that, for a finite channel impulse response, this complex process could be simplified. The resulting architecture is known as the *Viterbi equaliser* and is found in all GSM handsets and, increasingly, in multimedia UMTS mobiles where such equalisers give superior performance at high data rates.

For those who are interested, an overview of the Viterbi equaliser architecture and operation is provided in Appendix 5.1; it illustrates the way in which a dramatic reduction in complexity is achieved. Specifically, for data sequences of N m-ary symbols transmitted over channels where the impulse response is not significant beyond η symbol periods, the number of compare operations is reduced [5] from:

$$m^N \text{ to } N \times m^\eta.$$

In the case of GSM, the data sequences are 58 bits long and the worst case scenario defines multi-path to be insignificant beyond about four bit periods [6]. Then the number of compare operations (a convenient measure of implementation complexity) reduces to 928! It is this dramatic reduction in equaliser complexity that has made it practical to realise the rates of around 270k symbols required by GSM, in a heavy multi-path environment.

5.1.1.2 GSM equaliser performance

D'Avella *et al.* [7, 8] and Lopes [9] provide results of the performance of Viterbi-based MLSE equalisers both in general and, more comprehensively, for the detailed configuration applicable to GSM. They provide BER curves versus C/I or E_b/N_0 under a variety of conditions. Careful study of the specification for the GSM full rate speech codec [10] makes it clear that the GSM system is expected to operate down to raw BER levels of up to 5×10^{-2} (i.e., prior to the protection afforded by coding of the data that occurs before transmission) and this information can be used to establish the minimum acceptable C/I in particular scenarios.

Assuming that the mobile is properly synchronised both in time and frequency (which is assured by other means) the major factors that will affect equalised performance are the propagation environment and the mobile speed. To ensure that standardised conditions are used in the design and test of GSM equipment, the propagation conditions are defined in Annex 'C' of GSM 05.05 [6]. Key results from these papers are summarised in Table 5.1.

Looking at the results for C/N = 24 dB, when all the limitations can be considered to be due to multi-path with no significant contributions from thermal noise, it is possible to understand the remaining limitations of the GSM receiver after equalisation. The TU50 profile shows that the equaliser is operating under optimum conditions because there is significant multi-path but it is constrained to be 5 μs or less, which is well within the

Table 5.1. GSM equaliser performance

	BER in Rayleigh fading channel			
C/N (dB)	TU 50	RA 100	RA 250	HT 100
24	3×10^{-4}	2×10^{-3}	7×10^{-3}	1×10^{-3}
8		In the range 3×10^{-2} to 5×10^{-2}		

TU 50 – Typical urban @ 50 km/hr
RA 100 – Rural area @ 100 km/hr
RA 250 – Rural area @ 250 km/hr
HT 100 – Hilly terrain @ 100 km/hr

equaliser capability of 14 μs–15μs. Results for a two-ray model, which are not strictly comparable (in both cases for a BER of 2×10^{-3}), suggest that 3 dB to 4 dB is *lost* in a channel with no multi-path (reflecting the benefits of the fortuitous propagation channel 'coding' discussed earlier). The RA profiles have delay spreads constrained to be less than 1 μs and again reflect the loss of channel-coding benefit. Note that the move to RA 250 results in a further deterioration of the BER floor, but this time because of less accurate channel estimates (the channel is changing significantly during the reception of the transmitted burst). Finally, the HT100 estimate shows a reduced BER floor relative to RA100, again reflecting the channel coding benefit, even though it is not able to fully correct for multi-path in the region 14 μs to 17 μs, which is significant in this profile.

With the C/N reduced to 8 dB, thermal noise is now significant compared with the wanted signal and this constraint starts to swamp what are now second-order effects due to scenario-specific equaliser limitations. In summary, for fixed frequency transmission, it is probably sensible to use a C/N number in the range 10 dB to 11 dB for planning purposes on full-rate voice channels.

5.1.2 Evaluation of limiting system C/I scenarios

By the very nature of cellular networks, the frequency in the serving cell will be reused in another cell not very far away. This is true for both the downlink and the uplink. It is, therefore, necessary to examine both of these configurations to understand how frequency reuse affects system capacity. The downlink scenario will be addressed first as it is more straightforward, since the interference sources are the BTSs whose location is known and fixed.

5.1.2.1 Downlink interference scenario

For the downlink, the limiting scenario is clearly when the mobile is at the edge of the serving cell – farthest from the wanted signal and *not* proportionately farther from the potential interfering base stations. It also needs to be recognised that the worst-case interference will exist for the BCCH carriers – these are always required to transmit at full power, whereas other carriers' power will vary by time slot, depending upon the location of the mobile. The situation of the interfering sites is depicted in Figure 5.4. The

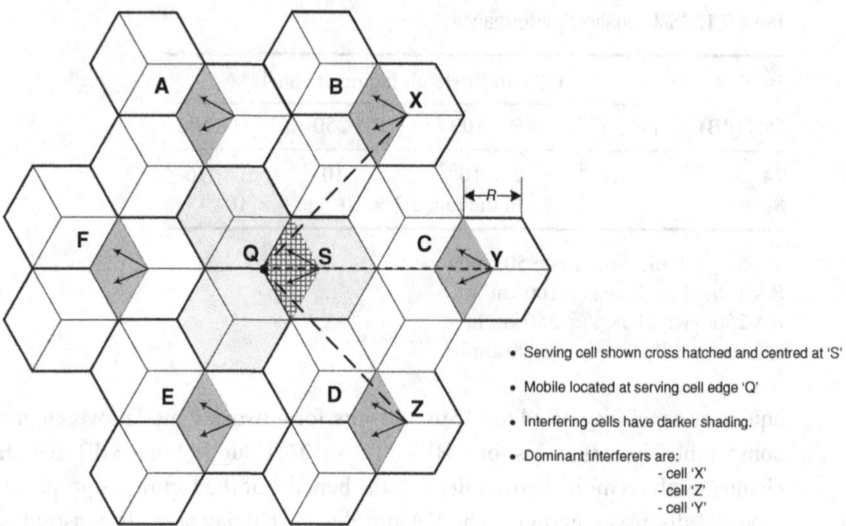

Figure 5.4 Downlink scenario for 3 × 9 reuse

diagram shows 3 × 9 reuse; that is, nine frequencies in total, deployed in a cluster of three tri-sectored sites and reused in all other clusters. The serving cell is the cross-hatched parallelogram in the cluster of three shaded hexagons labelled 'Q' at the centre of the diagram. The mobile is located at one of the points of the hexagon.

Surrounding the serving cluster are six otherwise identical clusters, labelled 'A' through to 'F'. The side of the hexagon is length R. These are the first tier of interfering clusters; second and higher-tier clusters are far less significant because of their greater range and the steep power fall-off with range and are ignored for this discussion. The cells using the same frequency as the serving cell in clusters 'A' through 'F' are shown shaded. Reuse cell 'F' can clearly be ignored, as the mobile 'Q' is directly in line with the antenna back lobe and will typically be 25 dB to 30 dB down on the main beam. For sites 'A' and 'E', the mobile will be between 90° and 100° off bore-sight. Although the cells notionally require 3 dB beam widths of ± 60°, operators typically deploy antennas with beam widths of ± 45° or even ± 40° to encourage crisper handover. Reference to typical antenna patterns shows that at 100° off bore-sight the gain will be −15 dB to −20 dB below the main lobe. For the purposes of this exercise, interference from these sites can also be ignored.

Clusters 'B', 'C' and 'D' can cause very significant interference. The reuse site for cluster 'C' presents main beam gain to the mobile whilst the reuse sites 'B' and 'D' are within 2 dB or so of main beam gain. For the purposes of this analysis all three sites will be assumed to offer main beam gain to the mobile, a slightly pessimistic view. Recall from Chapter 3 that in many deployments the propagation loss follows a fourth power law:

$$P_r(d) = \frac{P_t G_t G_r (h_t h_r)^2}{d^4} = \frac{{E_T}^2 K}{d^4}.$$

Trigonometry reveals that the distance XQ$(= \text{ZQ}) = \sqrt{13}\,R$ and YQ $= 4R$. If a pessimistic view is taken and the distance for all three significant interferers is set equal to $\sqrt{13}\,R$, the strength of the interferers relative to the wanted signal can now be evaluated.

It is also important to consider the absolute stability of the BTS signals to determine whether these will add coherently (on a vector basis) or on a power basis. Reference to the relevant ETSI specification [11] indicates that the BTS carrier must lie within ± 0.025 ppm, thus two base stations cannot be more than about 50 Hz apart; this amounts to a variation of $\pm 5°$ either side of the centre of each 0.577 ms burst and can be considered coherent over the time slot. Then, for carriers such as the BCCH, containing deterministic modulation for at least some time slots, coherent addition can occur for the interfering signals and this must be considered in evaluating the interfering power *cumulative distribution function* (CDF). In contrast, interfering carriers containing only traffic channels lose their mutual coherence since the voice traffic is generally uncorrelated. To assess the more challenging interference case, it is appropriate to add the three signals, assuming they are coherent, convert to power and then assess the probability distribution of this aggregate power.

Then for the coherent case, with each interfering signal of magnitude E_{T} and the other two vectors at angles α and β to the first, the resulting vector can be written:

$$V = E_{\mathrm{T}}(1 + \mathrm{e}^{\mathrm{j}\alpha} + \mathrm{e}^{\mathrm{j}\beta}),$$

and the vector magnitude squared, a measure of interference power may be written:

$$|V|^2 = V \times V^* = P, \quad \text{where } V^* \text{ is the complex conjugate of } V.$$

Some manipulation of exponentials reveals that:

$$P = E_{\mathrm{T}}^2(3 + 2\cos(\alpha) + 2\cos(\beta) + 2\cos(\alpha - \beta)).$$

For convenience, an interference multiplier, Q, is defined as the ratio of the sum of the interference powers, P, normalised to ${E_{\mathrm{T}}}^2$. Some numerical analysis of this expression for large numbers of random combinations of α and β reveals that values of Q for confidence percentiles of 50%, 90% and 95% are 2.2, 7.1 and 8.0, respectively. Then the ratio C/I can be written:

$$\text{C/I} = \frac{{E_{\mathrm{T}}}^2 \cdot K/R^4}{Q \cdot {E_{\mathrm{T}}}^2 \cdot K/(\sqrt{13}R)^4}, \quad \text{where}$$

K represents a constant common to the numerator and denominator,
Q is an interference multiplier dependent on required coverage confidence limits.

The corresponding C/I ratios for confidence percentiles of 50%, 90% and 95% are 18.9 dB, 13.7 dB and 13.2 dB, respectively. This appears to exceed the required 11 dB threshold for GSM at all times. However, further margins have to be allowed for Rayleigh fading and log–normal shadowing as, although the effects are experienced by both wanted and interfering paths, the probability of these effects manifesting themselves simultaneously on all three interfering paths is low. A factor not included in this analysis, which can significantly *reduce* the level of interference, is the use of antenna down-tilt. Vertical

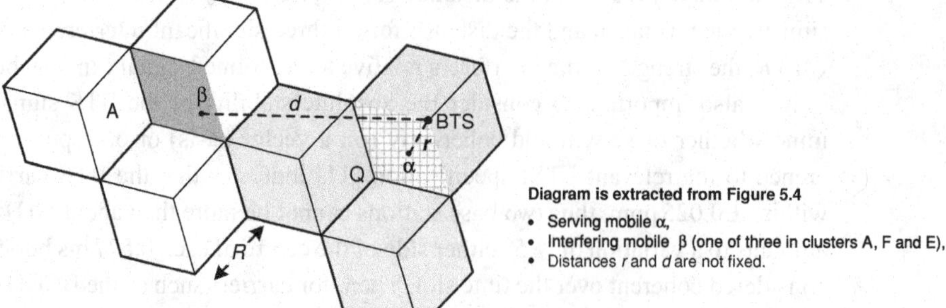

Figure 5.5 Detail of the uplink interference scenario

antenna main beam gains of up to 17 dB are readily available from vendors. Antenna down-tilt can be used to maximise this gain at the edge of the serving cell causing interference from the reuse cells, at shallower angles, to be significantly attenuated. This benefit, in reality, arises from a combination of the antenna tilt, the geography of the particular cell site and the characteristic localised signal 'lobing' in deployment due to the varying phase of ground-reflected multi-path.

In practice both 3 × 9 and 4 × 12 repeat patterns are used by operators for their BCCH plans. The 4 × 12 pattern clearly offers a higher confidence of coverage because the distance between wanted and unwanted BTSs is greater, but the particular pattern chosen will depend on the amount of spectrum available, the height of the BTS antennas (relative to the prevailing building height) and the confidence limits on coverage that need to be achieved. Additionally, non-uniform frequency plans can be used for the BCCH, which take advantage of geography and subscriber distributions to deliver an even tighter frequency reuse.

The above discussion illustrates the difficulties of precisely evaluating C/I. Nevertheless it can be seen in broad terms that adequate C/I ratios can be achieved and that it is generally sensible to operate the BCCH carriers on a less tight reuse pattern than the voice channels to ensure that these can always be correctly decoded.

5.1.2.2 Uplink interference scenario

Uplink interference arises when the S/N ratio at the BTS in the serving cell for a particular *physical channel* is degraded by interference from the same physical channel in the reuse cells. Examining Figure 5.4 once more, the desired uplink is that between the mobile Q and the serving base station, in the cross-hatched cell. Potential interferers are dictated by the antenna pattern of the *serving* cell and comprise active mobiles that share a common frequency and time slot with the served mobile *and* that can be seen by the serving BTS. For the 3 × 9 repeat pattern, the diagram shows that there are three potentially interfering mobiles in cells A, F and E.

Figure 5.5 shows one of the interfering mobiles, β, and the served mobile, α, in more detail. There are a number of differences from the downlink scenario, which need to be recognised before the uplink interference can be accurately modelled.

- The sources of co-channel interference, along with the served user are all *mobile*,
- GSM contains a *power control* feature, which allows the BTS to manage the mobile transmit power to a level that is just able to provide adequate call quality at the BTS, and thus minimise co-channel interference,
- In contrast to the downlink model, where the BCCH channel has always to be on and transmitted at maximum power, not all mobiles need to be active at any one time.

All of the above means that it is not practicable to construct a simple static model with the attendant ease of getting ballpark figures to address uplink interference. Instead, modelling is carried out using a series of snapshots, with the serving and interfering mobiles distributed with uniform probability – both in range and azimuth. For each 'drop' of the mobiles, log–normal shadowing and Rayleigh fading are introduced, along with the appropriate path loss, to establish the required transmitted power for the particular mobile and its associated serving BTS (e.g., for mobile β, this would be in cluster 'A'). Assessments of the co-channel interference at the serving BTS in cluster 'Q' can then be made.

Again, for each drop, an assessment will be made regarding the probability that each potentially interfering mobile is *actually* active. This probability will be calculated from a knowledge of the total system load and the number of frequencies available to the operator.

By now it will be apparent that uplink performance is evaluated on a statistical basis and the detailed methodology will not be discussed here; however this is well addressed by Koshi *et al.* [12] and McGuffin [13]. Suffice it to say that, in general, C/I on the uplink is rarely a problem. On the downlink, interference is dominated by BCCH carriers, which are required to transmit on full power (this interference is much greater than that from power-controlled mobile uplink). Even on non-BCCH carriers, the use of frequency hopping combined with the widely distributed and non-static location of uplink interferers means that in this configuration, too, the uplink is rarely the limiting link.

5.2 GSM system functionality

The GSM standard at Phase 2+ and beyond is rich with optional features designed to improve the link-level and system-level performance. Indeed, the benefit of some of these features is so great that many are activated on almost all deployed systems. The principles of operation of the key features and an indication of their advantages will be discussed here. The aggregated benefits of these features working in harmony will be seen in Section 5.2.3.

5.2.1 Frequency hopping

The GSM system includes a facility to change the frequency used for transmission 217 times/second (every burst period of 4.6 ms); this is known as slow frequency hopping (SFH), in contrast to some systems that change the frequency every symbol. Frequency hopping usually occurs across all frequencies available to the operator, other than those reserved for the (non-hopping) BCCH frequency plan.

Slow frequency hopping was introduced to ensure that fading between successive bursts of GSM was uncorrelated, even for slow moving mobiles. This feature, together with *convolutional coding* and the interleaving scheme [14], mitigates the effects of Rayleigh fading. With random frequency hopping across the GSM spectrum it is unlikely that more than one or two of the eight bursts, over which the GSM voice channels are interleaved, will be significantly affected and the channel coding can then recover any errors. This benefit of SFH arises from the *frequency diversity* introduced and benefits both the uplink and the downlink. This should be contrasted with *spatial receive diversity* at the BTS, which improves the uplink but would need to be introduced on most mobiles to benefit the downlink. Note that the same effect, reduction of Rayleigh fading, is at work here, so the introduction of SFH *and* receive diversity does not give twice the benefit. For *TU50* propagation and no receive diversity in use, about a 2 dB reduction in required C/I from 11 dB to 9 dB can be assumed [15] and this can be translated into capacity gain.

The second benefit of SFH is the introduction of *interference diversity*. Without SFH, and for a heavily loaded system with the corresponding incentive to adopt a tight reuse pattern, it is often the case that a small number of cells could suffer a higher level of interference than average because of unusual propagation conditions (co-channel interference) or neighbour cells transmitting on carriers close to the serving cell (adjacent-channel interference). Because these interference conditions are static, it might eventually demand reversion to the looser reuse pattern, to obtain acceptable performance, with the corresponding loss of maximum capacity. Slow frequency hopping is able to eliminate this problem substantially since different frequencies will interfere with the carrier at different times – often referred to as interference averaging; this can be better tolerated because of the channel coding and interleaving. There is also an additional benefit, which arises because the SFH makes it very unlikely that a significant number of bad C/I conditions will occur *in sequence*, which the coding would be unable to address.

The factors that influence the interference diversity benefit with SFH are the number of frequencies over which the system is hopped, the channel utilisation (the number of time slots occupied) and the difference in levels of the significant interferers [16]. At one extreme, if interferers are at the same levels in each cell and channel utilisation is 100%, then there is *no* interference diversity benefit! However, even in this situation, different reuse configurations can allow a considerable benefit. Consider an operator with 48 GSM frequencies. Assume that 12 frequencies have been withdrawn from the set to support a non-hopping 4 × 12 BCCH plan. Then 36 frequencies are available to support other traffic channels. Consider the potential capacities of a 3 × 9 and 1 × 3 reuse plan. For the 3 × 9 system, there are four frequencies available per sector, giving rise to a total of 32 physical channels. The corresponding number for the 1 × 3 system is 96. With reference to Erlang B tables and for 2% blocking this translates to a capacity of about 23 Erlangs and the channel loading is 23/32, 72%. For the 1 × 3 reuse case and supporting the same 23 Erlang/sector load, there *is* no hard blocking limit and channel loading is about 24%. For real systems, co-channel interference has to be considered. In the 3 × 9 system, hard blocking is reached before this is an issue; for the 1 × 3 system, capacity will be co-channel interference limited (*soft blocked*) but at capacities much higher than 23 Erlang, depending on the precise configuration. Some example figures are included in Section 5.2.3.

5.2.2 Power control and DTX

Minimising co-channel interference is a key goal in any cellular system, since it allows for either better user service quality or higher system capacity (through tighter frequency reuse), or a combination of these goals. In addition to SFH, GSM introduces two features specifically aimed at reducing co-channel interference; mobile and BTS power control and discontinuous transmission (DTX).

The power control feature recognises that it is not necessary for a mobile operating next door to the BTS to transmit the same amount of power as one at the cell edge, to achieve a specified link quality. In practice, there are adverse effects if power control is not adopted. The relative amount of interference from the reuse cells increases (mobiles there will also transmit at maximum power) and is particularly significant for mobiles at the cell edge. The mobile battery life is also shortened. When the mobile power is managed by the BTS to deliver just the required link quality, this translates to reduced co-channel interference, which can be used to support higher call quality or system capacity, depending upon the reuse configuration adopted. Although carriers supporting the BCCH in the downlink are *required* to operate at maximum power, other downlink carriers can operate power control to provide capacity benefits. Simulation suggests that a 3 dB improvement in C/I is possible compared with the non-power-controlled case, offering capacity improvements of 33% [17].

Discontinuous transmission (DTX) is a further enhancement aimed at co-channel interference reduction; it recognises that a person typically speaks less than 40% of the time in normal conversation. By turning the transmitter off during silence periods, co-channel interference can be reduced by about 50%. Discontinuous transmission is particularly beneficial in SFH systems. Without SFH, the frequency reuse plan must still be designed to address the worst case interference when the channel is active. With SFH, it is only necessary to design for the average interference level. A further benefit of DTX is that, again, power is conserved at the mobile unit.

5.2.3 Reuse patterns and fractionally loaded systems

GSM networks will usually employ two frequency reuse patterns to maximise system capacity. The repeat pattern for the BCCH will normally be 4×12, i.e., clusters of four tri-sectored sites. This is to ensure that the C/I for both the downlink and the uplink are such that the mobile can register if there is any possibility of viable commercial service; recall also that the BCCH carrier has to be transmitted at full power and without frequency hopping. However, almost invariably, all remaining spectrum will be employed in a much tighter reuse configuration, with SFH active. The prevalent configuration for non-BCCH carriers is a 1×3 pattern. As was discussed before, capacity in this configuration is subject to *soft blocking*; i.e., capacity has to be limited to a level that is less than 100% of the available number of GSM channels (the product of the number of carriers and the number of time slots (eight)). Why is a soft-blocking configuration so attractive? Here are some of the key reasons:

- Very tight reuse patterns maximise the number of frequencies available for SFH. The discussion on SFH highlighted that interference diversity is most efficient when the number of available hop frequencies is high and the channel loading is low, and this increases spectrum efficiency.
- Where the amount of spectrum is limited, tight reuse patterns allow efficient use of available spectrum whilst still providing uniform capacity and easy planning. For instance if 4 MHz of spectrum is available (= 20, 200-kHz channels), once 12 channels have been removed for BCCH, the remaining 8 channels cannot be used in a 4 × 12 or 3 × 9 plan. A 1 × 1 plan with soft blocking might be a good choice in this instance. Even where higher reuse plans can be used (e.g., if 4.2 MHz were available above) the trunking inefficiency of such single carrier cells would not be acceptable.
- The use of soft-blocking configurations and SFH mean that the results of any method of improving the system C/I margin can immediately be shared with each customer in the form of improved call quality or exchanged to support more users on the system. For instance, the introduction of *power control* and *DTX* can each provide about a 3 dB reduction in the level of co-channel interference. Soft-blocked systems can immediately increase the number of users until the interference in the system rises once more to the target C/I.
- Increasingly, as new methods are found of achieving satisfactory BTS and mobile performance at lower C/I ratios, soft-blocking systems can immediately realise the capacity or quality benefit. Features that allow reductions in target C/I which are already available include the *adaptive multi-rate codec* [18, 19] *interference rejection combining* [20] and, in prospect, forms of single antenna interference reduction (SAIC) [21].

The question arises: how much is the GSM capacity improved by the system features just discussed? It is impossible to give an absolute answer, as the capacity benefit will depend on a large number of factors including:

- The amount of available spectrum (trunking efficiency for hard blocked systems),
- Mobile velocity (C/I target),
- Whether receive diversity is deployed (C/I target),
- Percentage of mobiles below quality threshold (1%, 5%, 10%).

Nevertheless, there have been a number of papers that have attempted to address this question [22, 23, 16 and 24] and because the particular details of each study are not the same it is difficult to make comparisons. However, Wigard's figures [22] seem amongst the most conservative and some results from his work are shown in Table 5.2.

Wigard chooses to carry out his comparison for a network with 36 frequencies available in addition to the allocation needed for the BCCH overlay. Whereas the 3 × 9 hard-blocked plan provides a 40% increase in capacity, relative to the 4 × 12 reference configuration, the 1 × 3 soft-blocked system delivers a 73% capacity gain. Whilst the 1 × 3 solution demonstrates a significant benefit, in practice it is unnecessarily

Table 5.2. Benefits of soft-blocking networks

Reuse pattern	Frequencies/ cell	Channels/ cell	Blocking (2%)	Erlangs/ cell	% increase (vs. 4 × 12)	Number of frequencies
4 × 12	**3**	**24**	**Hard**	**16.6**	**Reference**	**36**
	2	16	Hard	9.8	−41	24
3 × 9	4	32	Hard	23.7	+42	36
	2	16	Hard	9.8	−41	24 (6 spare)
1 × 3	12/8	96/64	Soft	28.8	+73	36 **or** 24
1 × 1	36/24	288/192	Soft	20.2	+22	36 **or** 24

pessimistic. In most networks, the maximum operator frequency allocation is about 7.5 MHz (37 GSM frequencies) and from this must be subtracted the BCCH allocation of 12 frequencies and guard bands. It is, therefore, more likely that the non-BCCH allocation will be 24 frequencies. A row has been added, indicating the capacity of the same 4 × 12 configuration for this 24-channel case – which has significantly lower capacity due to higher Erlang blocking. The earlier capacity benefits normalised to this configuration are even more impressive! Note that in practice the BCCH carriers will carry some traffic but the benefits of soft-blocked systems, which can exploit the improved C/I of the advanced GSM features, are still impressive.

5.3 GPRS and EDGE

In Chapter 3, when planning for multimedia traffic was discussed, the approach for data systems was reviewed, but commentary on the GPRS and EDGE systems capacity was limited to the observation that these solutions would operate within the framework of the link budget dictated by speech. It is now appropriate to understand better the reality of these constraints.

In reviewing the limiting C/I system scenarios, it was apparent that the worst-case situation was usually to be found at the cell edge – for both the downlink and the uplink. Further thought makes it clear that a continuum of increasing C/I can be expected as the user moves towards the BTS, a consequence of the improving strength of the wanted signal and weakening reception of the interference. In voice systems this is interesting but of no consequence; once the transmission errors are insignificant, voice quality does not improve. For data systems, the opportunity is to operate at higher speeds when the C/I ratio permits (in practice, modern adaptive multi-rate codecs allow higher voice quality too). Then how can the improved C/I be exploited to increase the data rates?

There are three main classes of solutions to explore:

- Reduced channel coding,
- More bits per symbol,
- Incremental redundancy.

Table 5.3. GPRS and EDGE performance

Scheme	Coding scheme	Rate (kbits/s)	Coding family	Modulation	Coding rate	Payload unit (bits)	Number of payload units	Interleaving period (bursts)	Typical C/I (dB) (TU50)
GPRS	CS1	9.05	–	GMSK	1/2	181	1	4	−5 to 7
	CS2	13.4	–	GMSK	~2/3	268	1	4	7 to 11
	CS3	15.6	–	GMSK	~3/4	312	1	4	11 to 17
	CS4	21.4	–	GMSK	1	428	1	4	>17
EDGE	MCS1	8.8	C	GMSK	0.53	176	1	4	−5 to 3
	MCS2	11.2	B	GMSK	0.66	224	1	4	0 to 5
	MCS3	14.8	A	GMSK	0.85	296	1	4	2 to 7
	MCS4	17.6	C	GMSK	1	176	2	4	4 to 9
	MCS5	22.4	B	8PSK	0.37	224	2	4	5 to 10
	MCS6	29.6	A	8PSK	0.49	296	2	4	10 to 15
	MCS7	44.8	B	8PSK	0.76	224	4	4	13 to 20
	MCS8	54.4	A	8PSK	0.92	272	4	2	20 to 25
	MCS9	59.2	A	8PSK	1	296	4	2	>25

Reference to Table 5.3 will show that one or more of these approaches is used in the GPRS and EDGE standards.

Channel coding is introduced to detect and, wherever possible, correct bits that have been corrupted because of bursts of interference or Rayleigh fading of the wanted signal. These impairments can cause the C/I to fall below that required for error-free decoding of the signal. Table 5.3 shows that CS1 for GPRS employs a half-rate code, which would not be required away from the cell edge. Coding schemes CS2 to CS4 progressively weaken the protection offered by the coding, until in CS4 there is no protection at all. However in high C/I conditions, this is rewarded by a data rate for CS4 that is more than double that for CS1. As the user moves away from the cell edge, a process of *link adaptation* takes place to change to progressively weaker coding schemes. The mobile reports its received signal strength *RxLev* and an estimate of link quality *RxQual* every 480 ms and these are the key inputs to the BSC in making its judgements on when to change to a new coding scheme. There has been much analysis carried out on the required C/I needed to support a given BER or *BLER* (transmitted block erasure rate) [25, 26, 27]. This is found to be heavily influenced by the propagation environment, the user velocity, and whether or not SFH is operating, as well as the prevailing C/I. For information, an indication of typical C/I operating ranges for different coding schemes is included in Table 5.3 along with other pertinent information. The data are for TU50 conditions with SFH operating. Further information about the detail of GPRS and its coding schemes is available in the 3GPP specifications [28, 29, 30].

EDGE was conceived as an enhancement of GSM and of GPRS in particular. Enhancements usually labour under the constraint that they must not disturb the 'expensive' parts of the current solution and that they must demonstrate a key performance improvement;

EDGE was no exception to this constraint. For this reason, obvious methods of improving user throughput by increasing the symbol rate (with the corresponding need to seek new wider channel allocations and replace all power amplifier, filtering and combining equipment) could not be entertained. Instead, the basis of the EDGE solution [31] is to leave the symbol *rate* unchanged but move to a higher order modulation scheme *increasing the number of bits per symbol*. The GMSK solution used by GPRS is augmented by a number of transmission modes employing 8PSK (8 state phase shift keying), increasing the number of bits per symbol from one to three in these modes. EDGE also exploits different levels of channel coding to provide a total of nine *modulation and coding schemes* (MCS1 to MCS9). The same process of link adaptation detailed for GPRS can also be exploited to supervise movement between codes as the user C/I changes. As one would expect from Shannon's Law, additional C/I is required to decode the 8PSK modes at an acceptable BER; in TU50 conditions and with frequency hopping the additional C/I relative to GMSK is some 8 dB to 10 dB. The typical C/I operating ranges, under TU50 conditions, for the different EDGE MCS modes [32, 33] are also included in Table 5.3 together with other relevant information.

EDGE also provides one further level of flexibility particularly suited to RF propagation environments: *incremental redundancy* (IR). As has been mentioned on several occasions, the C/I for radio propagation can degrade quickly and exist only on a transient basis – either because of bursty interference or because of Rayleigh fading. The process of link adaptation is one that is averaged over some number of blocks, each block lasting 480 ms. What is ideally required is for one mechanism to address the random C/I drop of the type just described whilst link adaptation addresses the more sustained reduction in C/I that would be experienced as the user moves away from the BTS. Incremental redundancy is designed to deal with such a transient loss of C/I and the resulting errors. The basic idea is to choose a coding scheme suited to the average C/I and then, if transient corruption of some data occurs, resend a copy (or, if necessary, copies) of the corrupted block until the point that, when all received versions or the block are considered together, sufficient information is available to decode the originally transmitted information correctly. The EDGE format was designed with IR in mind. Further reference to Table 5.3 reveals that, depending upon the coding scheme, more than one 'payload' unit of a standardised size may be transmitted. MCS formats with the same payload size are distinguished by the same 'coding family' label; thus coding family 'A' can transmit one, two or four payload units, each of 296 bits. It is quite likely that corruption may only occur over one of these units and thus the retransmitted unit can be accompanied by one or three others, making a soft increase in coding redundancy. Note also that IR retransmission delay will normally be much less than waiting for link adaptation whilst avoiding the additional unnecessary adverse impact on user throughput for some period of time. These intuitive observations have been confirmed by detailed modelling [34].

To conclude this section on GPRS and EDGE performance it is useful to provide some quantitative information on system performance. The peak data rates that the user might experience in different areas of the cell are shown in Figure 5.6.

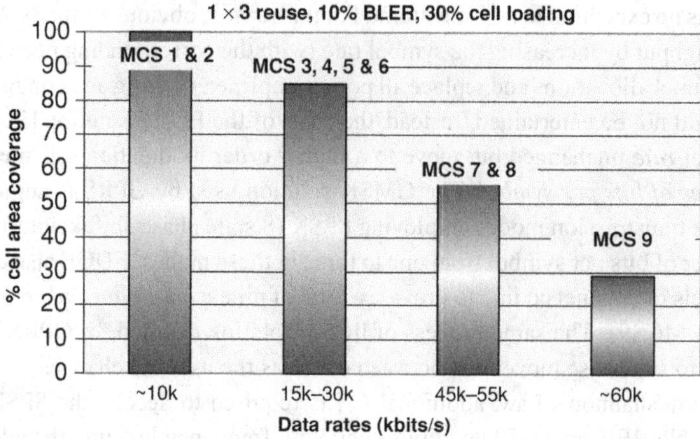

Figure 5.6 EDGE cell coverage

Questions on cell capacity and average user throughput, however, have to be specified very carefully. Factors to be considered include:

- Frequency reuse pattern,
- FER,
- Queuing delay,
- Number of users (load).

So for a useful median user rate assessment, the above parameters (and more) need to be specified; however, it is possible to make a qualitative observation, which is that because the soft-blocked cells, by definition, will have a lower average C/I level than cells with hard blocking, the latter will tend to have higher median user rates. However, when the aggregate traffic through a cell is considered, the soft-blocked cells, as expected, have higher capacity. By way of example, consider the cases shown in Table 5.4.

A more detailed discussion of the trade-off between cell load, median user data rates and spectrum efficiency can be found in [35, 36].

5.4 Access protocols and latency

For UMTS and 3G LTE, this section will be used to review the access protocol structure and develop estimates of system latency. For GSM and GPRS, this has already been covered in Chapter 2 and so all that remains is to 'rate' GSM against the metrics developed in that discussion. These ratings are shown in Table 5.5.

Table 5.4. GSM spectral efficiency

* 0.15 s delay	13 kbits/s full rate voice (4 × 12 reuse)	GPRS* (1 × 3 reuse)	EDGE* (1 × 3 reuse)
Bits/s per Hz per cell	0.043	0.16	0.45

Table 5.5. GSM performance against key metrics (see Section 2.3 for further discussion)

Ref.	Metric	Performance	Comments on GSM
1	Low access latency – 'push to happen'	**700 ms–2000 ms**	From registered but idle state
2	Higher peak data rates – 'push to happen'	**∼200 kbits/s**	Assumes four downlink time slots of EDGE and mobile close to BTS. [In theory, eight slots possible but unlikely with commercial operators]
3	Guaranteed QoS support – enable new applications	**Worst**	The arbitrary call set-up time and differentiated but not guaranteed QoS support mean that guaranteed QoS on packet bearers not practical
4	Enriched HLR functionality – support of new features	**Middle**	This is enabled by IMS and its successive releases. May in practice be limited by (3) above
5	Low-cost configurations – developing market access	**Best**	The very large radius cells possible with GSM and economies of scale for handsets and infrastructure mean that this will be the benchmark standard for low-cost configurations for some years to come
6	Self-configuring network	**Worst**	Some limited self-configuration is possible – e.g., picking up potential neighbour cells off air but much remains to be done
7	Self-optimising network	**Worst**	Relies heavily on operator staff skills and specialist third-party software
8	Common core network for all services	**Worst**	Although possible using IMS, the sheer legacy investment means that this is unlikely except for new 'green-field' networks
9	Convection air-cooled BTS	**Middle**	Pico/micro/minicells only. Macrocells invariably require fans or other forced air approaches
10	Common application support platform	**Worst**	As (8) above
11	Reduced back-haul costs	**Worst**	The most expensive part of back-haul is the BTS/BSC link and this remains circuit E1. Packet back-haul is possible with upgrades but is usually not possible to justify over remaining equipment life
12	Higher cell capacities – reduced capital	**Middle**	With soft-blocking cells, continuing evolution of EDGE and large cell capability this remains a more economic solution than UMTS for today's traffic
13	Reuse of currently deployed equipment	**N/A**	This is the reference for UMTS and 3G LTE
14	Ability to reuse existing licensed spectrum	**Best**	GSM has the narrowest channel bandwidth, and with soft-blocking solutions can be deployed in the narrowest band allocation. Product variants at 450 MHz, 900 MHz, 1800 MHz, 1900 MHz and possibly 700 MHz mean that it is most likely to have standard products for available spectrum
15	Is broadband data handover needed?	**N/A**	
16	Revised standards IPR policy	**N/A**	ETSI/3GPP policy developed around GSM

Wherever possible, absolute performance figures have been included. However in the majority of cases the metrics do not lend themselves to absolute performance figures (without a vast array of qualifications) and so the temptation is to degenerate into non-absolute terms, such as 'good' or 'poor'. For this reason and to provide some reference, the three solutions have been rated against each other by use of the terms 'worst', 'middle' and 'best'.

5.5 GSM RAN planning example

This section will address the detailed factors that need to be considered when undertaking GSM cell planning. Figure 5.7 illustrates a typical design flow. This will essentially be the flow followed in planning almost any network, although by no means all of the detailed activities are needed in every case. For instance, parts of process steps identified in (1), (2), (4) and (6) may not need to be considered, depending on the flexibility offered by the network solution and whether or not the new network will complement coverage already in place. This *detailed* design activity will usually only occur after the planning process of Section 3.3.6 has been completed and a clear business plan has been established and approved at company board level. The process will then take as its input the cell size and capacity requirements developed in the business plan. This same design flow will be adopted for the other worked examples included in Chapters 6, 7, 8 and 9. Note also that, in combination, the worked examples have been chosen to illustrate most aspects of the planning process, but that not all aspects (e.g., estimation of maximum cell size) are detailed in every case.

5.5.1 GSM features

The intention of this section is to focus on features that play a key role in shaping the performance of the particular air interface under discussion, even though some of the concepts may have been introduced earlier in the chapter. Worked examples in subsequent

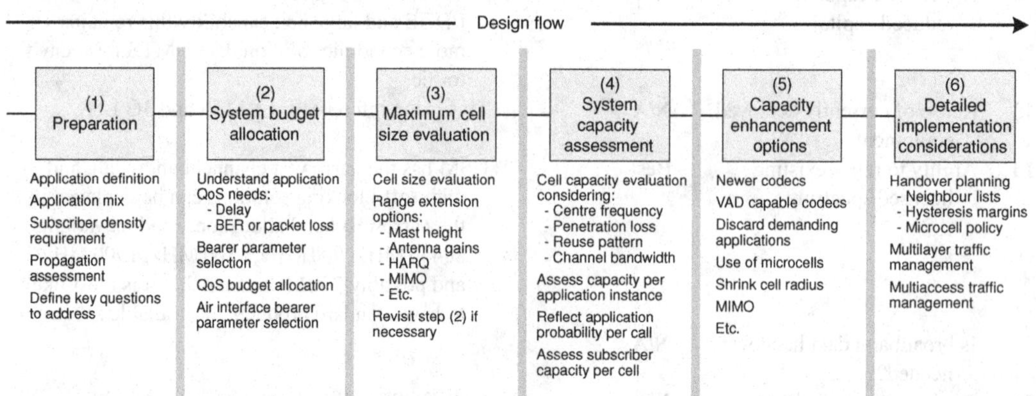

Figure 5.7 Wireless network design – a typical design flow

Table 5.6. Summary of GSM key features

Feature	Benefits			
	Capacity	Coverage	QoS	Deployment flexibility
Frequency band	✓	✓		✓
Sectorisation	✓	✓		
Frequency plan	✓		✓	
Slow frequency hopping	✓		✓	✓
Codecs	✓		✓	
VAD/DTX	✓			
Control channels			✓	
Handover	✓		✓	

(✓) Primary benefit (✓) Secondary benefit

chapters will contain similar sections, but will address those key features that are peculiar to that air interface alone. Thus, the worked example on UMTS will not, for instance, discuss new codecs (which are indeed used in UMTS) as they will be addressed in this GSM section.

Table 5.6 summarises the key features of GSM and indicates their primary benefits (in black) and in many cases also highlights secondary benefits (in grey) which can cause the feature to be deployed, even if the primary benefit is not a consideration in the particular case. Each of these features will now be discussed.

5.5.1.1 The GSM frequency bands

The frequency bands used by GSM systems are shown in Table 5.7. Separate bands are used for the uplink and downlink radio channels; in the original GSM 900 band, these channels are 45 MHz apart; in the DCS 1800 band, they are separated by 95 MHz. Each band is subdivided into a number of radio channels with a 200 kHz separation, with an additional guard band of 200 kHz at the beginning of each band. The individual channels in the band are given an absolute radio frequency channel number (ARFCN), which identifies a pair of channels allocated to a mobile for communication on the uplink and downlink. The original GSM band thus defines 124 channels with ARFCNs in the range 1 to 124 inclusive. The addition of 10 MHz to the extended GSM (EGSM) spectrum defines an additional 50 ARFCNs, numbered 975 to 1023 inclusive. The GSM 1800 frequency

Table 5.7. GSM frequency bands

	Downlink	Uplink	Number of channels
GSM 900	935–960 MHz	890–915 MHz	124
EGSM 900	925–960 MHz	880–915 MHz	174
GSM 1800	1805–1880 MHz	1710–1785 MHz	374

band at 1800 MHz defines a further 374 ARFCNs, numbered from 512 to 885 inclusive.

In most countries, national regulatory authorities subdivide the GSM frequency spectrum and provide network operators with licenses to operate within a certain frequency range. Consequently, the full range of ARFCNs is not normally available to the radio network designer, and careful planning is required to ensure that capacity and quality are maintained, and the effects of interference are minimised.

5.5.1.2 Frequency reuse

Because of the finite spectrum available, it is always desirable to use as tight a reuse pattern as the particular air interface allows so that capacity is maximised. Discussion in Section 5.1.1.2 (and reflected in the GSM recommendations) indicates that the C/I figure for co-channel interferers should be 9 dB or more. In practice, it is prudent to plan for at least 12 dB C/I, to allow for the effects of the *actual* propagation departing from the (usually) empirical law. There is also a requirement for the interference from 'adjacent' channels (separated by 200 kHz) to be less than −9 dB.

There are also other effects to consider. The radio planning process described earlier in Sections 5.1 and 5.2 allows the interference caused by frequency reuse to be estimated and various different reuse strategies to be evaluated. However, the chosen strategy cannot always be implemented in practice. Figure 5.8 illustrates the principles of cellular reuse for an arbitrary reuse pattern (in this case seven cells). The available radio channels are arranged into seven equal groupings (numbered 1–7), and each group of frequencies is assigned to one of the cells in a seven-cell cluster. The allocation is then repeated across the networks for each of the tessellating tiles of hexagonal cells.

The deployment strategy that hypothetically required a seven-cell reuse pattern assumed an ideal perfect hexagonal lattice for cell-site locations. In practice, there are often significant deviations from this pattern (owing to site availability) which result in non-ideal interference patterns for some cells, and adjustments to the rigid reuse plan will be required. Most commercial radio planning tools provide the option to use an automatic frequency planner (AFP), which examines a matrix of server-to-interferer estimates in

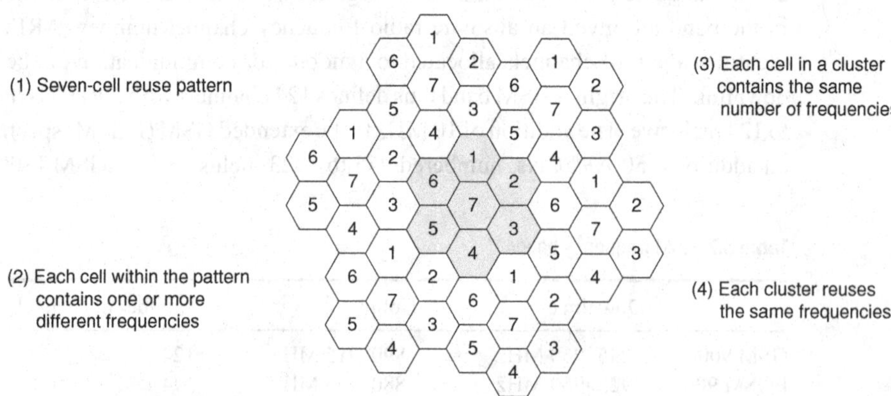
(1) Seven-cell reuse pattern
(2) Each cell within the pattern contains one or more different frequencies
(3) Each cell in a cluster contains the same number of frequencies
(4) Each cluster reuses the same frequencies

Figure 5.8 Ideal cellular reuse

the region under study, and then allocates frequencies to cells and calculates the overall cumulative system C/I. Search algorithms are then used to reallocate frequencies between cells in an attempt to minimise the interference, for a given number of available radio channels. During optimisation, even non-uniform BCCH frequency plans can be developed, based upon observed network performance on a loaded operational network.

5.5.1.3 Sectorisation

Initial network build usually focuses on providing coverage for the majority of the target population. To maximise coverage at the lowest cost, cells with omnidirectional antennas (360° azimuth) are usually deployed first. As the traffic on the system increases, these 'omni' sites can be subdivided by installing antennas that are directional in azimuth (as well as elevation) – albeit at the cost of providing additional transceivers. Usually, sites are tri-sectored, with each of the three cells covering approximately 120°, although other configurations (notably six-sector) are also possible. Note that, though the site may be tri-sectored, with each cell notionally covering 120°, the antenna −3 dB beam widths are typically narrower than this – in the region of 80° to 90° – to ensure decisive handover. This technique, called sectorisation or cell splitting, allows for tighter frequency reuse plans but at the cost of increased site complexity. Most manufacturers of GSM infrastructure allow a single BTS to control several 'cells', each of which provides coverage in one sector. Intra-cell handovers are performed either by the BSC, or in some cases by the BTS itself to minimise A_{bis} traffic due to handover control messages.

5.5.1.4 Frequency hopping

As an alternative to a fixed frequency plan, traffic channels can be deployed using frequency hopping (FH). In GSM the BCCH channel must always be transmitted on a fixed frequency, and so the BCCH carrier is invariably deployed according to a static frequency reuse plan. The remaining transceivers (TRXs) allocated to the site can, however, either be tuned to a fixed frequency, or frequency-hopped. In a frequency-hopping system, the TRX and MS retune to a new frequency in each time slot. The sequence of frequencies is known to the MS and TRX and so they are able to remain in communication. Nearby sites that reuse these frequencies also use FH, but with a different hopping sequence; consequently co-channel interference only occurs on the occasions when the serving cell and interferer happen to be transmitting on the same frequency – in other words, only on occasional time slots. The interference impairment caused to transmission is thus greatly reduced. An additional advantage is that the interference is effectively spread amongst all users of a site, which is not the case for fixed frequency reuse, resulting in improved overall system quality. The advantages of frequency hopping are denser reuse patterns and, hence, improved spectral efficiency and capacity gains.

There are two types of frequency hopping defined for GSM: baseband hopping (BBH) and synthesiser frequency hopping (SFH). In BBH, the number of unique frequencies in the hopping sequence is the same as the number of transceivers in the cell. Each TRX transmits on a fixed frequency, but baseband signals for each call are switched to a different transceiver every frame. In SFH, the non-BCCH transceivers retune to a new frequency every time slot, and the number of frequencies in the hopping set can be greater

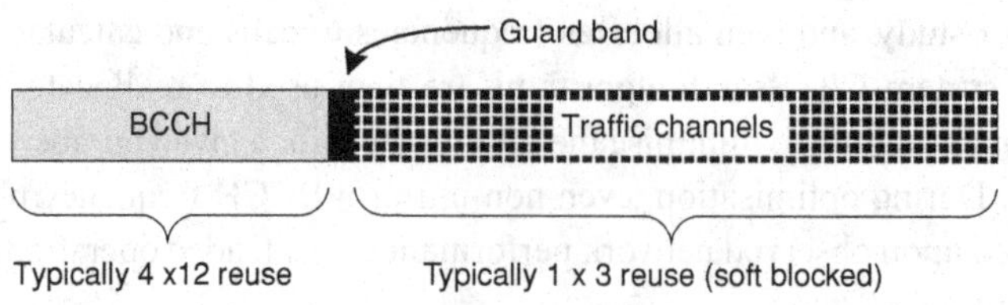

Figure 5.9 Separation of BCCH and TCH bands

than the number of transceivers, with the attendant benefits. The BCCH frequency may not be included in the list of hopping frequencies when SFH is used.

Synthesiser frequency hopping requires hybrid combining at the cell site; cavity combining is not possible. Hybrid combining results in some power loss and this must be taken into account when planning the cell-coverage requirements for the SFH system. In BBH systems, either cavity combining or hybrid combining may be used. For SFH, the BCCH channel remains fixed, and it is good practice to partition the available ARFCN between BCCH and TCH carriers, as shown in Figure 5.9; this makes for easier planning and better interference control.

The number of channels required for the BCCH band will vary, depending on the system capacity. The BCCH frequencies can then be planned using a 4×3 (4-site, 3-sector) or 5×3 reuse pattern. The reuse plan for the TCH frequencies can be much tighter, and a 1×3 plan is common. The available TCH channels are divided into three equal groups, and these are the ARFCNs which are defined in the MA for the SFH system. The number of channels required to support the planned traffic is determined by the design loading factor or carrier-to-frequency ratio. Synthesiser frequency hopping has a theoretical maximum loading factor of 50% but, in practice, loading will be limited to 30–40% to prevent unacceptable interference being encountered. In lightly loaded systems, for example sparsely populated rural areas, it is even possible to consider deploying a 1×1 plan for the TCH frequencies.

The frequencies to be used in the hopping sequence are defined for each cell in the mobile allocation (MA). Two types of hopping sequence are defined for GSM – cyclic hopping and pseudorandom hopping. In cyclic hopping, the frequencies follow a fixed sequence. However, in pseudorandom hopping, a randomly generated sequence is used to select frequencies from the MA. This process is controlled using the hopping sequence number (HSN), which takes a value in the range 0 to 64. A value of 0 indicates that cyclic hopping is to be used. A HSN in the range 1–64 indicates pseudorandom hopping; the HSN value is used as a seed to the random number sequence, consequently 64 possible sequences are possible. To avoid clashes between cells using the same HSN and MA, a mobile allocation index offset (MAIO) can be used; this defines the starting position in the pseudorandom hopping sequence.

In practical deployments, co-channel and adjacent-channel interference must be minimised through judicious allocation of HSN and MAIO. The HSN should be set to different values on different sites as far as possible (up to the maximum value of 64) as this will maximise the randomisation and possibility of clashes. On sectors on the same site, the same HSN should be used, and the MAIO value set differently on the

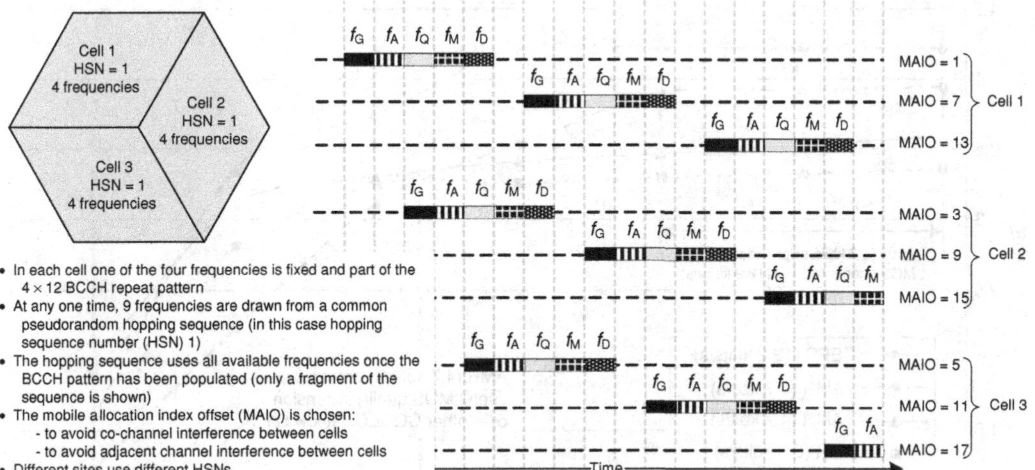

Figure 5.10 Using MAIO planning to avoid co-channel and adjacent-channel interference

different sectors to minimise the possibility of co-channel and adjacent-channel clashes on co-sited sectors. An example MAIO/HSN allocation for a 1 × 1 reuse plan is depicted in Figure 5.10. If the ARFCNs are allocated in a monotonically increasing sequence, then the MAIO numbers can be used to avoid adjacencies.

5.5.1.5 Improved codecs

In addition to the techniques just discussed, which seek ways of increasing capacity by reducing the reuse factor, an alternative approach finds ways to reduce the number of bits sent to convey the required information. For speech services, this approach relies on more modern codecs that compress the speech artefacts into a small number of 'tokens' for transmission across the air interface. On receipt, these same tokens are restored to their original form by the use of powerful signal processing.

When GSM was initially deployed, speech coding was carried out by a 13 kbits/second 'full rate' (FR) codec. Since then, a number of additional codecs have been developed for GSM including half rate (HR, 5.6 kbits/s), enhanced full rate (EFR, 12.2 kbits/s) and the AMR (adaptive multi-rate) codec [19], which provides a family of six codecs supporting speech at rates from 4.75 kbits/s to 7.95 kbits/s. By providing a family of codecs, it is no longer necessary to compromise the resulting speech quality by designing one codec to provide acceptable performance over the full range of C/I ratios encountered in the network. Instead, each codec member is optimised to deliver the best speech quality for a specific range of C/I values. As the user C/I varies, the network can deliver the best speech quality through dynamic adaptation of speech and channel coding rates. The AMR algorithms select the optimal channel rate (half rate or full rate) and speech codecs to provide the best combination of quality and capacity, according to the network designer's objectives. AMR half rate and its precursor, GSM half rate, allow the number of voice calls supported on the air interface to be doubled (although the capacity of signalling channels is unaffected).

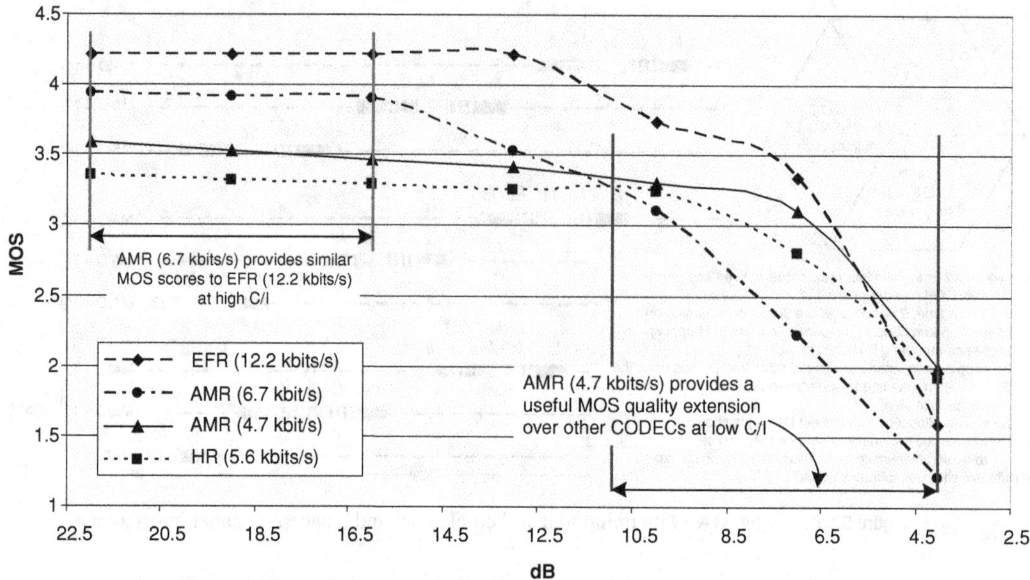

Figure 5.11 Mean opinion score vs. C/I for various codecs

GSM-HR was introduced earlier in the standards process and so the field population of mobiles that support it is quite high; it is thus a viable option for operators wishing to provide capacity at low cost, on a short timescale. AMR-HR was introduced more recently and has better speech quality than GSM-HR. In both cases, however, the gains in capacity are at the expense of a decrease in perceived speech quality, and an increase in background noise. Figure 5.11 shows how the perceived voice quality (MOS) varies as C/I reduces, for various codecs.

Enhanced full rate (EFR) is a modern codec with an encoded speech rate of 12.2 kbits/second. It is interesting to note that the AMR 6.7 kbits/s codec has almost the same speech quality at conditions of high C/I, before falling off significantly with lower C/I values. However, whilst the MOS is still greater than three, a second AMR codec (AMR 4.75) starts to provide superior performance to AMR 6.7 as C/I reduces further.

Figure 5.11 shows only two of the six AMR codecs for clarity. However, in practical deployments too, operators only allow network designers to select a subset of the AMR codecs to minimise signalling and the amount of rate adaptation, which in itself can be perceptible.

5.5.1.6 Voice activity detection, discontinuous transmission

Voice activity detection (VAD) and discontinuous transmission (DTX) are related techniques, which exploit the gaps in normal voice conversations to provide improved capacity, reduced interference and longer battery life for mobiles. In VAD, the codec on the transmitting device notices that the speaker is not talking and instead of encoding this silence as a 13 kbit/s stream (for example), it generates a lower bit-rate stream of

500 bits/s. As soon as the subscriber begins talking again, normal bit-rate encoding is resumed. DTX acts in a similar way: the data produced by the codec are assessed and if a suitable period of silence is found, it will insert a small number of 'silence descriptor' (SID) frames into the SACCH multi-frames transmitted to the air interface, instead of a larger number of normal speech frames. The background noise in the conversation is periodically sampled and sent in the SID; this is used at the receiving end as 'comfort noise' to avoid silence, which is otherwise found to be unsettling to the listener. The use of DTX can significantly reduce the amount of transmission on the radio uplink, which results in improved battery life for the mobile, and reduced uplink interference for other users of the system. DTX can be invoked on a call-by-call basis and is configurable by the network designer. The AMR codecs defined above can exploit VAD/DTX; the number of SID frames is higher than in traditional EFR systems, resulting in a more accurate representation of background noise.

5.5.1.7 Handovers in GSM

Cellular mobile communication systems are designed to support users who are moving around within the network coverage area. When a user who is making a call moves from one cell area to another, the GSM network performs a *handover* to ensure that there is no interruption in service. Generally, handovers are performed for one of the following reasons:

- To sustain a call – if the handover is not performed there is a high probability that it will terminate abnormally, due to poor signal strength or quality, or excessive distance from the cell,
- To ensure that each user is served by the nearest cell where possible,
- To manage traffic levels within the network and deal with congestion problems.

The network uses a number of indicators to determine how and when to perform a handover, including:

- Signal strength (uplink and downlink),
- Signal quality (uplink and downlink),
- Power budget,
- Distance (timing advance),
- Resource loading and availability.

Handover is a complex and costly procedure, and as such it is usually only undertaken when power management on the uplink or downlink has failed to fix any observed performance problems, or when the user can be better served by a neighbouring cell at lower power. Decisions on handovers are taken by the network rather than the MS, although the MS does supply information to the network about its current radio environment to aid the decision process, in the form of measurement reports (MRs). Such a technique is called MAHO (mobile-assisted handover). An alternative mechanism is MCHO (mobile-controlled handover) and is found in other systems, such as DECT cordless telephony. The decision to perform the handover can be taken in different places in the network, including the BTS, BSC and MSC.

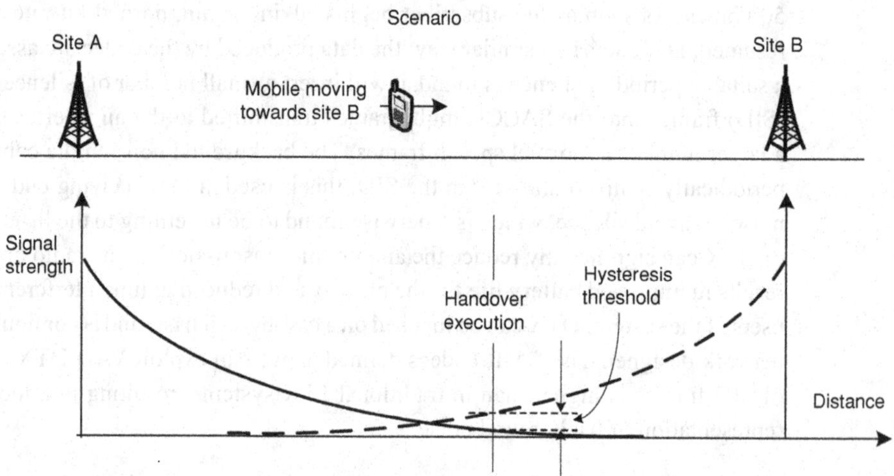

Figure 5.12 A simple GSM handover scenario

A simplified handover scenario is illustrated in Figure 5.12. The MS has originated a call on BTS A (the *serving* cell), and the user is moving away from the centre of the cell towards BTS B (the *neighbour* cell). As the MS moves further from BTS A, the received signal from BTS A ($RxLev_{serv}$) decreases, and the signal from BTS B ($RxLev_{neigh}$) increases. Typically, when the user is approximately equidistant from BTS A and BTS B, $RxLev_{serv}$ and $RxLev_{neigh}$ are very similar. As the user continues to move away from BTS A, the signal level continues to decrease, and if no action were taken it is likely that the call would drop. Throughout this journey, the MS is sending measurement reports back to the network (every 480 ms), and consequently the BSC responsible for BTS A and BTS B is able to observe and compare the relative RxLev values. When the difference ($RxLev_{neigh} - RxLev_{serv}$) becomes greater than the hysteresis threshold value, the BSC sets up a channel on BTS B, and instructs the MS to retune to this new channel. The communication between the MS and BTS B is established, and the link with BTS A is dropped. BTS B now becomes the serving cell (with BTS A now a neighbour cell) and the call can continue normally.

Manufacturers of GSM network equipment have implemented many different (and proprietary) algorithms for handovers. Note that the detailed implementations of these algorithms vary, as the manufacturers attempt to differentiate their network infrastructure, and consequently the manufacturer's operational documentation should be consulted when designing networks. Nevertheless, the general principles are common to all systems, and many of them are expressed in reference handover algorithms defined in GSM recommendations [37]. The principles behind these algorithms are now described.

5.5.1.8 Received signal level handover (RXLEV_HO)

Although the GSM specifications do not dictate how handover decisions are to be made, they do specify the input parameters to the decision process. The scenario shown in Figure 5.12 is an example of an RxLev handover, where the decision to perform the

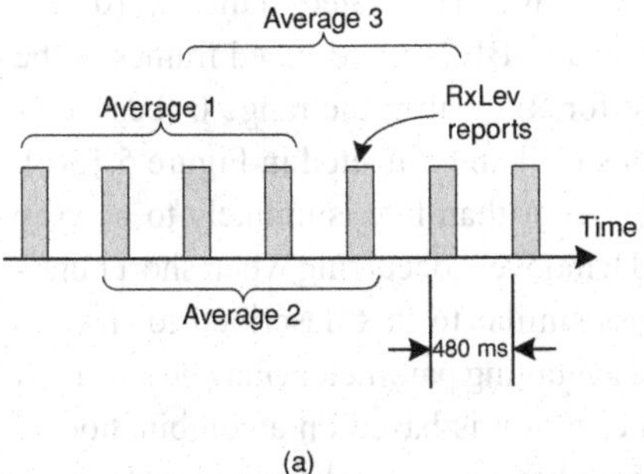

Signal (dBm)	*RxLev*
< −110	0
−110 to −109	1
−109 to −108	2
⋮	⋮
−49 to −48	62
> −48	63

(b)

BER (%)	Rxqual
<0.2	0
0.2 to 0.4	1
0.4 to 0.8	2
0.8 to 1.6	3
1.6 to 3.2	4
3.2 to 6.4	5
6.4 to 12.8	6
>12.8	7

(c)

Figure 5.13 RxLev averaging and RxLev and RxQual mapping

handover is taken on the basis of the relative strengths of the observed signal from the serving and neighbour cells. The RxLev figure is present in measurement reports according to the values in Figure 5.13(b). The MS constantly measures received signal strength on the downlink for both the serving cell and for the cells in the neighbour list, and sends an averaged figure for RxLev back to the serving BTS in the measurement report. The BTS measures RxLev on the uplink, and adds this to the measurement report, before sending the MR to the handover decision-making process. If the $\text{RxLev}_{\text{serv}}$ drops below some minimum threshold value, and $\text{RxLev}_{\text{neigh}} > \text{RxLev}_{\text{serv}}$, then a handover will take place.

Typically, the handover decision is not based on individual measurements, as these are liable to fluctuate quite rapidly, owing to multi-path fading and temporary obstructions on the radio path. Instead, the handover algorithm averages RxLev over a number of measurements – a so-called sliding window. This number may be defined by the operator. Figure 5.13(a) illustrates this averaging process for the case when averaging occurs over four measurement reports. It is important to realise that whilst the averaging mechanisms reduce the risk of short-term signal fluctuations producing an unnecessary handover, they can also introduce significant delays in the handover process. In the case where a mobile suddenly moves into an area that is poorly covered by its serving cell, for example by moving into the shadow of a large building, then it may take several measurement periods before the handover decision is triggered. In the worst case, the radio signal may continue to degrade such that the handover can no longer be executed and the call is dropped. Consequently, the parameters that define the averaging period can be configured on a cell-by-cell basis, or the averaging mechanism disabled completely.

5.5.1.9 Quality handover (RxQual_HO)

Received signal strength is by no means a guarantee of good call quality, as the signal may be subject to interference, which manifests itself as errors on the received frames. Consequently, a system that is designed to use only signal-level-based handovers may not have acceptable voice quality or data transfer rates. A preferred method is to use *quality* handovers, only handing over to a new cell when the quality of the received signal has degraded below some minimum acceptable threshold value.

RxQual is a measure of average received signal quality, calculated in the MS (downlink) and BTS (uplink) by measuring the bit-error rate (BER) on received frames in the channel decoder. The BER is mapped to a scale for RxQual in the range 0 (best) to 7 according to the values in GSM recommendations [37] and reflected in Figure 5.13(c). Clearly, the voice quality for values of RxQual greater than five is unlikely to be very good, and a handover may be necessary. To avoid handovers occurring when short bursts of interference are received, an averaging process similar to that described for RxLev handovers is employed; however the values of the averaging parameters may be specified independently. Typically, the handover decision criterion is based on a combination of RxQual and RxLev thresholds. Note that although a handover may be triggered by poor RxQual in the serving cell, there is no guarantee that the quality in the destination cell will be any better, because the measurement report from the mobile can only contain information about quality in the serving cell. Note also that the serving BTS may decide to move a user to a different radio channel if one is available rather than handing over to a neighbour cell, in an attempt to move the subscriber to a channel experiencing less interference – this is called an *intra-cell* handover.

5.5.1.10 Distance handover (Dist_HO)

The measurement report produced by the MS also contains the timing advance (TA) value, which is used to synchronise transmissions on the air interface as mobiles move away from the base station and the propagation delay increases. The TA can thus be used as a measure of distance from the base site. The TA takes a value in the range 0 to 63, representing time values from 0 to 233 microseconds, in steps of 3.69. Each step in timing advance corresponds to a distance of about 500 metres – a rather coarse indication, but nonetheless one that can be used as part of the handover decision-making process – for example to restrict coverage of a cell to a small geographical area or to eliminate spurious coverage 'splashes' at large distances from the base site.

Note that the range of TA values normally limits the effective radius of a standard GSM cell to about 35 km. However, extended-range cells, which dedicate two or more time slots to each user, allow the larger propagation delays associated with cells greater than 35 km to be accommodated, and are used to provide low-cost coverage in sparsely populated remote rural areas.

5.5.1.11 Interference handover (Int_HO)

The BTS is able to monitor unused time slots on the uplink, and it does this to measure the amount of interference or noise present on the radio link when the mobile is not transmitting. The measurement produced is called *interference on idle (IOI)*, and this value can be compared with a threshold defined for a particular cell as another potential trigger for a handover. Once again, the IOI measurement is subjected to the same averaging procedures previously described.

5.5.1.12 Power budget handover (PBgt_HO)

It is quite possible that an MS may be operating with acceptable quality and signal such that a handover is not triggered, however it may still be better served by a neighbouring cell. In situations where a number of possible cells are available to provide good service, the system will attempt to ensure that the MS is allocated to a cell such that the *power budget (PBgt)* is the most favourable, i.e., where the path loss is lower. This will result in the MS transmitting at lower powers, hence conserving battery life and reducing uplink interference for other users. In systems that are not interference or noise limited, power budget handovers are desirable, as this will mean that the network operates at high efficiency and delivers the highest service quality to its subscribers. The power budget is calculated by the BTS using knowledge of the transmitted power level and received signal strength.

5.5.1.13 Handovers for traffic management

The situation for operators is becoming increasingly complicated. Discussion in earlier chapters has indicated that an operator may deploy *microcells* and perhaps even *picocells* either to relieve the macrocell of high traffic volumes concentrated in a small area (such as a shopping mall) or to provide better in-building coverage. However, because the coverage area of these cells is smaller than for macrocells, traffic needs to be managed such that only stationary or slow-moving users are assigned to these cells if the risk of dropped calls due to a high-speed mobile quickly transiting the cell is to be avoided. Similarly, many operators now have access to GSM spectrum in more than one band. Depending upon the level of penetration of dual-band handsets in the particular market and the relative coverage of the most recently deployed network, the operator may wish to 'encourage' mobiles to 'camp' on a particular network so that traffic is balanced and congestion is avoided.

With the deployment of UMTS, further traffic management flexibility is afforded the operator. There may be a wish to 'direct' voice traffic towards UMTS for two reasons. When fully loaded, UMTS is a lower-cost solution for voice traffic than GSM; similarly, even some years after initial deployment, UMTS-coverage is far from ubiquitous and so it is desirable to reserve the limited capacity of GSM for those situations when either there is no UMTS coverage or the subscriber does not have a UMTS-capable handset.

All of the above traffic management requirements (and more besides) can be handled using a combination of new handover management algorithms and the introduction of some new parameters on existing procedures. There is insufficient space to go into detail here, but, by way of example, a parameter is available which is effectively *subtracted* from particular measurement report values, making a handover (or camping) candidate cell look more or less attractive. From the foregoing discussion, it is clear that a variety of scenarios can give rise to a decision to trigger a handover. The network typically uses a priority scheme when making decisions about handovers, executing the highest priority handover when more than one request is present.

5.5.2 Planning – preparation

5.5.2.1 Scenario sub-set selection

In this example, a GSM/GPRS system will be dimensioned to meet the requirements for the scenario summarised in Table 5.8.

Table 5.8. Scenario for the GSM/GPRS example

Chapter	Busy-hour call attempts	Application usage (%) ('Calls' in busy hour)							System characteristics			
		Voice	Mobile web (WAP)	Interactive gaming	Video download	VoIP	Web browsing	Messaging	Subscriber density (km^2)	Penetration (%)	Environment	Mobility (km/hr)
5	1	90	3	0	1.5	0	2.5	3	1000	70	Suburban	0–240

5.5.2.2 Key questions to address

At the start of any planning exercise, it is a good discipline to identify the key questions that need to be addressed. In the case of planning a network to support GSM voice and data subscribers, these would include:

- What propagation models are appropriate?
- What is the maximum cell size?
- What carrier dimensioning is required to meet the voice capacity?
- How many GPRS time slots should be configured to meet the data requirements?
- Can the subscriber density goals be met?

5.5.2.3 Propagation model selection and maximum cell size evaluation

Several propagation models are available for use in RF modelling tools for both 900 MHz and 1800 MHz systems. The Okumura–Hata model can be used for estimating path loss under the conditions given in Table 5.9. This model is commonly used for planning GSM900 systems. The COST 231 model is an extension of the Okumura–Hata model, to cover frequencies of 1500–2000 MHz and can be used for planning the GSM1800 frequency range. Both models have formulae to estimate path loss in open, suburban and urban environments. If microcells or picocells are being planned where the cell radius is less than 1 km, it may be necessary to use deterministic models, based on ray tracing, to provide high-resolution path-loss information. A less demanding, but less accurate, approach would be to use the Walfisch–Ikegami empirical model described in Chapter 4. In rare rural cases, employing very high masts, the maximum cell radius may, in addition, be limited by the timing advance to a theoretical maximum of 35 km (although this may be increased using extended-cell implementations). In most environments the range is limited by attenuation due to clutter and terrain to somewhat less than the 35 km limit.

In this example, suburban wide-area coverage is required and the Okumura–Hata model is appropriate. Indicative path loss values for GSM900 and GSM1800 are plotted in Figure 5.14. As expected (see Appendix 3.1) the maximum range of a GSM900 cell is greater than that for a GSM1800 cell of equivalent output power because of the longer wavelength and reduced attenuation due to clutter.

Usually, RF coverage planning is performed using a radio-propagation modelling tool of the type referenced in Chapter 4. The initial requirement to provide GSM coverage over the target area dictates that the co-channel C/I value at the cell edge on both uplink and downlink is at least 9 dB to enable the BCCH to be reliably received and decoded. In practical systems, it is necessary to plan for a margin of around 15 dB over 90% of

Table 5.9. Conditions for Okumura–Hata model

Frequency range	100–1500 MHz
Distance	1–20 km
Base station height	30–200 metres
MS antenna height	1–10 metres

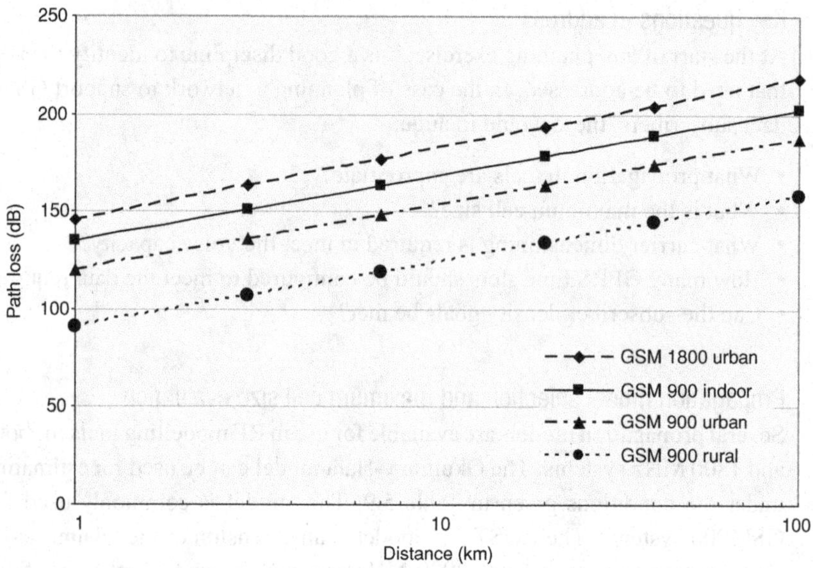

Figure 5.14 Path loss vs. distance for GSM900 and GSM1800

the system area, to allow for fading. Terrain, clutter and morphology data for the area under consideration should be obtained, with a spatial resolution of 50 m or better. Then, assuming a worst-case mobile sensitivity of −102 dBm, BTS power of +43 dBm and +17 dB antenna gain for a tri-sectored site:

$$\text{Allowable path loss } = (+43\text{ dBm} + 17\text{ dB} - (-102\text{ dBm}) - 15\text{ dB}) = 147\text{ dB}.$$

From Figure 5.14, and assuming that suburban propagation loss lies between urban and rural, this configuration will easily support cells with radii of up to 10 km; in reality, the cell radius will be dictated by subscriber density considerations.

5.5.3 Capacity planning for voice

The scenario shown in Table 5.8 calls for a GSM system covering a suburban area with a population density of 1000/km^2 and a traffic demand of 25 mE/sub. For 70% subscriber penetration, this gives:

$$\begin{aligned}\text{Traffic density} &= 1000 \times 0.7 \times 0.025\\ &= 17.5\text{ Erlangs/km}^2.\end{aligned}$$

The number of cells required to support this traffic will vary depending on the amount of spectrum allocated to the network operator, as this determines the number of unique frequencies available, and the feasible frequency reuse patterns. Table 5.2 summarises the Erlang capacities (for 2% blocking) for some fixed reuse patterns and spectrum allocations. If 10 MHz of spectrum is available (50 channels) and the system employs a 4 × 12 (four tri-sectored sites) reuse pattern, some 36 frequencies will remain once 12 frequencies have been reserved for the BCCH allocation. The capacity in this case is 16.6 Erlangs/cell or approximately 50 E/tri-sectored site. Then, the cell radius is defined by:

$$50/(\pi \times r^2) = 17.5, \text{ where } r \text{ is the cell radius.}$$

Thus, the apparent maximum cell radius is 0.95 km. In practice the cell radius might be larger or smaller than this. For example, Table 5.2 makes very clear the benefits of moving to a tighter (soft-blocking) reuse plan on the non-BCCH channels, especially if frequency hopping is used. In Chapter 10, the use of non-uniform frequency plans will be described, and this represents another method for increasing cell capacity and radius. On the downside, the operator may more typically have 7.5 MHz of spectrum at GSM900 and capacity is still required to support GPRS services. For this example, a 10 MHz hard-blocking plan will be assumed, with its associated 0.95 km radius cells.

Detailed planning of the GSM sites will also require dedication of capacity to support control channel traffic associated with activities, such as paging, location area updates, call originations and handovers. Calculations must be performed for the following control channels:

- Broadcast control channel (BCCH),
- Common control channel (CCCH),
- Stand-alone dedicated control channel (SDCCH),
- Cell broadcast channel (CBCH),
- Packet common control channels (PCCCH – for GPRS).

The dimensioning of these channels varies considerably from manufacturer to manufacturer, and detailed planning should be performed in accordance with the appropriate equipment planning guide. It will be necessary to specify typical call models to allow the calculations to be performed; an example call model for GSM is shown in Table 5.10.

5.5.4 Capacity planning for GPRS

GSM supports packet data using a new network element at the BSC, the packet control unit (PCU), and associated changes to the core network, as described in Chapter 2. At the cell level, the network designer has to determine how much radio capacity is required to support the estimated volumes of data traffic. More specifically, the number of time slots on the air interface that should be reserved for GPRS must be determined *in addition* to those required for circuit voice. As discussed in Chapter 3, designing for

Table 5.10. Typical GSM call model

Call duration	120 s
SMS ratio per call	0.1
Handovers per call	2.5
Paging rate	15/s
Location update ratio per call	2

bursty applications to guarantee low packet-loss ratios and low response times can be a complicated task, as this depends on the characteristics of the applications which the users are running – web browsing, email, VoIP, and so on. Nevertheless, planning in this way can be done and is done (see Chapter 7), but it is not normal practice for GPRS network dimensioning. This is because GPRS was not designed to support guaranteed QoS and a number of factors make support of real time services impractical, including:

- In most configurations GSM voice and GPRS contend for some shared resources, and thus actual capacity is unknown,
- PDP context activation time introduces large timing uncertainty,
- TBF timing adds further uncertainty and swamps queuing delays,
- Control messages pre-empt packet traffic,
- There are often insufficient GPRS network resources to support multi-slot mobiles and thus deliver maximum user rates.

Instead, design for GPRS systems is performed by estimating the required *mean* throughput data rates, in terms of kbits/s, which the cell must support, and allocating sufficient time slots to meet this average aggregate demand. It is also necessary to provide capacity needed for the control channels.

Before the number of time slots required can be calculated, the relative utilisation of the GPRS coding schemes must be estimated. As discussed in Section 5.3, GPRS uses an adaptive mechanism to select the most appropriate coding scheme for the prevailing radio environment. At high C/I, there is less need for error protection on the radio interface, and therefore CS3 or CS4 may be used, resulting in higher throughput on the link. At lower C/I, more protection is required and the more robust but lower throughput CS1 and CS2 coding schemes will be used. As the radio-link quality changes, the selected coding scheme changes as well, which means that the actual throughput experienced by the end user can vary considerably during a session. The theoretical user bit rate for each coding scheme is shown in Table 5.3.

The more recent enhanced GPRS (EGPRS) specification, sometimes referred to as EDGE (enhanced data rates for GSM evolution), introduced new modulation and coding schemes, enabling greater throughput and finer granularity in user bit-rate selection. The EGPRS modulation and coding schemes are labelled MCS-1 to MCS-9 and are also shown in Table 5.3. The four lowest bit-rate schemes use Gaussian minimum shift keying (GMSK) modulation; the higher bit-rate schemes, MCS-5 to MCS-9, utilise 8-phase shift keying (8PSK). Note that the higher bit-rate coding schemes for GPRS and EGPRS require a better-quality radio link if the BLER is not to be unacceptably high. Consequently, mobiles moving around within a cell's coverage area will need to change coding schemes to match the prevailing RF environment – this process is termed link adaptation (LA). Link adaptation is required by the standards but the actual detailed implementation of LA is left to the infrastructure manufacturers to define, and hence will vary. The LA mechanism can be controlled by defining the target operational block-error rate (BLER) at which the system is designed to perform – LA will switch to a more conservative coding scheme if the BLER rises above this value or to a more aggressive scheme if the BLER is lower. An alternative mechanism to LA (sometimes viewed as a

Table 5.11. Coding scheme utilisation in GPRS and EDGE

Air interface	Coding scheme	% utilisation
GPRS	CS1	15
	CS2	14
	CS3	20
	CS4	51
EDGE	MCS-1	2
	MCS-2	6
	MCS-3	2
	MCS-4	0
	MCS-5	5
	MCS-6	14
	MCS-7	12
	MCS-8	9
	MCS-9	50

Both GPRS and EDGE code utilisation are for 4 × 12 non-hopping repeat pattern and 20% BLER

part of LA) is incremental redundancy (IR), whereby unsuccessfully transmitted blocks are retransmitted without changing the MCS, but using a different puncturing scheme.

The radio planning tools referenced earlier in this chapter can provide coverage plots and cumulative distribution function plots, which show the estimated usage of each coding scheme across the network. The coding scheme utilisation on the uplink and downlink will differ. For planning purposes it is sensible to address the downlink, as this carries much more data traffic than the uplink in typical networks. Coding schemes CS-1 and MCS-1 have been designed so that their coverage footprint matches that of voice coverage; in other words, if a subscriber is successfully able to make a voice call at a particular location, he or she will be able to transmit packet data using CS-1 (or better). The actual coding scheme coverage and utilisation varies with operational BLER and LA/IR configurations, and it may be necessary to perform detailed simulations to obtain precise estimates of coding scheme usage and throughput. However, as a rule of thumb, a well-planned and maintained GSM network will experience an average C/I of about 15 dB on the BCCH, which will allow low and medium bit-rate EDGE coding schemes to be used over large parts of the network. Higher-rate coding schemes, such as CS-4, MCS-8 and MCS-9, may require good LOS coverage and may be selected near to the base site or in uncluttered rural areas.

Table 5.3 also includes the C/I ranges over which particular modulation and coding schemes (MCS) are likely to operate, although this will depend on the threshold BLER levels selected as discussed above. Using such data (perhaps from laboratory measurements, simulations, field studies or a combination thereof) along with system C/I plots and estimated subscriber geographical distributions, it is possible to estimate the percentage utilisation of the different MCSs, and hence to determine the throughput per cell.

As an example, Table 5.11 shows the percentage utilisation of different coding schemes measured for a 4 × 12 reuse system in an urban environment, operating at 20% BLER.

There is good utilisation of high-rate coding schemes (CS4 and MCS-9), which indicates that many users are close to the base site with a strong radio signal and low interference.

5.5.4.1 Calculation of required GPRS resources

It is now possible to estimate the number of GPRS/EGPRS time slots required in each cell. GPRS and EGPRS are considered separately, as not all mobiles support EGPRS, owing to its later introduction, and penetration of EDGE-capable mobiles is low in some networks. The first step is to calculate the time slot data rates. These are evaluated by forming the sum of data rates weighted by utilisation percentage:

$$\mathrm{TSR}_{\mathrm{AV(GPRS)}} = \frac{1}{100}\sum_{i}^{4}[\mathrm{CSU}_i \times \mathrm{CSR}_i] \times \frac{[100 - \mathrm{BLER}]}{100},$$

$$\mathrm{TSR}_{\mathrm{AV(EDGE)}} = \frac{1}{100}\sum_{i}^{9}[\mathrm{MCSU}_i \times \mathrm{MCSR}_i] \times \frac{[100 - \mathrm{BLER}]}{100},$$

where

- $\mathrm{TSR}_{\mathrm{AV(GPRS)}}$ and $\mathrm{TSR}_{\mathrm{AV(EDGE)}}$ are the average time slot throughputs for GPRS and EDGE mobiles in the cell, respectively,
- CSU_i and MCSU_i are the percentage utilisation of coding scheme i for GPRS and EDGE, respectively,
- CSR_i and MCSR_i are the user data rates for coding scheme i for GPRS and EDGE, respectively.

Note that the final terms in these equations reflect the need to account for data loss due to corruption of individual blocks. This will not be considered here. Using the data from Table 5.3 and Table 5.11, the mean time slot data rate for GPRS and EGPRS becomes:

$$\mathrm{TSR}_{\mathrm{AV(GPRS)}} = 0.15{*}9 + 0.14{*}13.4 + 0.20{*}15.6 + 0.51{*}21.4 \sim 16\ \text{kbits/s}$$

and the corresponding calculations for EDGE give:

$$\mathrm{TSR}_{\mathrm{AV(EDGE)}} \sim 46\ \text{kbits/s}.$$

Note that the average time slot data rate for the EDGE user is almost three times greater than for the GPRS, which illustrates the attractiveness of moving to EDGE.

Referring to the scenario in Table 5.8, the mean traffic load presented to each cell by GPRS and EDGE users can be evaluated, for the specified application mix. For this worked example, the aggregated system load per subscriber will be derived from the non-voice applications of Table 5.8 in conjunction with estimated session durations and mean data rates. Because of the low data rates and non-guaranteed delay performance with GPRS, applications such as gaming and VoIP sessions would be supported on UMTS. Accordingly, the dimensioning for GPRS is carried out only for the remaining applications, which are more tolerant of delay. For convenience, the relevant sections of Table 4.1 and Table 5.8 are brought together in Table 5.12.

Table 5.12. Evaluation of busy hour kbits/(hour subscriber)

		One 'call' / busy hour						
		Data rate (kbits/s)						
Application	%	Min	Max	Mean	Sessions/h	Duration (s)	Data/h	Comments
Voice	90	–	–	–	0.9	100	–	0.025 mE (GSM)
WAP	3	2	5	3	0.03	60	5.4	
Gaming	–	–	–	–	0	0	0	Not viable
Video	1.5	64	384	128	0.015	20	38.4	
VoIP	–	–	–	–	0	0	0	Not viable
WWW	2.5	64	2000	100	0.025	45	112.5	
Messaging	3	1	1	1	0.03	5	0.15	
Total							**156.45**	**kbits/(hour subscriber)**

It is necessary to determine the GPRS/EGPRS capacity required to support the data applications (i.e., excluding voice). The quantity of data traffic is not large; this is typical of many GPRS systems where uptake of data services has been rather slow and voice traffic dominates. To dimension the number of time slots required on a typical urban cell, we use the following system parameters from Table 5.8.

Mean traffic load per cell can be calculated using the 156.5 kbits/(hour subscriber) figure of Table 5.12 and the number of GPRS and EDGE subscribers from Table 5.13. Then

$$\text{GPRS load} = 156.5 \times 596 = 93.3\,\text{Mbits/hour or } 25.9\,\text{kbits/s}.$$

The corresponding EDGE load is

$$\text{EDGE load} = 156.5 \times 66 = 10.3\,\text{Mbits/hour or } 2.9\,\text{kbits/s}.$$

In principle, these loadings can be translated into the number of GPRS and EDGE time slots required merely by dividing by the average throughput/time slot figures for GPRS and EDGE, which we have evaluated. In practice, a 'rule of thumb' over-dimensioning

Table 5.13. System parameters for GPRS dimensioning example

Subscribers/km^2	700	70% penetration
%EDGE users	10%	Handset penetration dependent
%GPRS users	90%	
Cell radius	0.95 km	From voice calculation
Cell area	0.945 km^2	Each sector / tri-sectored site
Subscribers/cell	662	Sub density × (cell area)
GPRS subscribers/cell	596	Subs/cell × 0.9
EDGE subscribers/cell	66	Subs/cell × 0.1

by a factor of two is used to allow for peaks in demand and packet control plane traffic. The total number of time slots required is thus:

$$\text{Total data time slots} = (2 \times 25.9/16) + (2 \times 2.9/46)$$
$$= 3.3 + 0.13 \sim 4 \text{ slots.}$$

Most manufacturers provide the facility for some time slots to be reserved or dedicated to GPRS, whilst others can be dynamically switchable between GSM and GPRS, as data and voice traffic volumes fluctuate. This can reduce congestion at peak times but at the expense of longer delay as the switchable time slots are reconfigured. In this example, two time slots could be permanently allocated to GPRS and two further time slots made switchable.

Another consideration that will influence the number of time slots dedicated to data in an operator's network is the need to demonstrate competitive peak data rates to the customer base. Four-time-slot mobiles are almost *de rigueur* and this will be another consideration for deploying a minimum of four time slots for data in most networks.

Recall that, on completion of the voice dimensioning exercise, the cell radius was determined on the basis that the remaining 36 frequencies were dedicated to voice – so where are these 'data' time slots to be accommodated? In the capacity evaluation so far, the BCCH carrier has been neglected. Although some time slots will be needed for the BCCH data and control channels, in most configurations there will be at least four time slots available to carry this data traffic.

References

1 F. Kostedt and J. C. Kemerling, *Practical GMSK Data Transmission, Document # 20830067.002 – An Application Note* (MX.COM, 1998).

2 H. R. Anderson and J. P. McGeehan, Direct calculation of coherence bandwidth in urban micro-cells using a ray-tracing propagation model, *IEEE PIMRC'94*, **1** (1994) 20–24.

3 G. D. Forney, Maximum-likelihood sequence estimation of digital sequences in the presence of inter-symbol interference, *IEEE Transactions on Information Theory*, **IT 18** 3 (1972) 363–378.

4 G. Ungerboeck, Adaptive maximum-likelihood receiver for carrier-modulated data-transmission systems, *IEEE Transactions on Communications*, **COM 22** 5 (1974) 624–636.

5 J. M. Morton, *Adaptive Equalisation for Indoor Wireless Channels*, M.Sc. thesis (Virginia Polytechnic Institute and State University, 1998).

6 ETSI, *Digital Cellular Telecommunications System (Phase 2+); Radio Transmission and Reception*, GSM 05.05 version 7.1.0, Release 1998.

7 R. D'Avella, L. Moreno and M. Sant'Agostino, An adaptive MLSE receiver for TDMA digital mobile radio, *IEEE Journal on Selected Areas in Communications*, **1** 1 (1989) 122–129.

8 R. D'Avella, L. Moreno and M. Sant'Agostino, Adaptive equalization in TDMA mobile radio systems, *IEEE Vehicular Technology Conference*, **37** (1987) 385–392.

9 L. Lopes, Performance of Viterbi equalisers for the GSM system, *Second IEE National Conference on Telecommunications* (1989) 61–66.

10 ETSI, *Digital Cellular Telecommunications System (Phase 2+); Full Rate Speech Transcoding*, GSM 06.10 version 7.0.0, Release 1998.

11 ETSI, *Digital Cellular Telecommunications System (Phase 2+); Radio Sub-system Synchronization*, GSM 05.10 version 7.1.0, Release 1998.

12 V. Koshi, D. J. Edwards and A. M. Street, An approach to the calculation of the probability of co-channel interference for PCS/PCN cellular planning, *IEEE ICUPC'96*, **1** (1996) 91–95.

13 B. F. McGuffin, DCS1900 system configuration, *IEEE Vehicular Technology Conference*, **1** (1995) 112–117.

14 ETSI, *Digital Cellular Telecommunications System (Phase 2+); Channel Coding*, GSM 05.03 version 7.1.0, Release 1998.

15 P. E. Mogensen and J. Wigard, On antenna and frequency diversity in GSM related systems (GSM 900, DCS1800 and PCS1900), *IEEE PIMRC'96*, **3** (1996) 1272–1276.

16 H. Olofsson *et al.*, Interference diversity gain in frequency hopping GSM, *IEEE VTC'95*, **1** (1995) 102–106.

17 M. Almgren, H. Andersson and K. Wallstedt, Power control in a cellular system, *IEEE VTC 1994*, **2** (1994) 833–837.

18 H. Holma *et al.*, Performance of adaptive multirate (AMR) voice in GSM and WCDMA, *IEEE VTC 2003 Spring*, **4** (2003) 2177–2181.

19 3GPP, *Mandatory Speech codec Speech Processing Functions: AMR Speech codec; General Description*, Release 6, 3GPP TS 26.071 version 6.0.0 (2004–12).

20 S. Craig and J. Axnas, A system performance evaluation of 2 branch interference rejection combining, *IEEE VTC 2002 – Fall*, **3** (2002) 1887–1891.

21 M. Olsson, A. Brostrom, A. Craig and H. Arslan, Single antenna interference rejection, *IEEE VTC 2004 – Spring*, **3** (2004) 1698–1702.

22 J. Wigard *et al.*, Capacity of a GSM network with fractional frequency loading and random frequency hopping, *IEEE PIMRC'96*, **2** (1996) 723–727.

23 J. Naslund *et al.*, PCS based on DCS1800, *IEEE PIMRC'94*, **4** (1994) 1156–1161.

24 C. Carneheim *et al.*, FH-GSM frequency hopping GSM, *IEEE Vehicular Technology Conference* (1994) 1155–1159.

25 P. A. Gutierrez *et al.*, Performance of link adaptation in GPRS networks, *IEEE VTC Fall 2000*, **2** (2000) 492–499.

26 J. Gozalvez and J. Dunlop, High-speed simulation of the GPRS link layer, *IEEE PIMRC 2000*, **2** (2000) 989–993.

27 W. Featherstone and D. Molkdar, System level performance evaluation of GPRS for various traffic models, *IEEE VTC Fall 2000*, **6** (2000) 2648–2652.

28 3GPP, *General Packet Radio Service (GPRS); Overall Description of the GPRS Radio Interface; Stage 2*, Release 1999, 3GPP TS 03.64 version 8.12.0 (2004–04).

29 3GPP, *Physical Layer on the Radio Path; General Description*, Release 1999, 3GPP TS 05.01 version 8.9.0 (2004–11).

30 3GPP, *Channel coding*, Release 1999, 3GPP TS 05.03 version 8.9.0 (2005–01).

31 A. Furuskar *et al.*, System performance of EDGE, a proposal for enhanced data rates in existing digital cellular systems, *IEEE VTC'98*, **2** (1998) 1284–1289.

32 H. Arslan *et al.*, Evolution of EDGE to higher data rates using QAM, *IEEE VTC Fall 2001*, **4** (2001) 2267–2271.

33 D. Molkdar and W. Featherstone, System level performance evaluation of EGPRS in GSM macro-cellular environments, *IEEE VTC Fall 2000*, **6** (2000) 2653–2660.

34 S. Eriksson *et al.*, Comparison of link quality control strategies for packet data services in EDGE, *IEEE VTC Spring 1999*, **2** (1999) 938–942.
35 M. Eriksson *et al.*, System performance with higher level modulation in the GSM/EDGE radio access network, *IEEE Globecom'01*, **5** (2001) 3065–3069.
36 A. Furuskar *et al.*, Capacity evaluation of the EDGE concept for enhanced data rates in GSM and TDMA-136, *IEEE VTC Spring 1999*, **2** (1999) 1648–1652.
37 ETSI, *Digital Cellular Telecommunications System (Phase 2+); Radio Sub-system Link Control*, GSM 05.08 version 7.1.0, Release 1998.

Appendix 5.1 The MLSE equaliser

An excellent tutorial paper by Forney proposes and justifies a much simpler model to implement an equivalent of the sequence comparison envisaged by the MLSE process outlined in Section 5.1.1.1. The key assertions and assumptions are listed below and justified in [1, 2]. Nomenclature is based on that used in Figures 5.15, 5.16 and 5.17.

(i) Any real channel will have an impulse response that is essentially finite, being significant for some number of symbol periods ηT such that the impulse response $c(t) = 0$ for $t > \eta T$.

(ii) For a symbol sequence $X_N = (x_1, x_2 \ldots x_N)$ propagating as $x(t)$ on a radio carrier through a multi-path channel, the receiver response to a specific input symbol, x_i, can be completely defined (in the absence of thermal noise) provided comprehensive knowledge of the multi-path propagation characteristics is available. This signal is $r(t)$ (see Figure 5.15).

(iii) If the signal sampled was $r(t)$, this would give rise to a sequence R_N, the ith component of which would be given by $r_i = (x_i.c_0 + x_{i-1}.c_{-1} \ldots x_{i-\eta}.c_{-\eta})$ where $C_\eta = (c_0 \ldots c_{-\eta})$ is the set of complex coefficients of the actual channel impulse response at times t, $(t - T) \ldots (t - \eta T)$. Inevitably, perfect knowledge of this

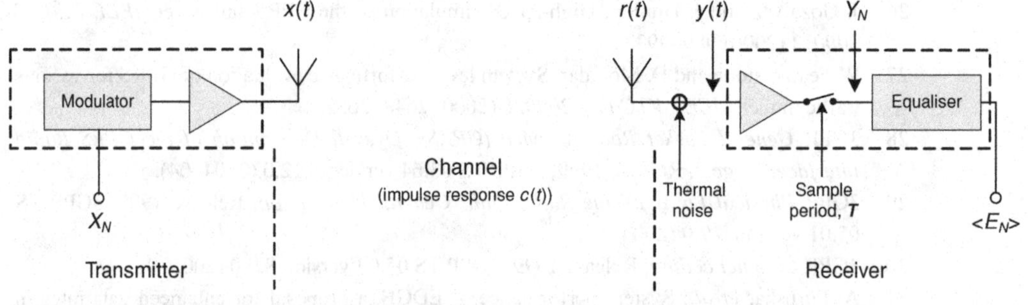

Figure 5.15 The role of the equaliser in GSM

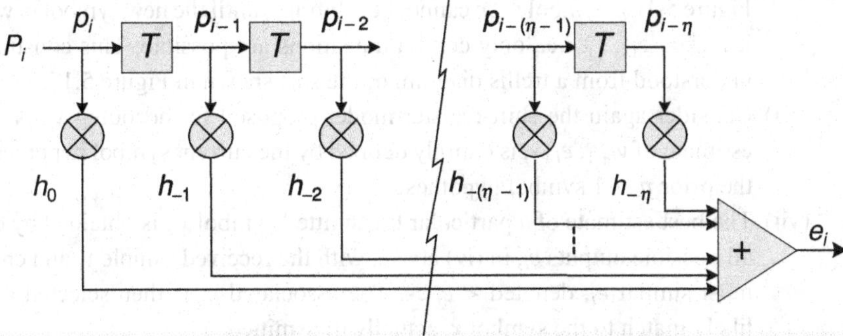

η is a measure of the significant channel impulse response duration, measured in bit periods T

P_i is a sub-sequence ($p_i, p_{i-1} \ldots p_1$) of one of the possible input sequences $\{P_N\}$ after arrival of the most recent character, p_i.

h_n is the weighting coefficient corresponding to the impulse response $h(t-nT)$

e_i is one estimate of the ith transmitted symbol (x_i) made using one of the possible sequences from $\{P_i\}$

Figure 5.16 Example MLSE symbol estimator

impulse response is not possible in practice, This equation can be recognised as a specific state of the convolution of the input symbol sequence X_N with the actual channel impulse response $C\eta$.

The signal actually sampled, $y(t)$, includes receiver thermal noise; this gives rise to a corresponding term y_i which is equal to r_i augmented by the noise.

(iv) Within the receiver, candidate responses to any possible transmitted sequence (drawn from $\{X_N\}$) can be constructed from a family of prototype sequences $\{P_N\}$, provided access to an imperfect but adequate estimate of the *actual* channel impulse response, C_η, is available in the form of $H_\eta = (h_0 \ldots h_{-\eta})$. Note that the family of sequences $\{P_N\}$ is, in fact, identical to the sequences $\{X_N\}$ at the transmitter from which X_N is drawn, but a different notation is used here to avoid confusion between the symbols *actually* transmitted and those candidates used in the receiver's estimation process.

Any possible estimate $e_i (= p_i.h_0 + p_{i-1}.h_{-1} \ldots p_{i-\eta}.h_{-\eta})$ of the signal received in response to x_i being transmitted, can be constructed by applying to the estimator, in turn, all possible subsets $P_i (= p_i, p_{i-1} \ldots p_{i-\eta})$. This process is shown in Figure 5.16 (Note that it is not possible to include the effects of receiver thermal noise to which the actual received signal, $r(t)$, is subject). The signal estimates formed in this way may then be compared with that actually received, to identify which is the most similar (more of this later). Each subset is drawn from a particular sequence $P_N (= p_1, p_2 \ldots p_N)$ which itself is one of the family of sequences $\{P_N\}$.

(v) In practice, it is not necessary to consider *all* possible combinations P_i. Because of the multi-path 'channel coding', only the most recent symbol is truly arbitrary. Consider the symbols contained in the shift-register model used to represent the multi-path channel just after the introduction of a new symbol p_i (see

Figure 5.16). Symbol $p_{i-}\eta$ cannot be arbitrary; until the new symbol p_i was entered it *was* $p_{i-(\eta-1)}$, i.e., only certain transitions are possible. This constraint is best understood from a trellis diagram of the sort shown in Figure 5.17.

(vi) Consider again the shift-register model proposed for decoding x_i. Note that the estimate of y_{i+1}, e_{i+1}, is entirely defined by the current symbol hypothesis, p_i, and the prior $\eta - 1$ symbol hypotheses.

(vii) The best estimate of a particular transmitted symbol x_i, is obtained by comparing all possible outputs e_i, in (iv) above, with the received sample y_i and choosing the most similar e_i, denoted $< e_i >$. The associated p_i is then selected as the most likely match to the symbol x_i actually transmitted.

(viii) 'Similarity is measured by calculating a difference term of the form $\delta = (y_i - e_i)^2$. The likelihood of a particular e_i being the most similar increases with reducing values of $\Sigma\delta$, the sum of δ over the η samples.

(ix) A state machine of the type illustrated in Figure 5.16 can be used recursively to decode Y_N, the received signal sequence, and arrive at the most likely estimate of X_N, the transmitted sequence. No more than m^η highest-scoring paths need be maintained at each symbol state – regardless of the number of symbols in the sequence X_N, where m is the alphabet of symbol states (e.g., 2 for binary symbols). Note also that the total number of 'compare' operations for an N symbol sequence must therefore be $N \times m^\eta$.

To illustrate this process, consider the trellis diagram of Figure 5.17. For simplicity, the multi-path is deemed to be insignificant for $\eta > 2$ and the modulation scheme is binary (symbol alphabet $m = 2$), so Figure 5.16 simplifies to a machine with two delay stages and three weighting coefficients. Then, at any point in time, $T = 0, 1, 2$, etc., the machine is in one of $m^{+\eta}$ states (00 through to 11). At $T = 0$, the equaliser has received no symbols and thus is known to be in the state '00' (this state is assured – it is either reset to this state on power-up or driven to this state by a sequence of zeros transmitted prior to the information bits in each frame). For clarity, it is convenient to start the description from this condition.

Between times $T = 0$ and $T = 1$, a symbol y_1 is received by the equaliser; its identity is clearly unknown and so the equaliser establishes two paths (in two separate shift-register sets). For path 1, where the received symbol hypothesis p_1 is a 1, the noise-free estimate of the received symbol $e^1{}_1$ can be calculated from the weighted combinations of p_1 and the register contents for path 1 (0 and 0) and compared with the noisy received symbol, y_1. The proposed decoded value for $y_1(1)$ and the squared difference between the noise-free and actual received symbols (δ) are recorded. The hypothesised symbol is entered into the register for path 1 and the machine register contents become 10. For path 2, where the received symbol hypothesis p_1 is a 0, the same process is followed and the corresponding register contents become 00. Note that there is no possibility of either path being in the states 01 or 11. Figure 5.17 shows example values of $\Sigma\delta$, which are included to show how the survivor paths are calculated.

After the second symbol y_2 has been received, *each* path has to allow for the received symbol hypothesis, p_2, being a 1 or a 0, thus a total of four machines now needs to be

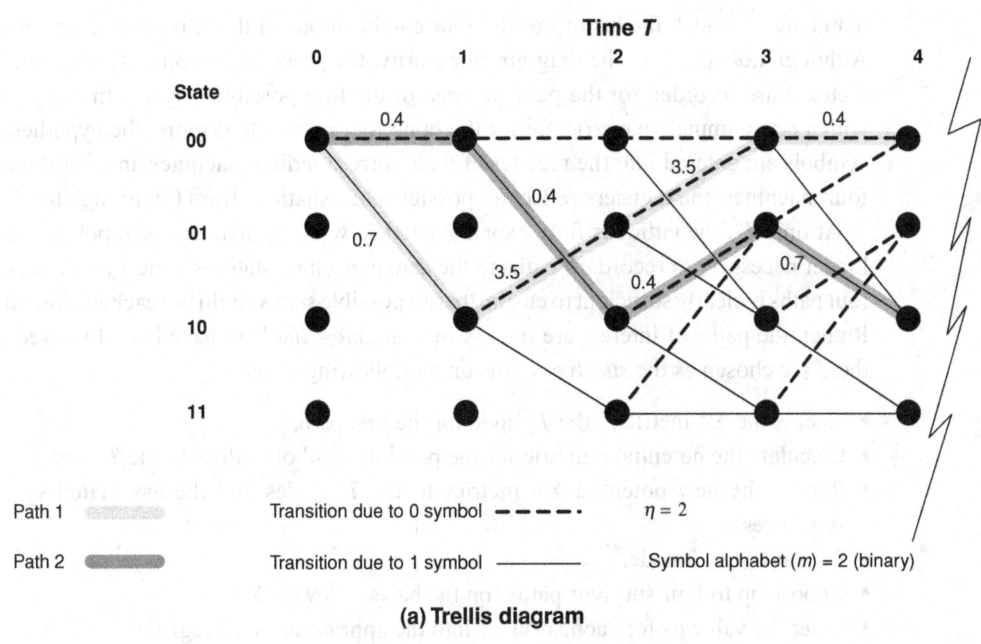

(a) Trellis diagram

Path 1	**Latest symbol (p_i) hypothesis**	N/A		p_1^1 1		p_2^1 0		p_3^1 0		p_4^1 0	
	Register contents		00		10		01		00		00
	δ		0.0		0.7		3.5		3.5		0.4
	$\Sigma\delta$		0.0		0.7		4.2		7.7		8.1
	Noise-free estimate of y_i (e_i^1)			e_1^1		e_2^1		e_3^1		e_4^1	
	Latest received symbol (y_i)			y_1		y_2		y_3		y_4	
Path 2	**Latest symbol (p_i) hypothesis**	N/A		p_1^2 0		p_2^2 1		p_3^2 0		p_4^2 1	
	Register contents		00		00		10		01		10
	δ		0.0		0.4		0.4		0.4		0.7
	$\Sigma\delta$		0.0		0.4		0.8		1.2		1.9
	Noise-free estimate of y_i (e_i^2)			e_1^2		e_2^2		e_3^2		e_4^2	

(b) Survivor calculation

Key: p_3^2 — Path 2 (superscript), Symbol 3 (subscript)

Figure 5.17 MLSE equaliser – trellis diagram and survivor calculation

maintained to track the history of the four combinations of the two most recent inputs. Although not shown in the diagram, for clarity, the proposed decoded symbol and the metric δ are recorded for the paths to *each* of the four possible states at time T_2 along with a new cumulative metric $\Sigma\delta$ for the same four paths. Once more, the hypothesised symbols are entered into the register of their corresponding machines and, between the four machines, the registers reflect all possible combinations from 00 through to 11.

At time T_2, the lattice is fully expanded and so with the arrival of symbol y_3 it is no longer necessary to record *all* paths to the active machine states at time T_3 – preserving four paths is clearly sufficient to ensure that all possible states could be reached if required. Rather, the paths of interest are those which are *most likely* to have been followed and these are chosen as the *survivor* paths on the following basis:

- Access the $\Sigma\delta$ metric to the T_2 node for the first path,
- Calculate the potential δ metric for the possible symbols values to the T_3 nodes,
- Record the new potential $\Sigma\delta$ metrics to the T_3 nodes and the associated symbol sequences,
- Repeat for each node,
- Choose up to four survivor paths (on the basis of lowest $\Sigma\delta$),
- Enter the value p_3 for each survivor into the appropriate shift register.

With the lattice fully expanded, the above process will be followed on the arrival of each new symbol. It should be noted that the presence of multi-path results in the amplitudes of both the received signal and the modelled estimates having a continuous range of values rather than discrete digital amplitudes. This naturally leads to a 'soft-decision' process being carried out, in which the signal amplitude differences are recorded as decimal numbers rather than quantised to the nearest state.

To summarise, the initially divergent hypothesis for the received symbol y_i, the symbol p_i (originally proposed as a 1 or a 0 with equal probability), will gradually converge to a common value (either a 1 or a 0) in all surviving paths as the cumulative geometric distance, $\Sigma\delta$, is found to be consistently lower for paths containing the correct hypothesis for y_i. This is usually true even if there was a noise or interference spike present at the time the sample y_i was formed, because the multi-path propagation 'coding' ensures that the effects of the actually transmitted symbol, x_i, are also seen in the received samples of later symbols, y_{i+1}, y_{i+2}, etc. It is likely, therefore, that only sequences containing the correct estimate of x_i will end up as a surviving path even though the error metric, δ, could be large for an individual case of $(e_i - y_i)^2$. This convergence is generally achieved within 5η received symbols [3].

In practical equalisers, a few concessions need to be made in the content of the transmitted signal. Firstly, it is important to force the equaliser into a known state before information is sent. Examination of the GSM transmit format in GSM 05.01 [4] reveals that three 'tail' bits (000) are added before and after the information burst. These are necessary to set the equaliser to a known state (prior to the start of data transmission) and to cause the equaliser to converge to one (000) state at the end of transmission to allow the most likely path to be evaluated. Also, as mentioned earlier, perfect estimates of the channel coefficients are not available. Instead, accurate estimates of the channel have to be made, and this information does not come for free! An additional processing block is necessary for this purpose, similar in structure to the previously maligned linear transversal equaliser but used instead as a channel estimator, as discussed in [5]. It operates in advance of the Viterbi decoder and is able to 'learn' the characteristic impulse response of the channel by means of a known 'training sequence' of 26 bits inserted in the middle of the transmitted data sequence. As this training sequence is repeated approximately every 0.6 ms, the equaliser is able to track rapidly varying propagation conditions.

To conclude this section on equalisation, it is appropriate to highlight some of the key benefits of these structures:

- Perhaps the key benefit of the Viterbi equaliser structure is that for data sequences of N m-ary symbols transmitted over channels where the impulse response is not significant beyond η symbol periods, the number of compare operations is reduced from:

  $$m^N \text{ to } N \times m^\eta, \text{ where } m = 2 \text{ for GMSK-based GSM.}$$

  In the case of GSM, the data sequences are 58 bits long and so the number of compares reduces from 288×10^{15} to 928! It is worth noting that if the impulse response extends for many symbol periods or – to put it another way – attempts are made to transmit a very narrow pulse through a narrow band channel, the equaliser complexity starts to rise exponentially as the impulse response, expressed in symbol periods, increases.
- The error probability for a Viterbi equaliser is much reduced in poor signal-to-noise conditions compared with decoding decisions made on a symbol-by-symbol basis. Interference affecting the amplitude of the current symbol can be offset by the (undisturbed) contributions of prior symbols, thanks to the unintentional 'coding' generated by the channel multi-path. This can still allow the composite current symbol to be correctly decoded.
- The Viterbi equaliser is particularly suited to 'partial response' modulation (PRM) schemes, such as GMSK. In such schemes, the phase trajectory is continuous (containing no abrupt transitions, as in FSK) and allows most of the transmitted energy to be constrained within a narrow bandwidth. However, when conventional detectors are used with PRM systems, there is typically a 3 dB loss in sensitivity compared with 'normal' modulation at the same bit rate; almost all of this 3 dB is recovered with the Viterbi-based MLSE approach.
- Although the MLSE approach was first developed to address pulse-amplitude modulation systems, it is readily extensible to phase modulation schemes by implementing in-phase and quadrature channels.

If justification were needed for including such an interesting topic as equalisers in a book essentially dealing with system performance, it would only be necessary to point to their use in almost all current and future systems. Introduced in wire-line systems to allow increased data rates, they found true volume application in the GSM system. As will be seen later in Chapter 6, UMTS was conceived to use rake receivers as a means of dealing with multi-path. In practice, most modern handsets employ equalisers to gain improved performance.

References

1 G. D. Forney, Maximum-likelihood sequence estimation of digital sequences in the presence of inter-symbol interference, *IEEE Transactions on Information Theory*, **18** 3 (1972) 363–378.

2 G. D. Forney, The Viterbi algorithm, *Proceedings of the IEEE*, **61** 3 (1973) 268–278.

3 J. M. Morton, *Adaptive Equalisation for Indoor Wireless Channels*, M.Sc. thesis (Virginia Polytechnic Institute and State University, 1998).

4 ETSI, *Digital Cellular Telecommunications System (Phase 2+); Physical Layer on the Radio Path, General Description*, GSM 05.01 version 7.1.0, Release 1998.

5 F. R. Magee and J. G. Proakis, Adaptive maximum-likelihood sequence estimation for digital signalling in the presence of inter-symbol interference, *IEEE Transactions on Information Theory*, **1** (1973) 120–124.

6 UMTS RAN planning and design

In the early 1990s, the need for a 'third-generation' cellular standard was recognised by many agencies worldwide. The European Union had funded a series of research programmes since the late 1980s, such as RACE [1], aimed at putting in place the enabling technology for 3G, and similar work was underway in Japan, the USA and other countries. The runaway success of GSM, however, had fortuitously put ETSI in the driving seat, as most countries wanted a GSM-compatible evolution for 3G. Recognising this need, and to reduce the risk of fragmented cellular standards that characterised the world before GSM, ETSI proposed that a partnership programme should be established between the leading national bodies to develop the new standard. The result was the formation of 3GPP (the 3G Partnership Programme) between national standards bodies representing China, Europe, Japan, Korea and the USA. It was agreed that ETSI would continue to provide the infrastructure and technical support for 3GPP, ensuring that the detailed technical knowledge of GSM, resident in the staff of the ETSI secretariat, was not lost.

Although not explicitly written down in the form of detailed 3G requirements at the time, the consensus amongst the operator, vendor and administration delegates that drove standards evolution might best be summarised as:

- Provide better support for the expected demand for rich multimedia services,
- Provide lower-cost voice services (through higher voice capacity),
- Reuse GSM infrastructure wherever possible to facilitate smooth evolution.

This was clearly a very sensible set of ambitions, but limited experience of multimedia in the fixed network, and of the specific needs of packet-based services in particular, meant that some aspects of the resulting UMTS standard were not ideal. These will be highlighted in Section 6.6. The initial release of UMTS, known as 'Release 99', will be used as the baseline to introduce the functionality of UMTS, although the key air interface improvements introduced in 'Release 5' (HSDPA) and 'Release 6' (HSUPA) will also be summarised.

6.1 UMTS system overview

In line with the goals above, UMTS can, at a high level, be summarised as the introduction of a new radio access solution on top of the GSM core network. This is reflected in

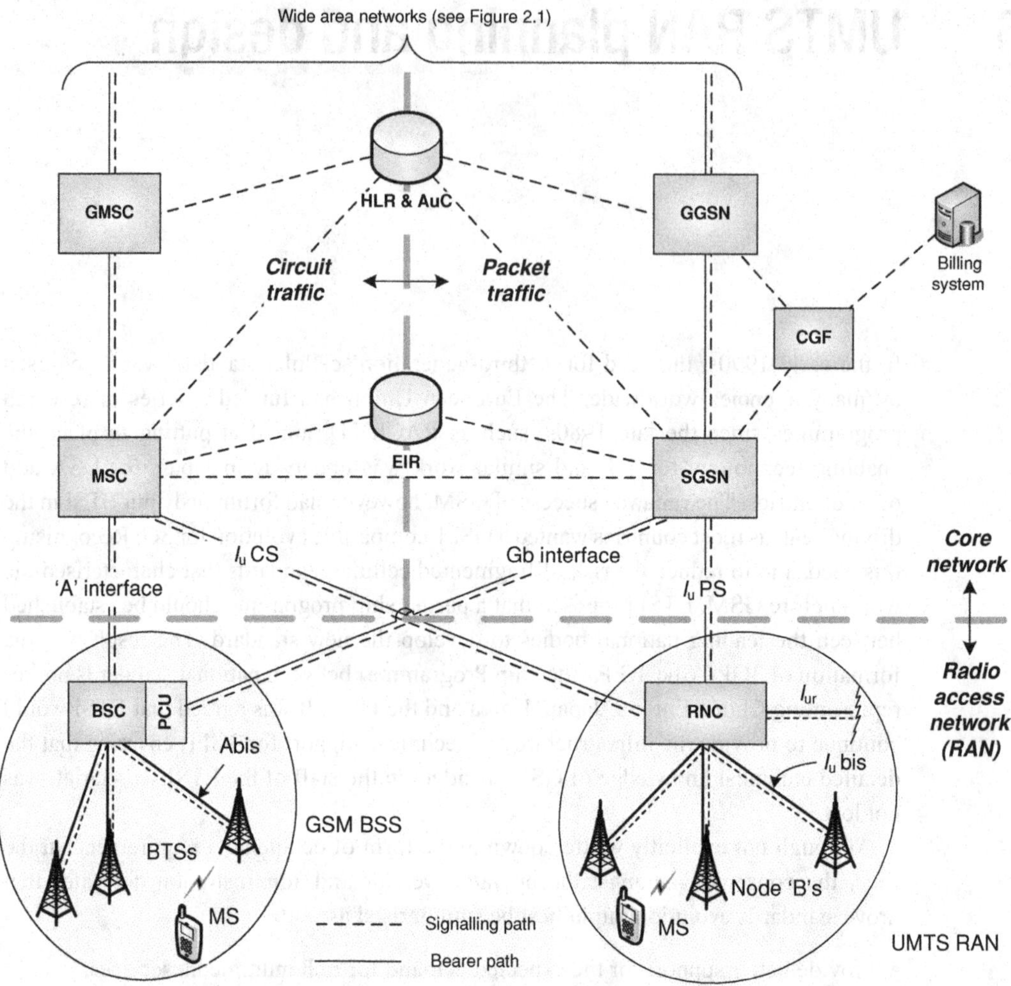

Figure 6.1 Partial view of 3GPP Release 99 network architecture with UMTS

Figure 6.1, where a new RAN (radio access network) is introduced comprising an RNC (radio network controller) and 'node Bs' (the UMTS base stations). The RAN reuses the existing MSCs and GSNs to support UMTS circuit and packet traffic respectively. In practice, these elements require upgrades to enable the variety of individual user rates that the access network now supports, along with other detailed changes. The new UMTS RAN is deployed alongside the existing GSM BSS, with each radio access network utilising the circuit or packet facilities of the common core network for wide area interconnect.

As will be discussed later, the limited transmit power available to the mobile means that in most deployments UMTS will be uplink limited; cells will also be smaller for higher data rates. This is particularly likely to be the case for cells supporting the low traffic levels to be expected early in network rollout. Higher network costs were foreseen for UMTS (relative to GSM) associated with the larger number of these smaller cells needed to cover a given area. The standard was also therefore developed to support deployments where UMTS would be limited to islands of coverage, until the use of applications requiring these higher data rates became more widespread. In such deployments, UMTS would be rolled out as an overlay to GSM (rather than a replacement) and the new standard would need to support seamless handover between UMTS and GSM, to maintain call continuity as users move out of UMTS coverage. For their part, UMTS subscribers would need a mobile capable of working on GSM and UMTS, as well as a subscription to both services. To facilitate handover to and from the UMTS network, a feature known as 'compressed mode' operation was introduced. This mechanism creates a 'gap' in the otherwise continuous UMTS transmissions, usually by increasing the transmitted data rate above the average level immediately prior to the planned gap. During this gap, the receiver can be reconfigured from its current mode – e.g., UMTS – to receive GSM BCCH transmissions, which can be used to identify handover candidates. Normally occurring breaks in GSM transmissions can be used for the corresponding GSM to UMTS handover. Compressed mode is important because it avoids the need for two complete receivers, reducing the size, weight and cost of the mobile as well as increasing its battery life.

As for GPRS, an imperative is the need to efficiently utilise access and core network resources for bursty traffic and this has led to the retention of the mobility management (MM) context (between the MS and the SGSN) and session management (SM) (through the use of PDP contexts between the SGSN and GGSN). Although there are some differences at a detailed level, particularly in the underlying transport protocols, the concept remains that introduced for GPRS. Indeed, within the MM context, the same three states are preserved. GPRS 'idle', 'standby' and 'ready' map to the UMTS packet mobility management (PMM) states of 'PMM detached', 'PMM idle' and 'PMM connected', respectively. In summary, even though a quite different physical layer is introduced for the access network, the MM and SM context structure is largely unchanged; further details may be found in [2].

One of the major differences between UMTS and GSM Release 98 circuit and packet services is the sheer variety of bearer rates and QoS attributes possible with UMTS – a consequence of targeting the support of multimedia services. However, the richness of the offering brings with it considerable complexity for the network designer and the need to introduce many additional parameters in signalling transactions during call set-up and handover. Figure 6.2 shows a high-level schematic of a UMTS network, with the defined interfaces (U_u, $I_u$bis, etc.) and the components of the *UMTS bearer service* superimposed. The UMTS specification addresses the *radio bearer service* only; one of four components that affect the application running between the mobile and application server. Messaging *is* defined between all UMTS network elements but the operator has full responsibility for determining transport architectures to deliver the I_u and core network bearer services.

The key to satisfactory operation of applications running between the mobile device and an application server, perhaps located within the operator's intranet, is the process shown on the left of Figure 6.2; 'application QoS to UMTS bearer service mapping'. If this has a familiar ring to it, it is because it is the same problem addressed in

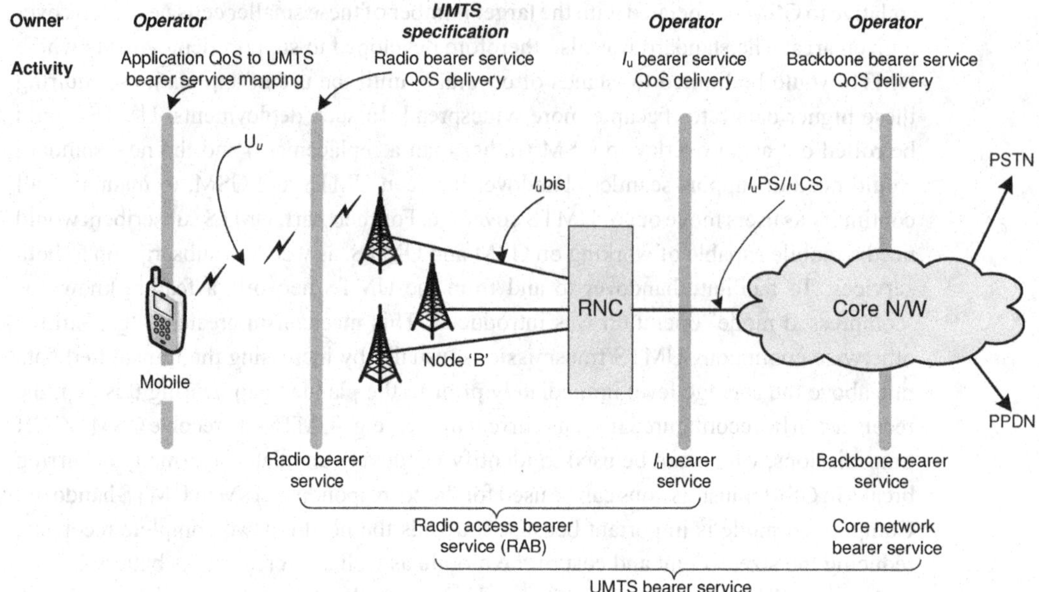

Figure 6.2 The UMTS bearer service

Chapter 3, using the ARC tool. Use of ARC or similar tools will enable the most economic bearer rate to be selected for the application and identify the absolute values of packet loss, delay, etc., that will be 'consumed' by the application buffering process in the mobile device and, by implication, specify the guaranteed bit rate, delay and packet loss ratio that must be delivered by the UMTS bearer service.

The quality of service concept and architecture for UMTS is specified in [3], although it is perhaps better considered a framework, as the almost infinite number of combinations of bearer rate, BER, delay, etc., mean that no attempt is made to define these comprehensively. The intent is rather to define four broad classes of UMTS bearer service attributes to meet the needs of corresponding application groupings and leave the operator to define *specific* bearers matched to the applications that will be offered by their network. Table 6.1 is based on information in [3] and indicates the attributes that are available to define specific UMTS bearer performance. Note that central to the efficient use of UMTS resources is the ability to specify *different* UMTS bearer service requirements for the uplink and downlink.

To illustrate the UMTS QoS concept, a simple example may be useful. Suppose that an operator wishes to support a web-browsing service. The first step would be to understand the application characteristics. The average user rate of 'mouse clicking' would be specified (defining the burst arrival rate) and, together with parameters of the specific distribution (known to characterise the uplink web-browsing burst-size distribution), would be entered into a modelling tool. This would determine the optimum bearer rate and mobile buffer size to meet the chosen delay and packet-loss ratios appropriate to this application. This specifies the requirements of the radio bearer. The delay and packet

Table 6.1. QoS attributes for UMTS

Attribute	Traffic class			
	Conversational	Streaming	Interactive	Background
Maximum bit rate	✓	✓	✓	✓
Delivery order	✓	✓	✓	✓
Maximum SDU size	✓	✓	✓	✓
SDU format	✓	✓		
SDU BER	✓	✓	✓	✓
Residual BER	✓	✓	✓	✓
Delivery of SDUs in error	✓	✓	✓	✓
Transfer delay	✓	✓		
Guaranteed bit rate	✓	✓		
Traffic-handling priority			✓	
Allocation and retention priority	✓	✓	✓	✓
Source statistics descriptor	✓	✓		

losses arising from the queuing process in the mobile and transmission across the UMTS air interface are, of course, only one component of the uplink QoS budget; to this needs to be added the I_u bearer and backbone bearer service QoS impacts which also form part of the link to the application server. Further, what the end user is interested in, of course, is the *end-to-end* delay between mouse click and receipt of the web page, so in putting together the overall QoS budget for the application, not only does the uplink have to be considered, but also the web-server delay, and the corresponding delays in the downlink. Because of the known *Pareto* distribution of downlink burst sizes, with its probability of very large bursts, it is likely that for this application, by far the largest part of the end-to-end QoS budget would be allocated to the downlink. Similarly, the single most expensive part of the end-to-end link is usually the radio bearer component; so in a typical uplink or downlink QoS budget, by far the largest part will be reserved for this element of the link.

Having established a set of radio, I_u and backbone bearers for both the uplink and downlink associated with web-browsing, this particular combination of bearers would always be used for that operator's web-browsing service and represents a *specific* UMTS bearer service supported by that operator. Mobiles supported by the particular operator will be configured to associate *this* UMTS bearer service with the web-browsing application and when the user selects the web application *this* bearer service will be requested in the 'QoS information element' of the PDP context activation request passed to the GGSN [4]. Had this been a circuit call set-up, the corresponding information would have been included in the 'bearer capability information element' (BCIE) of the call set-up request message sent to the MSC. Note that, like the application-to-QoS mapping process, the engineering of the I_u and backbone bearer services is left to the operator to carry out.

The QoS request is conveyed to the network during the PDP context activation phase, known as session management (SM); a similar SM phase occurs in GPRS (see

Table 6.2. UMTS channel structure

	GSM	GPRS	UMTS		
Channel class	–	–	*Logical*	*Transport*	Comment
Broadcast	BCCH	PBCCH	BCCH	BCH/FACH	BCH = broadcast channel FACH = Forward access channel PCH = Paging channel PCCH = Paging control channel RACH = Random access channel CCCH = Common control channel DCH = Dedicated channel DCCH = Dedicated control channel CPCH = Common packet channel DTCH = Dedicated traffic channel DSCH = Downlink shared channel
Paging	PCH	PPCH	PCCH	PCH	
Common control	RACH AGCH	PRACH PAGCH	CCCH	RACH FACH	
Dedicated[a] control	SDCCH FACCH SACCH	– PDTCH	DCCH	DCH/DCH RACH/FACH CPCH/FACH	
Dedicated[b] traffic	TCH-F TCH-H	PDTCH	DTCH	DCH/DSCH RACH/DSCH	

[a] Note that the separate 'associated' control channels of GSM are not needed, as the data and control information are multiplexed on to the same dedicated connection – for both the uplink and downlink

[b] Although the DCH/DCH uplink/downlink combination was expected to be the dominant bearer configuration at the time of design, other transport bearers can be used and may be more efficient for light, bursty traffic

Section 2.1.2). However it is at this phase that some of the additional complexity of UMTS is introduced. Before an 'activate PDP context accept' message can be transmitted to the mobile following its 'activate PDP context request', the UMTS bearer service manager has to check that resources are available. This involves signalling between the bearer managers in the mobile, RAN (radio bearer and I_u) and core network (backbone bearer) to determine whether the specific UMTS bearer service request can be met in full and, if not, what service *can* be offered. If the full service cannot be offered, the network may respond to the mobile with a 'modify PDP context request' message containing details of a reduced bearer service offer. The mobile then needs either to accept the offer or to terminate the PDP context activation request.

To conclude this overview of UMTS network architecture, it is helpful to examine the sequence of events that takes place once a user has decided to initiate a UMTS session. UMTS supports both circuit calls and packet calls; the commentary here relates to the establishment of a packet session, as these are likely to be of more interest in the future, but corresponding procedures exist for circuit call set-up. The channel structure for UMTS is shown in Table 6.2. It begins to be clear that the session set-up process for UMTS is very similar to that for GPRS. To avoid repetition, only the key differences will be discussed here. More detail on the common elements of the session initiation process may be found in Sections 2.1.1 and 2.1.2, and [2].

At power on, the mobile will carry out its own initialisation processes and then search the UMTS frequencies for cells transmitting a BCCH and 'camp' on a suitable cell. It will decode the information on the BCCH and locate the RACH and FACH channels for future use. When the user initiates a UMTS session from the handset, a UMTS attach request message will be sent on the RACH channel. Note that, because this is the first time (for a while) that the user has been connected to the network, authentication will take place as part of the attach process using information from the USIM. All signalling transactions relating to session set-up will take place on the RACH (uplink) and FACH (downlink). Once authentication has been successfully completed, serving RNC and routing area information will be held in the MM context and the status will be PMM connected. With a connection now in place between the mobile and the SGSN, the mobile will send a PDP context-activate request to the SGSN. This message, in addition to information to identify the PDP context uniquely, contains the requested QoS. The SGSN will send a 'create PDP context request' to the GGSN and, following PDP context establishment (with an appropriate core network bearer), will start a negotiation with the RAB manager to determine whether, with the current system load, the requested QoS can be granted. The RAB manager will then set-up the appropriate radio and I_u bearers to deliver the requested (or offered) QoS. Table 6.2 makes it clear that, depending on the requested QoS and expected call duration, dedicated channels (DCH) may not be used for transport, as these can be quite inefficient for low-rate bursty traffic and take some time to set up. Following this process, the SGSN sends a 'PDP context accept' message to the mobile and is able to route PDP PDUs between the mobile and GGSN and start charging. Note also that the mobile may choose to reject the PDP context (if the QoS offered is less than that requested) by sending a 'PDP context deactivate' message.

Although some of the considerable complexity of the UMTS system has been omitted for clarity, it is still possible to see that, to deliver the flexibility needed to support multimedia services, it is necessary to increase the number of fields both within the network elements and within the messaging for call set-up (some messages may be up to 30 bytes long). With limited bandwidth signalling channels, this must (and does) have an impact on session set-up time. Network design for the operator is also considerably more complex than for GPRS, where only limited attempts were made to deliver guaranteed QoS. Firstly, it is left to the operator to decide what QoS is needed to support new applications introduced to the network. It is also the operator's responsibility to engineer the I_u and backbone bearer families. At the present time, the number of packet-based services requiring specific levels of QoS is fairly low and thus it is still economic to address the problem by over-dimensioning the network to ensure that there is sufficient capacity. The industry acknowledges that there is much left to learn about the economic delivery of packet services with guaranteed QoS that will become essential when IP-based applications become the volume product.

6.2 The UMTS air interface

In Chapter 2, when GSM was introduced as a vehicle for discussing features common to most cellular systems, only a page or so was dedicated to introducing the physical layer of GSM. The transmit scheme was frequency-division duplex, with one frequency dedicated to transmissions to the base station, and a second used for relaying information in the reverse direction. The frequencies in each direction were shared in time by up to eight users; at any point in time, one user had complete use of the up- and downlink frequencies. UMTS adopts a scheme known as wideband code division multiple access (WCDMA); it too uses frequency-division duplex, but there the similarity ends! Every user assigned to a UMTS carrier shares that frequency *simultaneously*. A mobile receives data from the base station relating to *all* active users at a point in time but retrieves its specific information by multiplying the incoming signal by a unique code assigned only to that user and also used during the transmission process to *spread* the user data over the channel bandwidth. This scheme is less intuitive to understand but has a number of advantages in the context of the goals established for UMTS:

- From day one, UMTS was conceived as a wideband system and designed to operate in 5 MHz channels (compared with the 200 kHz chosen for GSM),
- The wide channel allocation is intrinsically capable of supporting the high user data rates needed for some multimedia services,
- Different user rates can be accommodated by using different spreading factors; for instance high user data rates would be delivered using lower rate spreading codes (and thus still contained within the 5 MHz channel),
- The user rates can be changed on demand and can be different for different users; each user is allocated capacity in 10 ms frames during which the rate is constant,
- The voice capacity of a single UMTS carrier can easily exceed 50 channels, compared with a maximum of eight for GSM. Even though there is some complexity overhead for UMTS, the intrinsic costs are not fundamentally different, especially for mature solutions, resulting in a much lower cost for supporting voice users *provided the carrier is fully loaded.*

The combination of the WCDMA air interface and 5 MHz channel structure can thus meet the goals set for UMTS. The remainder of this section will address the fundamental drivers of this solution. The factors that need to be considered in deploying an efficient UMTS network will be discussed and the coverage and capacity that can be delivered by such systems will be presented.

For completeness, it is appropriate to mention at this stage that the UMTS solution embraces *two* different air interfaces; the WCDMA FDD solution already introduced and a time-division duplex (TDD) solution. The TDD solution utilises one carrier frequency for both uplink and downlink transmission; as its name suggests, within a frame, a proportion of time is dedicated to the downlink and the remainder to the uplink with a short guard period between the two sections. The TDD frame is divided into 15 slots which, for a given deployment region, may be allocated in an almost arbitrary downlink-to-uplink ratio provided certain minimum constraints are met. The TDD solution has not been widely exploited commercially, with the exception of China, where trials of potential solutions continue. This has probably been influenced by the following considerations:

- Cellular traffic, currently, is still dominated by voice so the possible asymmetry between up and downlinks offered by TDD is not an advantage,
- The UMTS FDD solution was therefore commercially exploited first,

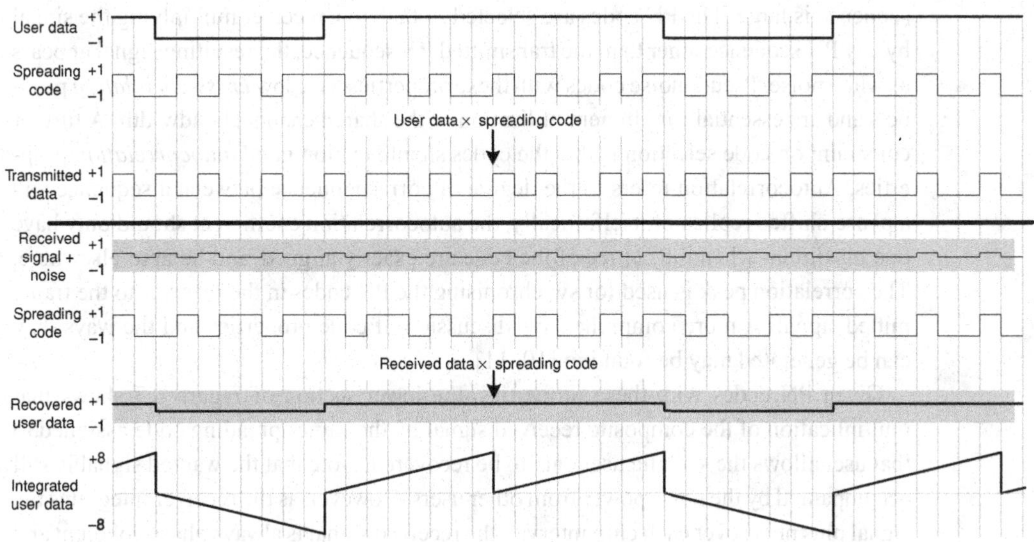

Figure 6.3 Principles of spread-spectrum coding

- The demand for 3G services has been limited to date (no need for further capacity),
- The spectrum allocated in Europe for TDD is usually close to that for the FDD system. Operators planning to secure additional capacity through deployment of TDD solutions would need to reuse their current sites and interference between co-sited FDD and TDD solutions would be difficult to address.

For these reasons the UMTS TDD solution will not be covered here, but the interested reader is referred to [5, 6, 7, 8, 9] for further information.

6.2.1 WCDMA air interface physical layer principles

The underlying concept of code division multiple access and the functionality that remains central to the operation of WCDMA is the way coding of the user data allows many users to share the same channel. Figure 6.3 introduces this concept. The user data are multiplied by a specific *pseudonoise* (PN) code (or spreading code) assigned to that user to spread the information across the whole of the UMTS channel. The transmitted signal is much attenuated through the normal propagation process and will usually be *below* the total noise power at the receiver input. This noise power will be made up of thermal noise along with other user interference. In a CDMA system with many active users, the noise-like signals from other users will be greater than thermal noise.

The properties of the PN codes chosen for CDMA are very important. Pseudorandom sequences or 'PN codes' are sequences of 1s and 0s generated by an algorithm so that the resulting numbers look statistically independent and uniformly distributed. In fact, a pseudorandom signal is not random at all; it is a deterministic, periodic signal that is known to both the transmitter and the receiver. Even though the signal *is* deterministic, it appears to have the statistical properties of sampled white noise, provided the length of the sequence is large. The PN codes are selected so that, to a receiver multiplying the signal by any PN sequence other than the transmitted PN sequence, the resulting signal appears as white noise. Pseudonoise codes with these properties have low *cross-correlation* products and are essential for efficient utilisation of the shared channel bandwidth. A further constraint on code selection is that the codes should exhibit good *autocorrelation* properties. Autocorrelation refers to the degree of correspondence between a sequence and a phase-shifted replica of itself. Ideally, the autocorrelation coefficient should only have one maximum when the copies of the code are exactly aligned, and be zero elsewhere. The correlation peak is used for synchronising the PN codes in the receiver to the transmitted signal. A more comprehensive discussion of code properties and the ways they can be generated may be found in [10, 11].

Given PN codes with these properties, the lower section of Figure 6.3 shows that multiplication of the composite received signal by the same spreading code assigned to that user allows the specific user data to be recovered. Note that the wanted signal is still accompanied by the noise power from other users. However, as the receiver integrates the signal plus noise over each chip interval, the received signal is always phase coherent and adds in proportion to the number of chips, N, whereas the interference is not coherent and will add in a noise-like way, proportional to $\sqrt{N}$. In the example shown in the diagram, there are 8 chips/user data symbol and thus $N = 8$. The ratio of the wanted user signal power to the sum of thermal noise plus user interference power will thus improve by $(N/\sqrt{N})^2 = 8$, or more generally by the spreading factor N. Using UMTS parameters, if the user data originates from an 8 kbit/s voice codec, the *spreading factor* can be calculated as:

$$\text{Spreading factor}\,(N) = \frac{\text{WCDMA chip rate}}{\text{User data rate}} = \frac{3.84 \times 10^6}{8 \times 10^3} = 4.8 \times 10^2.$$

This suppression of other user interference by the spreading factor means that more than one user can share the channel simultaneously. For example, in the uplink and for 8 kbit/s speech, if (say) 5 dB is allowed as the required energy per bit relative to the sum of thermal and interfering noise spectral densities $[E_b/N_0]$, it is very apparent that many tens of users can be supported in *one* channel, even allowing for other system overheads and effects (provided the noise contributions from each user are similar). This should not be surprising when it is recognised that 5 MHz is being used to transmit 8 kbit/s data. In fact, we are seeing another example of Shannon's Law at work (see Chapter 1), where it will be recalled that the maximum possible information rate in a channel is a function of the communication channel bandwidth and the carrier-to-noise ratio. In this case the excess channel bandwidth is utilised to code the user signal redundantly (using the spreading code) and thus the required carrier-to-noise ratio drops dramatically. This is the key mechanism at work in all CDMA systems and predicts that, for variable user rate solutions, such as WCDMA, the number of users that can be supported for higher-rate services will be much lower, as is indeed the case. With this central tenet of CDMA operation understood, the other key aspects of WCDMA will be discussed next.

6.2.1.1 Power control

A working assumption made in demonstrating the way excess channel bandwidth is exchanged for capacity in CDMA systems was that 'noise contributions from each user are similar'. This does not come for free. Consider a user close to the cell edge and another not that far from the base station. Both mobiles have the same maximum transmit power but the mobile at the cell edge will suffer much larger transmission loss. If both transmit at maximum power the mobile at the cell edge (and indeed most mobiles not close to the cell site) will not be 'heard' at the base station because of the excess interference caused by the mobile close to the cell site. This is unacceptable and a power control system is clearly needed for the uplink, and in practice also offers benefits for the downlink – even though there is no near–far problem. Power control on the downlink allows some additional power to be allocated to mobiles close to the cell edge (to mitigate the effects of higher out-of-cell interference) or for mobiles that are moving slowly (where channel coding and interleaving offer less benefit).

Then, to maximise cell capacity, it is important that at all times the received E_b/N_0 is kept as close as possible to the target value for the particular bearer rate. Any power above this level causes additional interference and will reduce the number of mobiles that the cell can support. How frequently does the control loop need to operate? Power control in GSM is managed by the BSC and averaged over about 100 frames; adjustments are therefore made about twice a second. However, deviation from the optimum received power levels is much less important in GSM because the base station is largely protected from other user in-cell interference through time and frequency orthogonality. Power control is, therefore, primarily introduced on the uplink to extend battery lifetime. In WCDMA, because of the impact on capacity, power control is much more important and is designed to eliminate substantially all transient propagation effects. Propagation loss above that which may be expected from the typical r^{-4} law can largely be attributed to shadowing or multi-path. The update rate necessary to counter the effects of range variation and shadowing is quite low, and is of the order of 100 Hz or so. The impact of multi-path can be much more severe.

In introducing the topic of multi-path in Section 3.1.1, the phenomenon was explained by the presence of a few major reflecting surfaces within the beam width of the transmit and receive antenna systems and (implicitly) treated as a series of point reflections. In practice, returns from most objects giving rise to significant multi-path (large buildings, the ground, vegetation, etc.) are the result of contributions from a large number of reflecting surfaces very closely spaced in time about the apparent peak of the reflection. This behaviour is due to secondary reflecting structures, perhaps windows of buildings or roughness in the surface of the ground. The magnitude of the return from the composite object, at a point in time, is thus the result of adding all of the reflections in the vicinity of the object projected in the direction of the mobile. Because all of the reflecting sub-surfaces of the composite reflector are physically close together, very large variations in returned power can be caused by movements of the mobile as small as one or two wavelengths. The actual multi-path return for typical reflecting structures thus comprises an average return accompanied by fast Rayleigh fading about the mean. Rayleigh fading at typical mobile speeds can sometimes give rise to 20–30 dB variation below the mean value in the space of 10 ms to 20 ms [12]. Therefore, WCDMA introduces *fast power control* to counter the effects of Rayleigh fading as well. The fast power control loop, managed by the RNC, is designed to measure the received signal strength, issue instructions to the mobile and execute them at a 1500 Hz rate. Note that this means that the mobile must maintain up and down control channels *at all times* that it is in the 'PMM connected' state – even when it is not transmitting data.

In addition to fast power control there is *outer loop power control.* So far it has been assumed that all mobiles are managed by fast power control to the same E_b/N_0 target. This is not the case. The goal is usually to deliver a specified block erasure rate (BLER) for the chosen bearer. The actual E_b/N_0 target will vary according to the radio bearer service selected; configurations interleaved over longer periods of time (up to 80 ms compared with the minimum 10 ms) and employing more redundant coding will usually have lower target values. Outer loop power control also allows these target values to be altered as propagation changes (e.g., different mobile speeds or changing multi-path environment). The RNC will monitor reported BLER values and, through the node 'B', increment or decrement the mobile power transmitted until the desired BLER is restored.

6.2.1.2 Rake receivers

In Chapter 5, it was shown that equalisers are used in GSM to eliminate the worst effects of multi-path delay spread. It will be recalled that such equalisers use a 'training sequence' to estimate the multi-path channel coefficients and that these values are then used to pre-distort all possible input sequences, which can then be compared with the received signal. The most similar candidate sequence is then chosen as the best estimate of the transmitted signal sequence. WCDMA adopts a different solution to this same problem, known as 'rake reception'.

By definition, multi-path comprises energy from a direct ray, plus some number of delayed copies of the transmitted signal, where the delay is dependent on the additional distance to and from the reflecting object. In an equaliser, these delayed copies are treated as a problem that distorts the wanted signal. *Rake reception* recognises that there is energy in each of these reflected rays and seeks to add the indirect rays coherently to the direct ray, at the same time eliminating the multi-path distortion and improving the magnitude of the wanted signal.

Figure 6.4 illustrates this process and shows the key elements of a rake receiver together with thumbnail sketches of the signal at the input to the receiver and at points YY and ZZ. The signal present at the input to the receiver comprises a direct ray (AB) and two multi-path rays (ADB and ACB). The matched filter, which immediately follows the filtering and conversion process, detects the timing of the signal peaks and allocates correlation receivers or 'fingers' to each peak (at times t, $t+\tau_1$ and $t+\tau_2$). The de-spreading and integration process illustrated in Figure 6.3 is carried out by the correlator–code generator sub-system. The magnitude, phase and timing of the vector at point YY in finger 3 corresponding to multi-path ray ACB is shown in the second sketch, along with vectors representing the second multi-path ray ADB (finger 2) and the direct ray AB

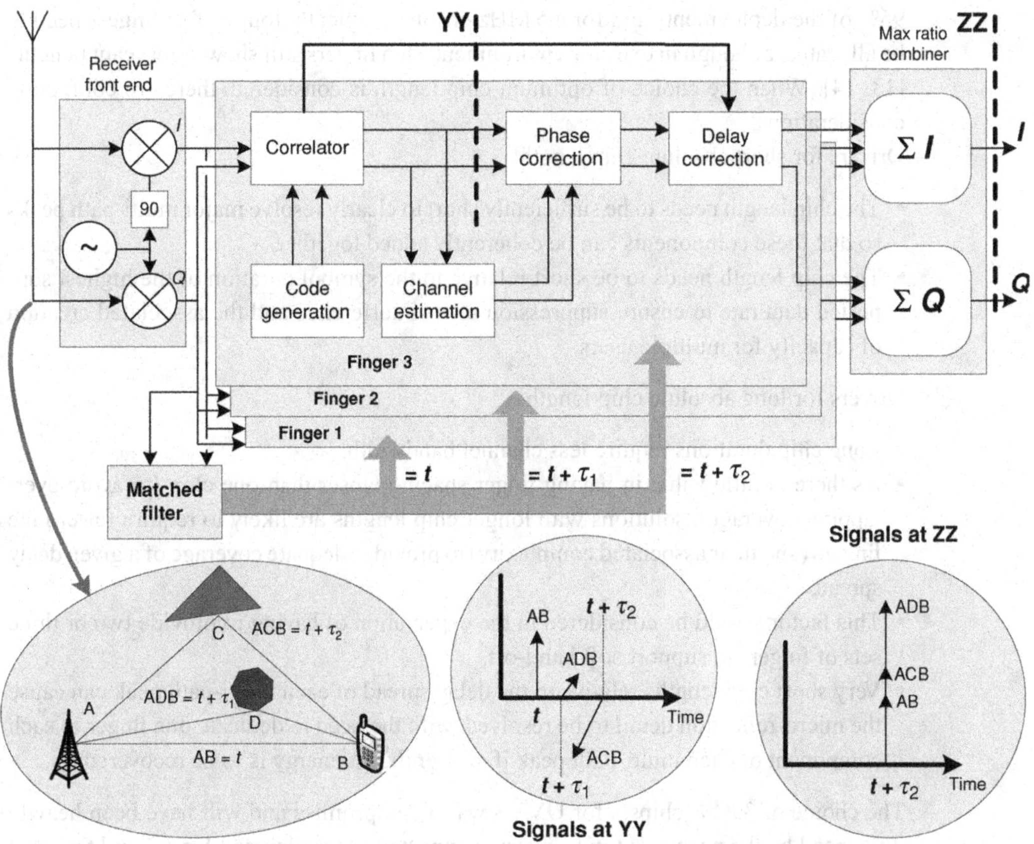

Figure 6.4 Principles of rake reception

(finger 1). Over a period of time, the vectors at point YY are samples of the integrated user data and reflect the amplitude and phase variation of each ray.

Along with the user data, the base station also transmits 'pilot channel' information (analogous to the GSM training sequence), which allows the phase and timing of the multi-path rays, relative to the direct ray, to be determined. The phase-correction and delay-correction blocks add delays and phase rotations to the samples from fingers 1 and 2 so that the three signals can add coherently in the maximum ratio combiner. The term 'rake reception' originates from the magnitudes of the reflected returns, which appear as impulses along the time axis and might be likened to the tines (or fingers) of a rake.

In CDMA cellular solutions, the chip rate chosen for a particular implementation turns out to be a key factor that ultimately limits the performance of the system and drives cost and complexity. It is, therefore, useful to explore some of these trade-offs. The aim of the equaliser is to eliminate substantially the effects of multi-path without loss of sensitivity. This required sufficient rake fingers to be available to collect most of the energy in the multi-path components. Extensive measurements of multi-path were carried out in the 1990s at channel bandwidths foreseen for 3G systems. These show that, if no more than 1 dB of the available multi-path power is to be lost in

95% of the deployments and for a 5 MHz channel, typically four or five fingers need to be allocated, although in extreme environments, ten fingers still show significant benefits [13, 14]. When the choice of optimum chip length is considered, there are conflicting considerations.

Drivers for short absolute chip length:

- The chip length needs to be sufficiently short to clearly resolve major multi-path peaks so that these components can be coherently added together,
- The chip length needs to be short relative to the symbol duration of the highest supported data rate to ensure suppression of its interference and the associated creation of capacity for multiple users.

Drivers for long absolute chip length:

- Long chip durations require less channel bandwidth,
- As there is little value in having finger spacing closer than one chip (to avoid overlapping coverage), solutions with longer chip lengths are likely to require fewer rake fingers (and their associated complexity) to provide adequate coverage of a given delay spread,
- This factor should be considered in the expectation of having to provide two or three sets of fingers to support soft hand-off,
- Very short chip lengths relative to the delay spread of each multi-path peak can cause the micro-reflection detail to be resolved, with the need to dedicate one finger to each component of each multi-path peak if all significant energy is to be recovered.

The choice of 3.84 Mchips/s for UMTS was a compromise and will have been heavily influenced by the proposed total spectrum expected to be allocated for 3G and the need to support multiple operators for competition reasons. The result is that, in order to accommodate user data rates much above about 200 kbit/s, spreading factors as low as four and eight are used. This means that other user interference is poorly suppressed and only a few users at this rate can be accommodated. Additionally, because the chip length is now close to the symbol length, the rake receiver can no longer perfectly align in time the multi-path-delayed versions of the wanted signal with that from the direct path. The timing imperfection due to the now significant chip length is such that there can be inter-symbol overlap (interference) of 25% or more. This distortion of the wanted signal cannot now be eliminated and thus represents a floor to the BER that can be achieved – regardless of the signal-to-interference ratio. The rake equaliser is starting to fail to do its job in this situation.

6.2.1.3 Soft handover

At the beginning of Chapter 2, one of the overheads of adopting cellular reuse as the basis of mobility in modern mobile systems was identified as the need to support and manage the process of handover. This was discussed further in Chapter 5 and will feature again in Chapter 10, when the optimisation of mobile systems is addressed. However, because WCDMA introduces an additional form of handover, it is useful to cover briefly the principles of both mechanisms here.

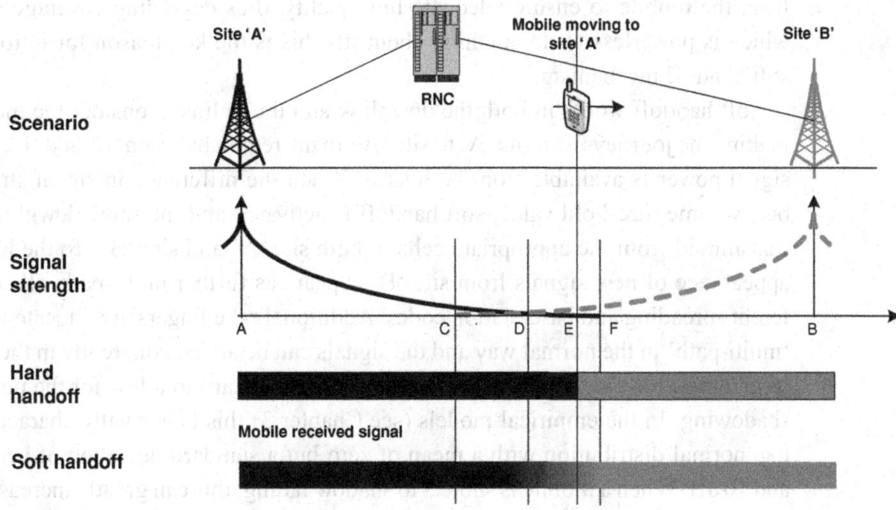

Figure 6.5 Soft handoff in WCDMA

Figure 6.5 shows a mobile user moving between coverage provided by site 'A' and site 'B'. In GSM, the mobile reports the received signal strength and quality of the serving BTS and also reports the signal strength of the BCCH allocation, using the broadcast channels of other base stations in the vicinity of the serving BTS. When the signal strength or call quality starts to deteriorate as the user moves from point 'C' to point 'D', the BSC will decide to which cell (from the BCCH allocation) the mobile should handoff. It will then cause the mobile to drop the connection from the serving cell in site 'A', and in this case connect to a cell in site 'B'. Note that this does not happen until point 'E' by which time the signal from site 'B' is significantly greater than that from the serving cell. This is quite deliberate and the difference in signal strengths is known as the hysteresis margin. If the handover had taken place at point 'D', where the signals are roughly equal, small fluctuations in signal strength would cause oscillation of the connection between the old and new cells. This oscillation is often referred to as the 'ping-pong' effect for obvious reasons.

A fundamental difference between WCDMA and GSM is that there is no time orthogonality or frequency reuse to protect cells from adjacent cell interference. On the uplink, because there is no code synchronisation between mobiles, all intra-cell and inter-cell interference is seen by the base station. On the downlink, the mobile sees some proportion of interference from other users in its own cell and all of the interference power from users in other cells. Although in the downlink all users start out orthogonal through careful code selection, some proportion of other user interference becomes visible, owing to multi-path degrading the initial orthogonality. In the absence of any other mechanism, fast power control (on both the downlink and the uplink) means that there is the possibility of a mobile – still connected to a cell in site 'A' – moving into the planned coverage area of site 'B'. Site 'A' still controls downlink and uplink power to and from the mobile to ensure adequate link quality, thus degrading coverage in site 'B', which is powerless to do anything about it! This is the key reason for introducing the soft handoff mechanism.

Soft handoff works in both the downlink and the uplink. Consider the mobile again making the journey from site 'A' to site 'B'. In the region between 'C' and 'F', significant signal power is available from both sites. When the difference in signal strength falls below some threshold value, soft handoff is activated and the same downlink signal is transmitted from the appropriate cells in both site 'A' and site 'B'. To the handset, the appearance of new signals from site 'B' appears as further multi-path, albeit with different spreading and base-station codes. Additional rake fingers are allocated to the new 'multi-path' in the normal way and the signals can be added coherently in the combiner. In planning for a single BTS–mobile link, it is necessary to allow for the possibility of shadowing. In the empirical models (see Chapter 3) this is normally characterised as a log–normal distribution with a mean of zero but a standard deviation of between 6 dB and 10 dB. When a mobile is subject to shadow fading, this can greatly increase the interference level in the serving cell, as the mobile and BTS increase power to preserve link quality, with a corresponding loss of capacity. However, it is far less likely that the path to *several* base stations will be subject to the same degree of shadowing and the power distribution of the combined signal from two or more base stations is much more benign. This greatly reduces the capacity loss from interference. This increase in capacity has to be offset against the need for one or more additional downlinks in the soft handoff cells, which would otherwise be available for use. Additional back-haul resources are also needed from sites to the RNC for the same reason. Depending upon the propagation characteristics and the thresholds chosen to determine the number of cells in the *active set*, the percentage of calls in soft handoff for UMTS will be in the range 20% to 50%.

On the uplink, no additional facilities are required and several base stations may receive transmissions from the same mobile, depending upon its location. The base station tags frames from each mobile with a link-frame quality indicator and the RNC selects the best frame. A special case of soft handover is *softer* handover. This occurs when the soft handoff links are from different sectors of the same base station.

Viterbi made an extensive analysis to quantify the benefits of soft handoff [15] as a function of the number of base stations in the *active set* (base stations participating in soft handoff) and standard deviation of the log–normal shadowing. For the single-service IS-95 case, and for standard deviations up to 6 dB, he showed that most of the possible benefits of soft handoff could be delivered with a maximum of three base stations supporting soft handoff. Under these conditions, he showed that capacity gains of up to twice or improvements in cell area of slightly more than double were possible. For practical reasons largely associated with the location of users and the service rate of the users in soft handoff, the benefits of soft handoff for multi-service networks are expected to be smaller.

The choice of soft handover threshold – the point at which data transmission from more than one cell is enabled – is also important. From the earlier discussion and in extreme cases, where the best soft handoff candidate is significantly lower than the serving cell (e.g., 10 dB), were soft handoff to be enabled, it is clear that the system cost of transmission from two cells would negate any marginal benefit. More comprehensive modelling [16] suggests that soft handover should be enabled when the signals are with in 2 dB to 3.5 dB of each other, depending upon whether the cells are sectored or omnidirectional.

6.2.1.4 Synchronisation and common pilot channels (SCH and CPICH)

In any cellular system, there are some activities that need to happen before communication can take place. In GSM, activities such as time and frequency synchronisation are fairly transparent and were omitted for clarity. In WCDMA, specific physical channels (SCH and CPICH) are necessary to allow the mobile to achieve time synchronisation (initially to time slot, then to frame level) and then identify the scrambling code for the detected cell. Configuration of these channels directly affects the cell coverage and capacity and the time taken for the mobile to 'acquire' the network. For these reasons, a brief discussion of the factors affecting performance is appropriate. Cell acquisition takes place in three stages:

- During the first step of the cell search procedure, the mobile (UE) uses the SCH's primary synchronisation code to acquire slot synchronisation to a cell. This is typically done with a single filter matched to the primary synchronisation code, which is common to all cells. The slot timing of the cell can be obtained by detecting peaks in the matched filter output.
- During the second step of the cell search procedure, the UE uses the SCH's secondary synchronisation code to find frame synchronisation and identify the code group of the cell found in the first step. This is done by correlating the received signal with all possible secondary synchronisation code sequences, and identifying the maximum correlation value. Since the cyclic shifts of the sequences are unique, the code group as well as the frame synchronisation is determined.
- During the third and last step of the cell search procedure, the UE determines the exact primary scrambling code used by the found cell. The primary scrambling code is typically identified through symbol-by-symbol correlation over the CPICH, with all codes within the code group identified in the second step. After the primary scrambling code has been identified, the primary CCPCH can be detected and the system- and cell-specific BCH information can be read.

Further details on this process may be found in [17].

A key parameter affecting the mobile's synchronisation performance is the power allocation between the SCH (comprising the primary synchronisation channel (PSCH) and secondary synchronisation channel (SSCH)) and CPICH, which are both concerned with the cell search procedure. It has been suggested that the optimum allocation scheme is for the ratio (SCH: CPICH) to be in the region of (0.65: 0.35), which should allow typical synchronisation times of about 160 ms [18]. Within the SCH, a ratio of (0.75: 0.25) is proposed for the (PSCH: SSCH) allocation. Although it may appear that power for the SCH may dominate the budget, because the SCH is only transmitted for 10% of each time slot [19], the continuously transmitted CPICH is by some margin the largest single allocation. In practice, all of the other mandatory downlink common channels (SCH, acquisition indication channel (AICH), and paging indication channel (PICH)) are normally allocated powers *relative* to the CPICH. Depending upon the load in the

Table 6.3. Management of system impairments in GSM and UMTS

System impairment	Features	
	GSM	UMTS
Interference	– Frequency reuse – Serving cell time orthogonality	– CDMA spreading gain – User code orthogonality
Interference diversity	– Frequency hopping	– Interference averaging (all users share the same channel)
Multi-path	– Equaliser	– Rake receiver and maximum ratio combining
Rayleigh fading	– Frequency hopping & channel coding – Receive diversity – Transmit diversity	– Fast power control – Rake diversity – Receive diversity – Transmit diversity
Shadowing	– Slow power control – Channel coding	– Slow power control – Soft handoff – Channel coding

cell, these other channels may in aggregate represent up to an additional 50% of the CPICH power. The question of how much power to allocate to the CPICH, therefore, begins to assume significant proportions, as power allocated to the downlink common channels cannot be allocated to carry commercial traffic and is in effect 'wasted'.

After switch-on, a mobile phone determines its serving cell by choosing the best CPICH signal. It has already been seen that the CPICH is also used by mobile phones to obtain initial system synchronisation. However, in addition, the CPICH is used to support channel estimation for the dedicated channel. Thus, in summary, CPICH power determines the cell coverage area. Because of this, the number of users can be balanced among neighbouring cells by appropriately adjusting the CPICH power amongst a cluster of base stations. This can reduce the inter-cell interference, stabilise network operation and facilitate radio resource management. On the other hand, there are some constraints on setting the CPICH power: Values which are too high will create interference called 'pilot pollution' to the neighbouring cells, and decrease the network capacity. Setting CPICH power too low will cause uncovered areas between cells, where the CPICH power is too weak for the mobile phone to decode the signal, making network access impossible.

Much effort has been spent on developing procedures for choosing the optimum amount of power that should be allocated to the CPICH [20, 21, 22] and the results generally mean that CPICH is typically in the region of 5% to 10% of the maximum BTS power. Further, once a network has been deployed on its final cell sites (as opposed to a generic planning model), significant improvements in capacity and other figures of merit can be obtained by altering the vertical angle of the cell-site main beam, to recognise the specific geography of the location.

In concluding this section, it is useful to contrast the way GSM and WCDMA manage the various system impairments that are encountered in mobile systems. Table 6.3 lists

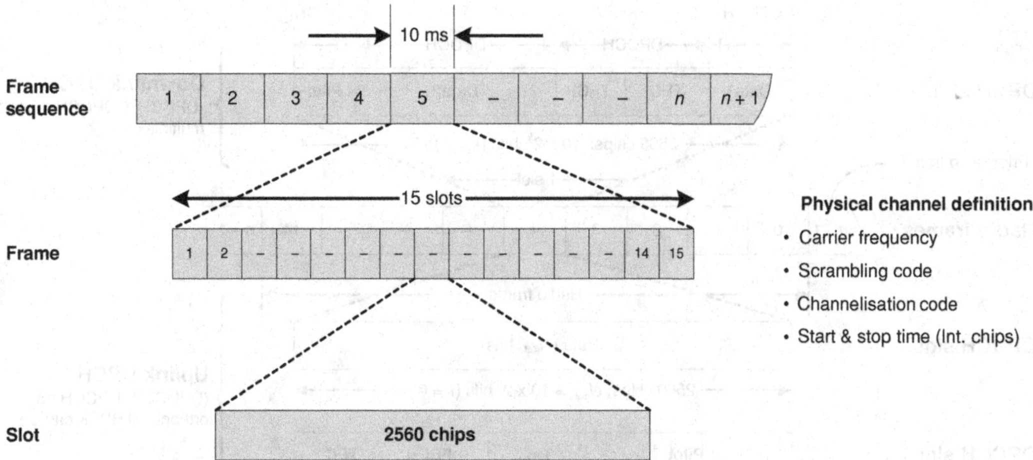

Figure 6.6 UMTS WCDMA Release 99 physical channels

the key system impairments and identifies the functionality that manages the impact of each in GSM and WCDMA.

6.3 UMTS physical channels and transport channel multiplexing

In this section, the discussion of WCDMA principles introduced above will be extended to the implementation that is specified in the 3GPP UMTS Release 99 FDD solution. In addition to coverage of the spreading and modulation process, which gives rise to the UMTS *physical channels*, attention will be paid to the way in which *transport channels* are multiplexed on to *physical channels*. It is the concept of *transport channels*, introduced earlier in this chapter, which provides operators great flexibility in the way services can be supported, potentially enabling more efficient solutions.

6.3.1 Physical channel structure

Figure 6.6 provides a top-level view of the physical channel structure. Physical channels are defined by a specific carrier frequency, scrambling code and channelisation code, together with a start and stop time. The role of channelisation and scrambling codes is discussed further in Section 6.3.2. Start and stop instants are measured in integer multiples of chips. Multiples of chips used in the physical channels are a time slot, comprising 2560 chips, and a radio frame, comprising 15 time slots. In principle, a physical channel is continuous, from the instant when it is started to the instant when it is stopped. In practice it is convenient to identify 'slots', which represent intervals at which a channel format is repeated, and 'radio frames', within which the transmitted data rate is kept constant.

Figure 6.7 shows the slot format for the dedicated physical channel (DPCH) [19]. It will be recalled that UMTS WCDMA requires omnipresent fast power control in the

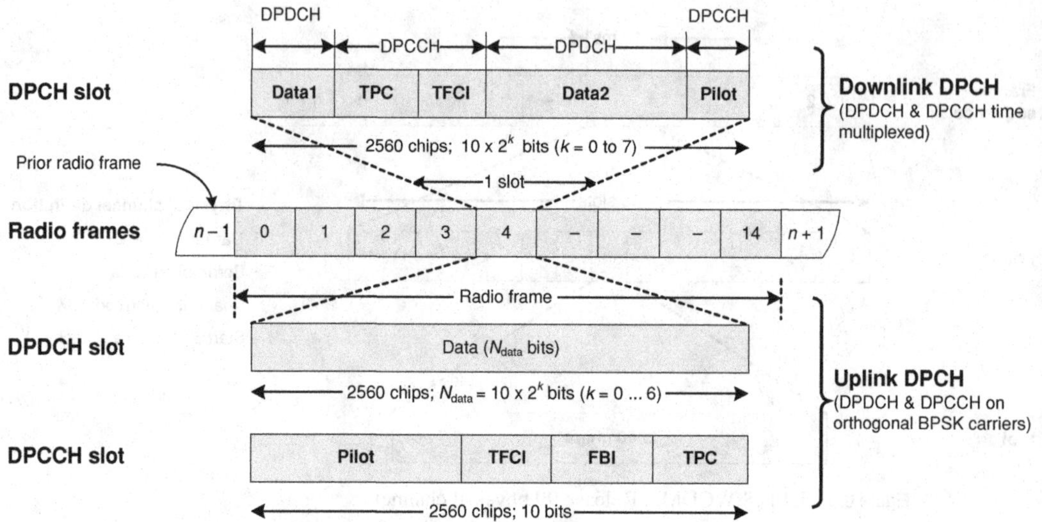

Figure 6.7 Slot format for the downlink and uplink dedicated physical channel

DPCH to ensure correct system operation and thus there will always be the need for control as well as data channels. Indeed, for the DPCH, fast power-control information carried in the TPC (transmit power-control) field is transmitted *whether or not* data are being carried and for as long as the channel exists. Further, power control is the key driver in the choice of a slot period of 0.667 ms. This enables the 1.5 kHz fast power-control loop, which ensures that the carrier-to-interference ratio remains optimised for the selected BER, despite transient changes in propagation loss about its mean value. In both the downlink and the uplink, the underlying data rate for the DPCH is determined by the choice of spreading factor and will be dictated by the service the operator wishes to support. Figure 6.7 defines an exponent k, which indirectly specifies the number of data bits. Small values of k correspond to large spreading factors and lower data rates; at all times, the number of chips in the DPCH slot remains constant.

In the downlink, the control channel (DPCCH) is time multiplexed with the data channel (DPDCH) and, as well as TPC, supports *pilot* and *TFCI* (transmit format combination indicator) bits. Modulation on the downlink is QPSK. The dedicated pilot bits are used for channel estimation in situations where the CPICH channel may not be representative of the propagation channel in use – for instance, when adaptive antennas are deployed. The TFCI bits specify the transport format combination in use for the current slot and radio frame. There is a one-to-one correspondence between a specific value of the TFCI and a corresponding transport format combination. The TFCI is used by the mobile to determine how to decode, de-multiplex and deliver the received data on the appropriate transport channels. This will be discussed further, later in this section.

In the uplink, the control and data channels are modulated onto two separate BPSK carriers in phase quadrature, rather than being multiplexed on to a single carrier, as is the case for the downlink. This approach has been adopted to avoid the possibility of audible interference in the mobile that could arise in slots where there were no data to be

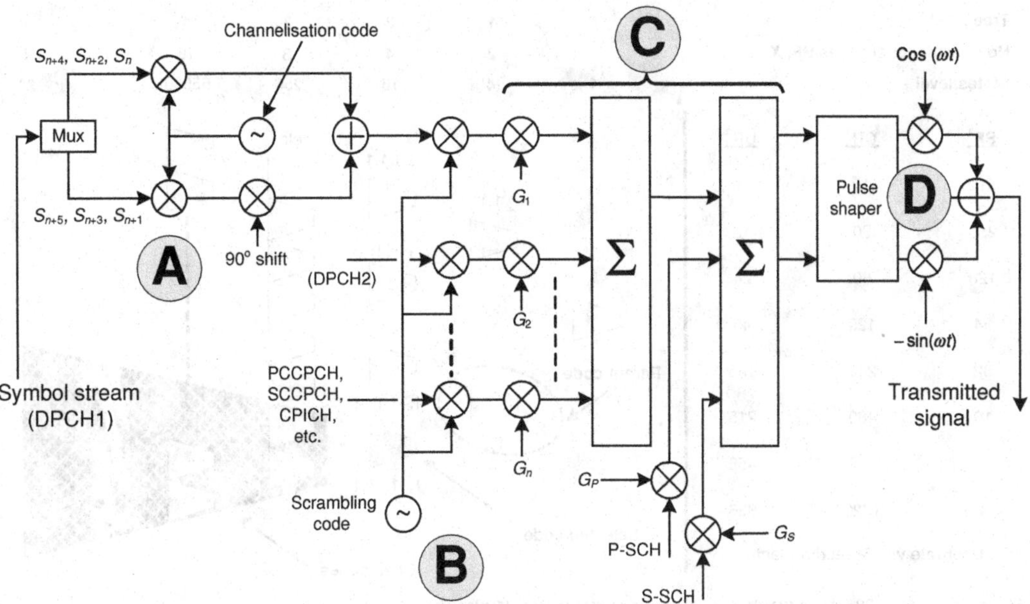

Figure 6.8 The spreading and modulation process

sent and thus where periods of transmit silence would alternate with transmitter activity (to support the control channel). Such interference would otherwise occur at a 1.5 kHz rate – right in the middle of the speech band. The spreading factor of the uplink DPCCH is fixed at 256, whereas that for the DPDCH may be selected by once again using the parameter *k*, to match the service to be supported. Because the data rates of the DPDCH and DPCCH will usually be different, there is a facility to apply an offset between the power levels of the two vectors and thus ensure a common BER for each channel. Finally, the control channel includes *FBI* bits in addition to information supported in the downlink. These bits may be used by the mobile to provide feedback to the base station in support of various forms of transmit diversity.

Although this discussion of the physical channel structure has focused on the DPCH, the same underlying granularity of time slots and radio frames is common to all physical channels, though the actual format varies significantly from channel to channel.

6.3.2 Spreading and modulation

The process of spreading the data symbols and modulating the resulting wideband signal onto an RF carrier is depicted in Figure 6.8 and is carried out in four distinct stages, identified as A to D. In stage A, the incoming stream of symbols is split, alternate symbols being categorised as 'odd' or 'even' and fed to multipliers, where they are spread by in-phase and 90°-shifted samples of the channelisation code. The benefits of spreading, along with some of the desirable attributes of the code, such as auto- and cross-correlation performance, were discussed earlier but there is still one further desirable attribute to be considered: code orthogonality.

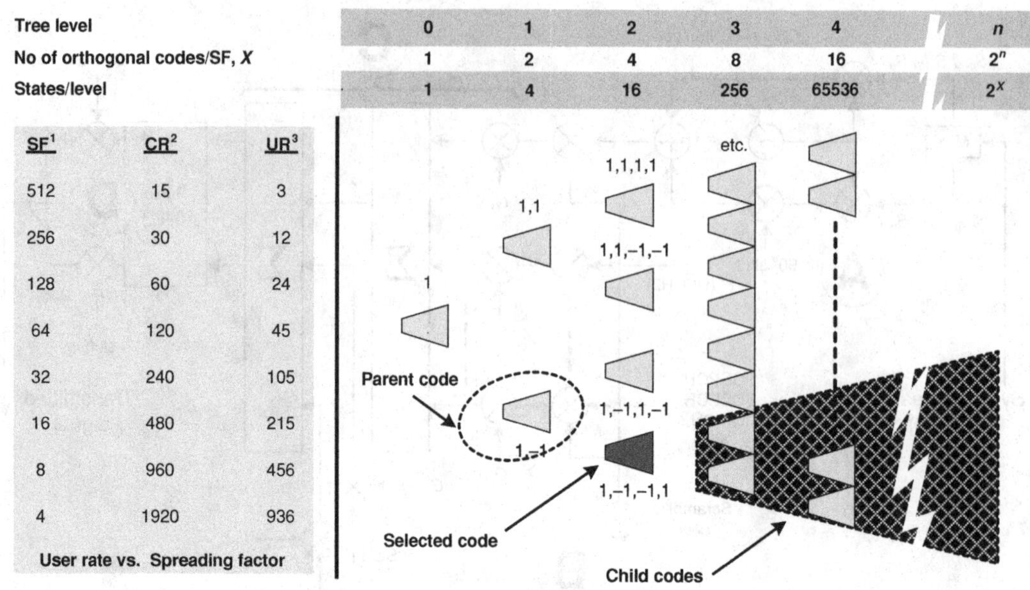

Figure 6.9 User spreading codes – The OVSF code tree

Recall that the receive process of de-spreading (using the spreading code) is carried out in the presence of interference power from other users of the same spectrum. If a further criterion is used in selecting the spreading code so that each code is orthogonal to each other code, the power from each of the other codes will sum to zero [23, 24]. In practice, multi-path causes perfect orthogonality to be lost but, even so, the fraction of in-cell interference seen by a single user is reduced to values typically in the range 0.3 to 0.6, depending upon the level of multi-path propagation [25]. This translates into improved system capacity. Note that this reduction in in-cell interference can only be achieved in the downlink, where all transmissions are synchronised in time. In the uplink, imperfect synchronisation of mobiles means that orthogonality of signals received at the base station cannot be preserved and the full multi-user interference is seen. For similar reasons, inter-cell interference is seen in full.

Figure 6.9 shows the orthogonal variable spreading factor (OVSF) code tree. This is an important resource and is managed by the radio network controller (RNC). Each code at a given level is orthogonal with all others on the same level. In the WCDMA downlink, the number of levels n is 9, giving rise to a maximum of 512 orthogonal codes at the highest spreading factor. Whilst this may seem a large number, these codes in practice can get consumed quite quickly. Suppose a high-rate service is to be deployed, requiring a code from level 2. If orthogonality is to be preserved, neither the 'parent' code at level 1 nor any of the 'child' codes at level 3 below the selected code may be used concurrently.

Following the spreading process using the OVSF channelisation codes, downlink transmissions from one cell to individual users can now be resolved. However, the OVSF codes do not possess good cross-correlation properties, exhibiting significant correlation compared with white noise sources [26, 27]. Additionally, the autocorrelation performance is non-ideal, with significant peaks outside of true code alignment. So to resolve transmissions from more than one cell it is necessary to follow the spreading process with multiplication by another code, the *scrambling* code shown in step B of Figure 6.8. This code is a pseudonoise (PN) sequence known as a Gold code and is unique for each BS. These codes have very distinctive peaks in their autocorrelation function at zero and mask the multiple peaks of the OVSF codes. The excellent cross-correlation characteristic of these codes also reduces multiple-access interference (MAI).

For completeness, it is useful to note that a similar process is followed on the uplink. The channelisation codes in this case merely serve to resolve the DPDCH and DPCCH channels. However, the polynomials generating the Gold codes used for scrambling are longer and are able uniquely to resolve several millions of handset transmissions [28].

Step C recognises that in the downlink, in addition to code tree management, the RNC must allocate transmit power to the various physical channels. At any time, the total transmit power must equal the sum of the power allocated to the common control channels plus the user traffic channels. The RNC manages this dynamically using an admission control process. The power that needs to be allocated depends on the bit rate, the intended cell size and the particular bearer in use (the channel format – interleaving, coding rate, etc.) reflected in E_b/N_0. Once the power to be allocated to the CPICH has been established, the majority of the other common control channel powers are set as a proportion of the pilot channel power; this is because their channel rates are fixed in relation to CPICH. The only downlink common-control channel that is not fixed is the SCCPCH. This is because this channel conveys the FACH and PCH transport channels and their data rate will be influenced by the amount of packet *traffic* carried by the FACH and the level of paging anticipated. These are operator choices and will be determined by the needs of the particular deployment. A summary of the main physical channels for the uplink and the downlink is included in Table 6.4. More comprehensive information on the format of the physical channels may be found in [19] and the channels themselves will be discussed further in Section 6.3.3 in the context of their relationship to *transport channels*.

Once the power to be allocated to each of the control channels is known, these maximum power limits will be implemented by means of the gain factors G_1 to G_n, and G_p, G_s for the synchronisation channels. Note that the synchronisation channels are introduced after the scrambling process and are therefore transmitted unscrambled. These channels may thus be read by the mobile to find the scrambling code employed by a particular cell, and used to demodulate all other channels.

The only dedicated-user physical data channel is the DPDCH (present in the uplink and downlink). For the downlink, more than 40 combinations of spreading factor, number of active slots and ratios of data to control information are specified. However all of these are predetermined and so the required power to be allocated on selection of a particular format is immediately known. In addition to the dedicated channels on the downlink, data

Table 6.4. Mapping of Transport Channels to Physical Channels

Transport channel	Physical channel, abbreviation and title ('fixed' and 'variable' refer to physical channel bearer rate)	
BCH	PCCPCH	Primary common-control physical channel – **(fixed)**
FACH	SCCPCH	Secondary common-control physical channel – **(variable)**
PCH		
RACH	PRACH	Physical random-access channel – **(variable)**
DCH	DPDCH	Dedicated physical data channel – **(variable)**
	DPCCH	Dedicated physical-control channel – **(variable)**
DSCH	PDSCH	Physical downlink shared channel – **(variable)**
CPCH	PCPCH	Physical common packet channel – **(variable)**
None	SCH	Synchronisation channel – **(fixed)**
(the Channels only	CPICH	Common pilot channel – **(fixed)**
exist as physical	AICH	Acquisition indication channel – **(fixed)**
channels)	PICH	Paging indication channel – **(fixed)**
	CSICH	CPCH status indication channel – **(fixed)**
	CD/CA-ICH	Collision detection and channel assignment indicator channel – **(fixed)**

may be sent on SCCPCH (FACH) and PDSCH (DSCH). The SCCPCH needs to cover the whole cell and cannot be subject to fast power control, as it is used in the call set-up process. In the absence of fast power control, additional margin has to be provided to allow for Rayleigh fading and this, together with the need for coverage of the whole cell, will mean that channel rates for the SCCPCH will be constrained. The same factors that constrain the data rate mean that transport channels supported on the SCCPCH are not very efficient for sustained traffic, such as circuit voice and video. However, the FACH is well suited to light bursty traffic as it is much faster to access compared with the PDCH, which needs negotiation to establish.

The PDSCH and its associated transport channel, the DSCH, is a shared channel and is also intended to support bursty traffic. The support of high-rate traffic causes particular problems in CDMA systems as it means that the parent code above the selected rate may not be used along with all the child codes. This is illustrated in Figure 6.9. Thus if one user wishes to transmit occasional high-rate bursts of data on a DPCH (which would far from fully utilise the cell radio resources) it might mean that only a few additional users could be admitted because of code limitations. Whilst this could be addressed using additional scrambling codes, interference from these users would not be orthogonal and would be fully visible, with the consequent loss of capacity. The shared channel instead allows part of one code tree to be shared on a time-division basis. Each radio frame may be allocated to a different mobile user; the spreading factor may also vary on each radio frame to match the needs of the particular application supported. To support power control and to indicate which frames the mobile must decode, each mobile sharing the PDSCH must concurrently support a downlink DPCH. The appropriate power allocation for the DPCH and the PDSCH will again be implemented by means of the gain factors

G_1 to G_n. Further useful discussion of power allocation to physical channels and its role in cell planning can be found in [29].

With appropriate powers allocated, all physical channels are summed to generate a composite chip stream resolved into real and imaginary components. This forms the input to the final process – translation of the complex baseband signal to the desired carrier frequency, shown as step D. The input rectangular chips are subject to a root raised cosine filter to ensure that the transmitted power remains within the allocated channel bandwidth [30]. Similar activities to those depicted in Figure 6.8 are carried out in the mobile, albeit focused on uplink code resources and management of mobile transmit power.

To conclude this section on spreading and modulation, it is appropriate briefly to mention radio resource management (RRM) and, in particular, *admission control.* At a top level, radio resource management is responsible for the efficient use of all radio resources using the processes described in Section 6.2.1, together with another activity, admission control.

In the system coverage and capacity analysis of WCDMA, addressed in Section 6.4.2, it becomes clear that, as the number of users in the system increases, various radio resources are consumed:

- Code resource has to be allocated from the node B,
- Transmitter power has to be allocated from the node B,
- Code resource has to be allocated in the mobile,
- Transmitter power has to be allocated from the mobile,
- Total interference noise power at the node B and mobile has to be acceptable.

The radio bearer service requested is analysed and the incremental resources required on the uplink and downlink (in terms of power and code resource) assessed. Only if these resources are available and the resulting increase in received noise power is acceptable is the new user admitted. Admission control is applied separately for the uplink and downlink.

The dynamic nature of the transmitter power management process can be seen in Figure 6.10. Resources are first reserved for the pilot channel (CPICH), the channel carrying the BCH (PCCPCH) and the channels carrying the synchronisation (SCH) and FACH/PCH bearers (SCCPCH). Without these physical channels, the system cannot operate correctly. This typically takes about 15% of the available power. The scheduling process within RRM then allocates power for services with the tightest QoS requirements; these are normally the dedicated channels. The diagram shows power allocated to a large number of AMR voice users and one high-rate DCH user (note that the latter call terminates on a radio-frame boundary, as required). Finally, power is allocated to a downlink-shared channel (PDSCH). The rapid power variation between frames reflects time-multiplexed allocation of the channel to different users, with different spreading factors and different physical locations in the cell. A more detailed description of the protocols and processes of RRM and the broader function of radio resource control is provided in [31].

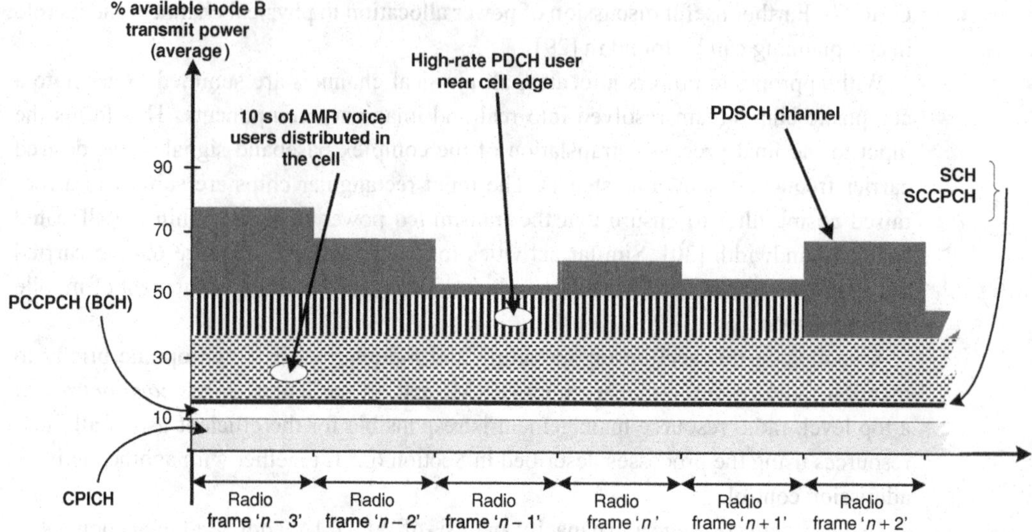

Figure 6.10 An example of RNC admission control transmitter power management

6.3.3 Transport channels

When GSM was designed, only a limited number of services were foreseen; a speech service that offered acceptable voice quality, and a number of circuit data services of different rates. At a later date, these basic services were extended by the provision of packet-data support, again with a variety of predetermined coding and data-rate schemes. For GSM, the interleaving and channel coding were chosen to match the BER tolerance of different classes of bits from the GSM full rate codec. With the availability of the enhanced full rate speech codec, further coding rules were devised *specific to the new codec*. In GSM, therefore, the details of the physical layer were dictated by the service or application. In contrast, a key goal for UMTS was the ability, flexibly and quickly, to support a rich array of multimedia services throughout its life, the majority of which, by definition, had not been invented. To address this requirement, the concept of *transport channels* was introduced.

The role of transport channels in the context of the whole radio resource management process can be seen in Figure 6.11. As discussed in the introduction to this chapter and reflected in Figure 6.2, specific radio bearers are requested at call set-up. It is then the responsibility of the radio resource control (RRC) function to determine whether a new user can be admitted into the cell and, if so, what applications may be allotted to it [31].

Once RRC has allocated resources to the new 'call', the layer-2 entities, radio-link control (RLC) and media-access control, are used to pre-process control and user plane data from higher layers to form inputs to the transport channels. Specifically, the RLC is responsible for:

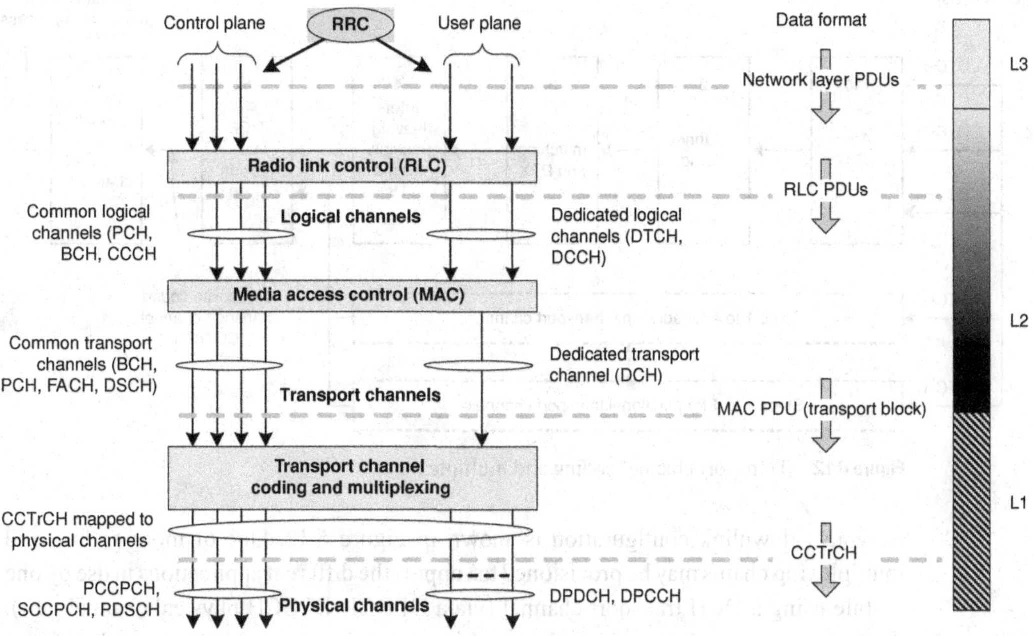

Figure 6.11 The radio resource management process

- Segmentation of potentially large network layer protocol data units (PDUs) into RLC PDUs,
- Allocation of higher-layer messaging to *logical channels*,
- Provision of transparent, unacknowledged and acknowledged mode transmission [32],
- Error correction through retransmission of data in acknowledged mode.

The MAC function executes radio-resource allocation on behalf of the RRC entity. Services provided by the MAC include:

- Segmentation of RLC PDUs into *transport blocks* for use by the transport channels,
- Mapping of data received on logical channels to transport channels (see Table 6.2),
- Priority handling for multiple traffic streams on one transport channel,
- Selection of transport formats for each active transport channel to deliver the short-term needs of the service.

The transport channel coding and multiplexing entity (that identified as special to UMTS) is the last element of the chain and maps the requested radio-bearer service to specific physical channels. The key functionality provided by this process is:

- Addition of cyclic redundancy checks to transport blocks,
- Coding of transport blocks,
- Rate matching,
- Interleaving,
- Mapping to physical channels.

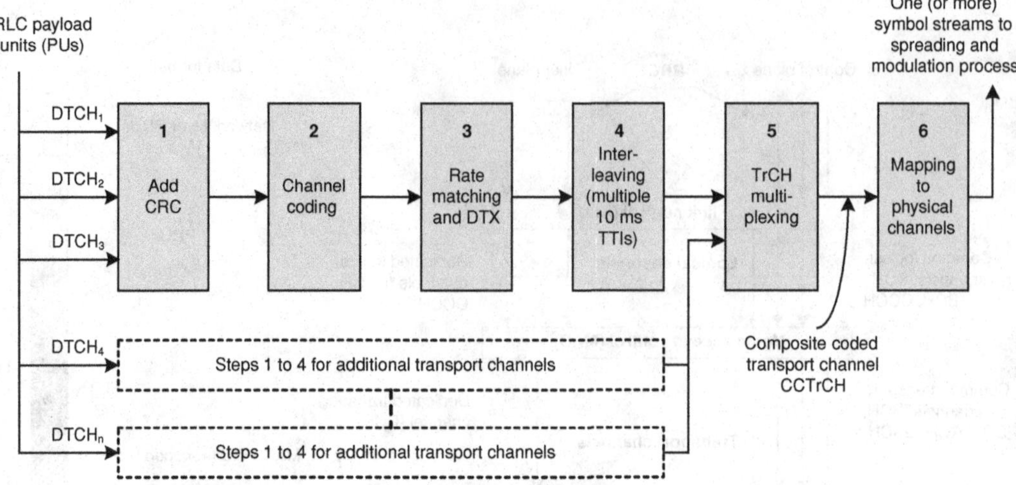

Figure 6.12 Transport channel coding and multiplexing

A typical downlink configuration is shown in Figure 6.12. One or more coding and multiplexing chains may be provisioned to support the different applications in use by one mobile using a DCH transport channel (via a DPDCH/DPCCH physical channel pair). Alternatively a similar configuration could support common control channels using a FACH transport channel (via an SCCPCH physical channel). Further details of transport channel to physical channel mapping are provided in Table 6.4.

Table 6.5 illustrates the flexibility available within a transport channel. The parameters are categorised into 'dynamic' and 'semi-static', the latter remaining fixed for the duration of the transport channel. The transmit-time interval (TTI) is an integer number of 10 ms radio frames, over which the transport block set is interleaved. Interleaving over longer periods will usually reduce the required E_b/N_0 for a given BER if the additional delay can be tolerated. Each transport block has appended to it a CRC word; the word size is also programmable. Two coding modes are available, convolutional and turbo-coding; the coding rates may also be varied.

The rate-matching parameter is a positive integer and influences the way in which transport channels are aggregated into one composite data stream when the data rate of the physical channel does not equal the sum of the data rates of the individual streams.

Table 6.5. Transport channel attributes

Attribute	Parameter
Dynamic (from TTI to TTI)	Transport block size Transport block set size
Semi-static (for all TTIs during the life of the traffic channel)	Transmit time interval ($n \times 10$ ms, the radio frame duration) Coding type (convolutional or turbo-coded) Coding rate (1/2, 1/3, etc.) Rate-matching parameter CRC size

In some cases, there may be too little capacity on the physical channel, and in others there may be excess capacity. Puncturing or repetition of the traffic channels is used to ensure that these parameters match. The higher the priority of the rate matching attribute, the greater the access given to the physical channel (i.e., a higher bit rate). Finally, the transport block size and the transport block set size are variable on a TTI to TTI basis and are used by the MAC to vary the instantaneous transport channel data rate when required. Further details of the MAC and transport channel coding and multiplexing process can be found in [33, 34]. The transport channel, in concert with physical channel parameters, can thus provide a very wide variety of throughput, BER and delay combinations.

To illustrate the way this flexibility might be utilised, consider the case of a user who has a speech conversation underway along with three concurrent (and separate) web-browsing sessions, with varying levels of activity. The first point to recognise is that all transport channels within a CCTrCH, will experience the same carrier-to-interference conditions. Therefore, if the level of BER that can be tolerated differs between the services, then this must be addressed by coding or other means. So for a modern speech codec with three classes of bits with different levels of BER tolerance, three separate transport channels would be set up, with coding and CRC levels adjusted as required. However, three web-browsing sessions, which by definition would have the same BER and delay requirements, could be multiplexed at the MAC layer on to a *single* (different) transport channel. It is unlikely that data would be transported at peak rate on all three sessions simultaneously and, therefore, variation of block size and block set size from TTI to TTI across the three multiplexed channels would enable the QoS needs to be met. With an appropriately chosen physical bearer rate, it should be possible to deliver all four services across one physical channel, subject to the particular application requirements. Further discussion of the optimisation process can be found in [35, 36].

Although a DCH transport channel has implicitly been chosen in the above illustration, much the same process would have been carried out if other transport channels had been used for the sessions. Similarly, the downlink has been chosen throughout this discussion of transport channel to physical channel mapping as the more complex aspect to address. A similar process is followed in the uplink management.

6.4 UMTS coverage and capacity

This section will address the factors influencing the performance of WCDMA and outline how these same considerations may be used to develop estimates of coverage and capacity, which will form a part of the cell planning process. The limitations of these estimates will be discussed and the underlying principles of 'static' network simulators, used for detailed cell-site planning, introduced. In Chapter 5, when developing coverage and capacity estimates for GSM, it was found convenient to establish a target carrier-to-interference (C/I) ratio, which would deliver the required BER performance. This C/I ratio was then used to determine the system configurations (e.g., maximum range or capacity), which would allow these requirements to be met with the required level of confidence. This same two-step process needs to be followed in WCDMA but, to make

Table 6.6. IMT2000 propagation environment

	Pedestrian A		Pedestrian B		Doppler spectrum	Vehicular A		Vehicular B		
Tap	Delay (ns)	Power (db)	Delay (ns)	Power (db)		Delay (ns)	Power (db)	Delay (ns)	Power (db)	Doppler spectrum
1	0	0	0	0	Classic	0	0	0	−2.5	Classic
2	110	−9.7	200	−0.9	Classic	310	−1.0	300	0	Classic
3	190	−19.2	800	−4.9	Classic	710	−9.0	8900	−12.8	Classic
4	410	−22.8	1200	−8.0	Classic	1090	−10.0	12900	−10.0	Classic
5	–	–	2300	−7.8	Classic	1730	−15.0	17100	−25.2	Classic
6	–	–	3700	−23.9	Classic	2510	−20.0	20000	−16.0	Classic

life more interesting, analysis of system deployments will reveal that cell coverage is impacted by the amount of traffic in the cells, and vice versa.

Another consideration is that because a variety of user data rates can now be expected in most deployments – in line with Shannon's Law – a corresponding family of C/I thresholds will be needed. Further complication is introduced by the very flexibility of the transport channel. How much coding should be added to the payload data? Over what period should data be interleaved for a given mobile speed and propagation profile? It turns out that a more useful figure of merit is a normalised term, E_b/N_0, where E_b is the energy per bit and N_0 is the thermal noise power spectral density. Appendix 6.1 discusses E_b/N_0 further in the context of spreading gain and also illustrates that E_b/N_0 is directly equivalent to S/N (or C/I). However, E_b/N_0 is generally more useful since it is normalised and, therefore, does not need to make explicit reference to the bandwidth or bit rate. This normalised terminology will be used in the following discussions.

6.4.1 Link-level simulation

ETSI specified a series of propagation environments in which UMTS is expected to operate [37]. These are shown in Table 6.6 and reflect measurements taken at 2 GHz, the primary frequency deployment band for UMTS. The 'taps', numbered one to six, reflect up to six significant peaks of multi-path energy with the 'delay' and 'power' headings reflecting the time delay relative to the first tap and average power referenced to the strongest tap. The presence of 'A' and 'B' environments reflects a low delay scenario, A, and a second scenario, B, with much greater delay, both of which are encountered frequently.

This environment is discussed further in [38], where Buehrer develops a model that may be used in simulations comprising:

- Time delay spread (causing frequency-selective fading),
- Doppler spread (causing temporal fading),
- Angular spread (causing spatial-dependent fading).

Interestingly, Buehrer's model introduces a means of representing the dependency of the level of reflected energy as a function of the *angle* of the return. This is particularly important in systems such as UMTS where adaptive antennas may be deployed.

The process of link-level simulation can conveniently be captured in the following steps [39]:

- Generation of a user signal corresponding to the specific *physical channel* for which E_b/N_0 information is to be determined. The key activities in this process have been discussed in Section 6.3.
- The transmitted waveform is then subjected to the effects of the time-varying propagation channel. A simulator is used to introduce propagation effects, which potentially includes multi-path, fast fading and Doppler frequency shifts. The precise effect of propagation can vary greatly with the particular multi-path channel and mobile speed.
- A programmable level of 'interference' (relative to the wanted signal) is then added. This is often approximated as white Gaussian noise, although, in some simulations, multiple independent user signals are generated to approximate the real world more closely.
- The final stage of the simulation is to demodulate the received signal, which now comprises an attenuated and distorted version of the transmitted signal, accompanied by a noise-like signal representing other user interference. The demodulation is accomplished by a comprehensive model reflecting the *actual* processes and algorithms that run in the mobile or base station receiver under characterisation. A high-level view of these processes is reflected in Figure 6.4 and was discussed earlier; key functions to be modelled must include:
  - Channel estimation,
  - Fast power control,
  - Number of rake fingers and allocation algorithm,
  - Phase and delay correction,
  - Maximum ratio combining algorithms (algorithms used for diversity combining),
  - De-interleaving and decoding.

  The whole of the above simulation sequence is carried out for each slot of each radio frame as the mobile moves on its predetermined trajectory.

The BER performance will reflect the aggregated statistical likelihood, at points in time during the simulation run, of encountering phase and amplitude distortion of the received signal together with significant levels of interference which, after de-interleaving, are sufficient to cause an incorrect symbol-decoding decision. The level of interfering signals is varied until the desired mean BER is achieved over a variety of mobile trajectories.

In practice the number of simulation scenarios is, in principle, infinite and in reality very large. The E_b/N_0 versus BER performance will vary as a function of:

- Mobile velocity (how well does power control offset fast fading),
- Bearer rate (and control channel overhead),

- Propagation channel,
- Interleaving period (how many radio frames per TTI),
- Channel coding scheme.

Most equipment vendors, therefore, establish a set of physical channels that they will implement in their equipment (reflecting a specific bearer rate, control channel overhead, interleaving period and channel coding scheme, etc.) and characterise these over a variety of propagation scenarios.

It is important to recognise that the resulting E_b/N_0 figures will be specific to the particular realisation of the receiver under test. For instance, how good is the particular rake finger allocation algorithm and how many fingers are employed? (this will determine how much of the available energy is recovered). Similarly, how good is the channel estimation process? (this will affect the quality of the coherent integration). They effectively represent figures of merit for a given vendor's equipment, and so tend to be commercially sensitive. Table 6.7, therefore, summarises uplink performance for dedicated physical channels, provided during the early specification stages of WCDMA [40]. These data certainly do not represent the state of the art today. Also, they represent a particular implementation of the bearer rates tabulated, in terms of coding schemes, interleaving, etc. Nevertheless, they still allow some interesting observations to be made.

Counter-intuitively, the E_b/N_0 figure *reduces* with increasing data rate. However, when this is considered from the perspective of the received *power* ($E_b/N_0 \times$ bit rate) the C/I requirement for higher rate services will increase as Shannon's Law would suggest. There is still some effect, which may be divorced from the bearer rate, which indicates that higher rate bearers are more efficient than lower rate bearers. This, in fact, is largely an artefact of the need to provide power to support the dedicated physical control channel that must always accompany a dedicated physical data channel. This power becomes increasingly significant at low user data rates when the power needed to support the control channel can equal or exceed that required for the data channel. A further factor that increases the E_b/N_0 for low-rate channels, such as voice, is that 'turbo-coding' is not effective for small block sizes and so less efficient convolutional coding is used.

The major column headings, 'static', 'case 1', 'case 2', etc., reflect the impact of propagation on the required E_b/N_0 figure. 'Static' represents a situation with line-of-sight and no multi-path; the other cases are for different mobile speeds and multi-path environments; these are defined in [41]. Information is also provided as a function of block erasure rate (BLER). At first sight, it may not seem obvious that it may be advantageous to operate at other than the target BLER for the service; indeed for delay intolerant services, choosing E_b/N_0 on the basis of the target BLER *is* the right answer. However, if the requested QoS allows for a longer transmission delay, the target BLER can be met using a lower E_b/N_0 and retransmission of any corrupted data. If delay is not an issue, the question then arises, what is the optimum number of retransmissions? There are two opposing factors. Increasing the allowed BLER reduces the E_b/N_0 required (and thus creates additional cell capacity) *but* requires retransmission of data (that erodes cell capacity). This optimum point will depend on the target BLER and the

Table 6.7. E_b/N_0 versus bearer rate

	Propagation conditions (see [41])								
	Static (dB)		Case 1 (dB)		Case 2 (dB)		Case 3 (dB)		
Rate	BLER $= 10^{-1}$	BLER $= 10^{-2}$	BLER $= 10^{-1}$	BLER $= 10^{-2}$	BLER $= 10^{-1}$	BLER $= 10^{-2}$	BLER $= 10^{-1}$	BLER $= 10^{-2}$	BLER $= 10^{-3}$
12.2 kbps	1.6	2.5	5.5	8.8	3.5	6.0	1.8	2.9	–
64 kbps	−1.0	−0.8	3.1	6.0	1.0	3.3	−0.7	−0.3	−0.1
144 kbps	−1.8	−1.6	2.5	5.4	0.6	2.6	−1.4	−1.1	−0.8
384 kbps	−1.7	−1.5	2.6	5.8	0.8	3.0	−0.9	−0.6	−0.3

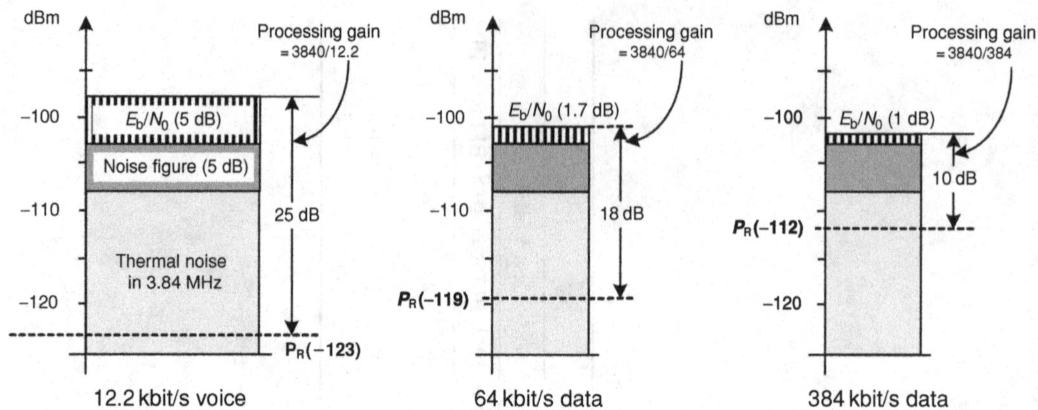

Figure 6.13 Single-user base-station sensitivities

particular physical dedicated channel but is usually in the range 10^{-2} to 10^{-1}. This is discussed further in [42].

The interested reader can find further information on the current sensitivities that the base station and mobile must meet, along with the associated test conditions, such as multi-path, interference levels, etc., in [43] and [41], respectively. Information and insight into deployment and simulation scenarios is available in [44] and [45].

6.4.2 Evaluation of system coverage and capacity

Armed with E_b/N_0 figures for particular services it is now possible to see how this translates into a 'sensitivity' figure for the base station. Figure 6.13 illustrates the way this is evaluated. Thermal noise is present at the input to the receiver in a 3.84 MHz bandwidth. To this must be added the noise figure of the RF pre-amplifier, which for the base station will be in the region of 5 dB once filter and other losses are considered. Finally, the E_b/N_0 margin required for the particular service to operate at the required BLER must be included. All of these factors so far serve to increase the required level of the wanted signal. However, the benefit of WCDMA processing gain, capturing the difference between coherent integration of the wanted signal and incoherent addition of other signals, has yet to be included. For 12.2 kbits/s AMR voice, the processing gain is about 25 dB and reduces the required level of the wanted signal to −123 dBm – well below thermal noise. Similar calculations for 64 kbits/s and 384 kbits/s data services are also shown. Note that although the required E_b/N_0 reduces for higher rate services, as discussed earlier, this minor benefit is more than offset by the reduced processing gain, giving rise to much lower sensitivities in these cases, as expected.

In the discussion above, it has implicitly been assumed that only one mobile is operating in the cell and so this is the single-user sensitivity. Figure 6.14 shows a more representative scenario, with many mobiles sharing the same spectrum in the same cell. Each of these users will transmit its information on a unique scrambling code and this transmitted power will appear as noise to receivers not de-modulating the signal with the appropriate

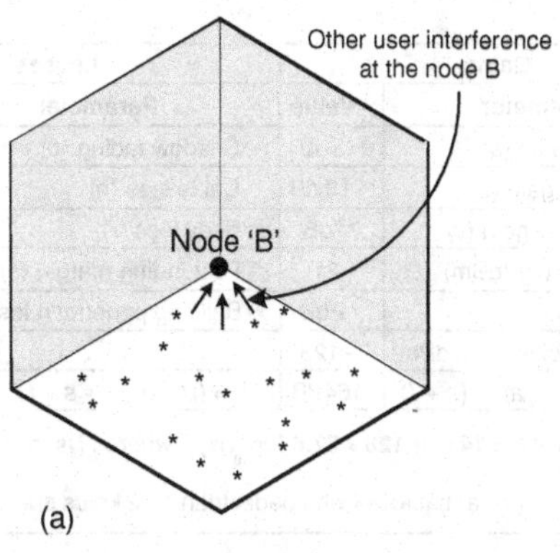

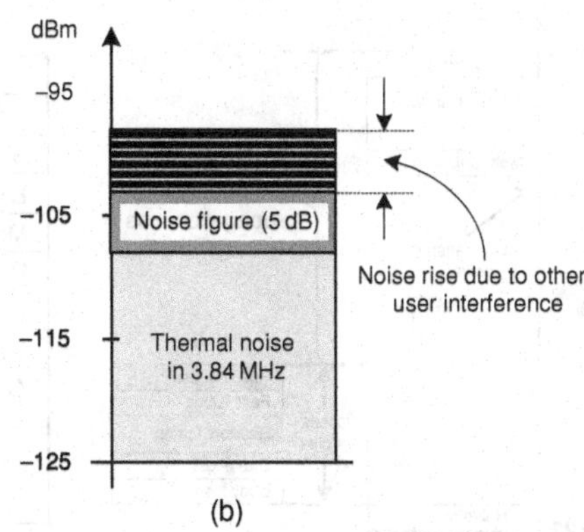

Figure 6.14 Uplink interference scenario

code. Each user of the same service will add an equal amount of noise power at the base site, as it will be power controlled by the base station to about the same C/I target. Users of higher-rate services will generate correspondingly larger incremental noise powers. The maximum range in the uplink will, therefore, reduce as more users are admitted to the cell. This effect is known as *cell breathing* and to ensure expected coverage under the planned load conditions, an *interference margin* must be added to the normal link budget parameters. When the total system is considered, because the base station will be able to transmit 10 watts or more to support an individual user on the downlink, it is very apparent that the uplink, with maximum mobile powers of 0.125 watts, will determine the maximum cell radius. Figure 6.15 shows the maximum cell radius for the uplink (and thus the system) for a number of different user services; it also shows how the cell radius changes for the cases of (1) a single user and (2) multiple users causing a 6 dB noise rise at the receiver.

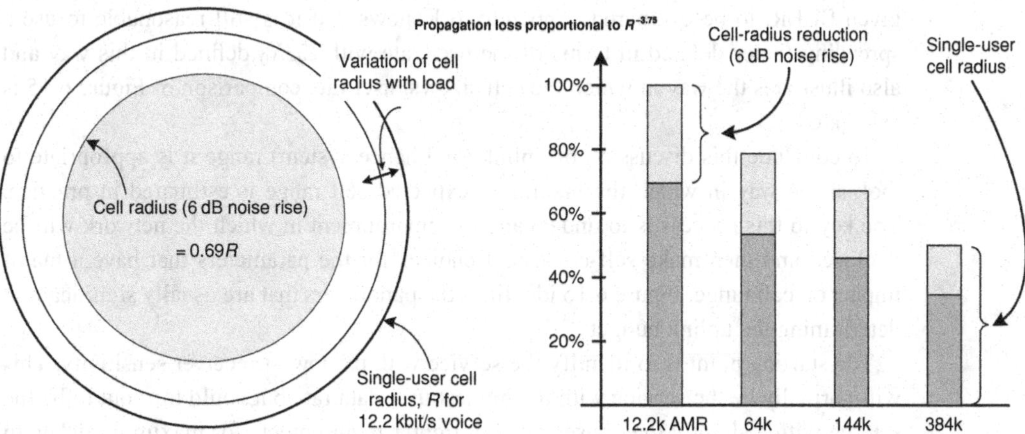

Figure 6.15 Impact of multi-user interference and user data rate on cell radius

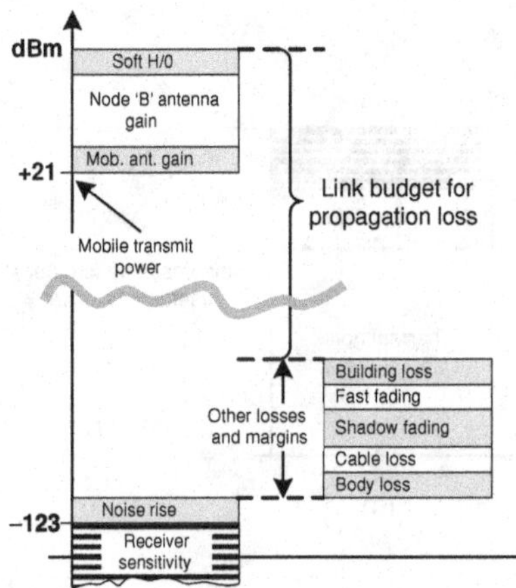

Gains		Losses	
Parameter	**Value**	**Parameter**	**Value**
Soft handoff gain (*a*)	3 dB	Shadow fading (*p*)	8 dB
BTS antenna gain (*b*)	18 dB	Cable loss (*q*)	2 dB
Mobile antenna gain (*c*)	1 dB	Body loss (*r*)	2 dB
Mobile power (*d*) (dBm)	21	Fast fading margin (*s*)	5 dB
Noise rise (*e*)	2 dB	Building penetrat'n loss (*t*)	0
Receiver sensitivity(*f*) (dBm)	–123		
$G = (d + c + b + a) - (e + f)$	164 dB	$L = (p + q + r + s + t)$	17 dB
$\mathbf{G - L = 147 = [128 + 37.6 \log_{10}(R)]}$ where *R* is in km			
$\mathbf{R = 3.2}$ **km** (for a macrocell with pedestrian 12.2 kbit/s speech users only)			

Figure 6.16 Estimation of maximum cell radius

In reality of course, the 12.2 kbits/s data stream does not directly form the symbol sequence, which is spread by the 3.84 Mchip/s code. As has been seen, there are CRC bits, channel coding and possibly repetition and puncturing, which significantly modify the symbol rate that is subject to spreading – the *channel rate*. The channel rate for 12.2 kbits/s voice is, in fact, 60 ksymbols/s, that for 64 kbits/s data is 240 ksymbols/s and that for 384 kbits/s data is 960 ksymbols/s. Yet, in Figure 6.13, when single user sensitivities were calculated, processing gains reflecting a spreading factor in relation to the *user* data rate were included. Why was this? The answer lies in the way the channel rates for each service were selected. The relationship between the amount of channel coding and the reduction in E_b/N_0 for a given BLER is not linear or readily evaluated. Link-level simulations were therefore carried out for different coding rates and it was important to recognise the overhead of additional channel coding when comparing different implementations of the same service. E_b/N_0 was therefore defined in terms of the required receive power and the user bit rate, which directly allowed the merits of the different schemes, for a given BLER, to be compared. Appendix 6.1 shows that it is still reasonable to use a spreading factor defined in terms of the user rate with E_b/N_0 defined in this way and also illustrates the way in which the cell size vs. user rate comparison of Figure 6.15 is calculated.

To conclude this discussion of uplink (and hence system) range it is appropriate to look at the way in which the maximum expected cell range is estimated in practice. The key to this process is to understand the environment in which the network will be deployed and then make self-consistent choices for the parameters that have a major impact on cell range. Figure 6.16 identifies the parameters that are usually significant in determining the uplink budget.

The starting point is to identify the service with the lowest receiver sensitivity. This will normally be the service with the highest user data rate but could turn out to be the service with a slightly lower user rate but tighter requirements for maximum delay. In this example, 12.2 kbit voice for pedestrian users is the only service deployed and so the figure of −123 dBm estimated in Figure 6.13 is selected. As has been seen, to guarantee coverage under load, a margin needs to be introduced to allow for other user interference. In this case, a figure of 2 dB has been chosen, as this cell is located in a lightly populated rural area and maximum cell loading is anticipated to be low. Again, consistent with the deployment region, there is expected to be little vehicular traffic and so no losses are included for penetration into buildings or cars. In other networks, figures of 15 dB or 8 dB (whichever is relevant) might be included to ensure coverage in buildings and cars respectively.

There then follow a series of associated gains and losses that are often helpful to consider together. The mobile transmit power for handheld devices is usually 21 dBm and, for slow-moving pedestrian users, fast power control will be needed to counter the effects of Rayleigh fading since interleaving does not provide sufficient diversity gain. The effect of fast power control is to raise the mean transmit power, and thus some 'headroom' or back-off from the maximum mobile transmit power needs to be included in the link budget if power control is to work correctly. The fast fading margin has been studied [46] and found to vary as a function of the target set for E_b/N_0; a value in the range 4–5 dB is typical. Conversely, for cells covering a motorway, high vehicle speeds will mean that interleaving is effective and the power-control margin can be set to zero.

The next set of associated parameters relates to the target end user. Mobile handsets normally have very low antenna gain and voice users will usually have the handsets next to their heads. So in this case an allowance has to be included for body loss. However, if the target population is data users, both a higher antenna gain (2 dB) and 0 dB body loss can be used. The base-site antenna is a major contributor to the link budget. Although normally deployed for capacity reasons, in this case tri-sectored sites with the additional 5 dB of antenna gain are used to enhance cell range. Associated with the antenna, and slightly offsetting this benefit, is the allowance that needs to be included for cable losses between the node 'B' equipment and the terminals of the antenna. The final significant elements in the link budget are the allowance for shadow fading and the effect of soft handoff, which can help offset these losses. A discussion of shadow fading is available in [47] with parameters specific to 2 GHz propagation provided in [48] and a value of 8 dB is chosen in this case. The available link budget can now be calculated as $(G - L)$.

The link budget is then used with a propagation model appropriate to the deployment region, to evaluate the maximum cell size. In this example the propagation model defined in 3GPP RF System Scenarios [44] is used to solve for R, although an excellent and more comprehensive discussion on models for a variety of deployments is provided in [49].

6.4.2.1 Uplink 'pole' capacity

In the previous discussion of system coverage, it will have become very clear that a key parameter to manage in any UMTS (and CDMA) system is the noise rise at the BTS, since, in most cases, this will limit the maximum cell size. This noise rise in turn is clearly some function of the number of active mobiles in the system. This is extensively treated in [50] and [51] and reconciled with practical results in [52]. However, it is helpful to understand the principles at work in a CDMA uplink. A derivation of the uplink pole-capacity equation is, therefore, provided, which illustrates the key factors.

Over the next couple of pages, expressions will be developed linking the noise rise at the BTS to the number of active mobiles, *n*, and other key system parameters. It will be seen that as the number of mobiles in the system increases, and if there were no constraint on the available power from the mobiles, at some point the noise rise at the BTS would become infinite. The number of active mobiles that would give rise to this theoretically infinite noise rise at the base station is, unsurprisingly, known as the *pole* capacity and cell loading is often referenced to this theoretical value.

Consider a cell with a number of active voice users, each accurately power-managed so that an equal power, P, is incident on the BTS from each mobile. If there are n active mobiles being served by the cell, for a specific user, one of these signals will be the wanted signal, of power P, but for clarity designated P_S and the other $n-1$ (unwanted) users are interferers, each also of power P but designated P_I. If, on average, each mobile is active some proportion δ of the time a noise ratio, N_R, can be defined, which is a direct measure of the relative powers present at the input to the BTS when there are $n-1$ interfering users (active and passive). Then N_R can be defined as

$$N_R = \frac{\text{incident interfering power} + \text{receiver thermal noise power}}{\text{receiver thermal noise power}}$$

$$= \frac{P_N + P_I(n-1)\delta}{P_N}$$

and the signal-to-noise ratio can be defined as:

$$S/N = \frac{P_S}{P_N + P_I(n-1)\delta}. \tag{6.1}$$

Further, assume that each mobile is transporting user data at a rate R symbols/second and that after the introduction of coding to provide resilience against propagation impairments the coded symbol rate is C symbols/s. The coded user data are then spread by a pseudonoise code at a rate of W chips/second. In the introduction to Section 6.4, when discussing the approach to be used to establish system coverage and capacity, it was found useful to use the normalised term E_b/N_0 instead of the more widely used S/N or C/I terminology. An equivalent figure of merit E_{CH}/N_0 can be defined referenced to the energy/channel symbol, E_{CH}, and the 'noise' power spectral density as follows:

$$E_{CH} = \int_0^{\frac{1}{C}} P_S(t)\,dt = \frac{P_S}{C} \tag{6.2}$$

for the rectangular symbol, and N_0 is the (thermal noise + interference) power spectral density. Then, assuming a matched filter is used, the RF channel bandwidth will be set equal to W, the chip rate, and combining Equations 6.1 and 6.2 reveals:

$$\frac{E_{CH}}{N_0} = \frac{P_S}{C} \times \frac{W}{P_N + P_I(n-1)\delta}. \tag{6.3}$$

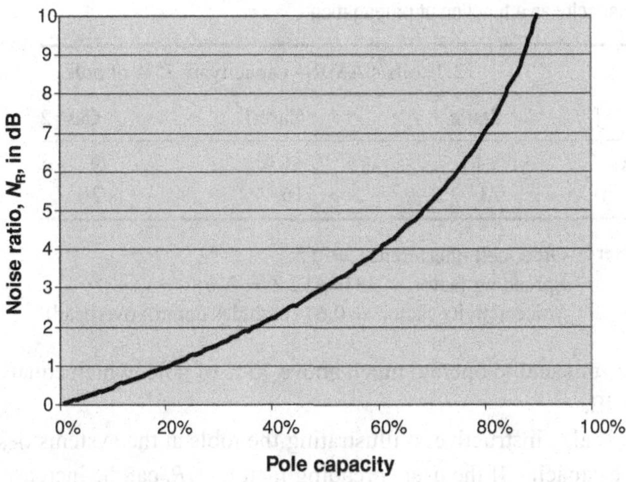

Figure 6.17 Uplink noise ratio as a function of pole capacity

In Appendix 6.1, E_b/N_0 is justified as a more useful figure of merit in that it reflects the energy required per user symbol and it is shown that:

$$\frac{E_b}{N_0} = \frac{C}{R} \times \frac{E_{CH}}{N_0}. \tag{6.4}$$

Then combining Equations 6.3 and 6.4, recognising that P_I equals P_S equals P and rearranging, an equation for P/P_N can be developed:

$$\frac{P}{P_N} = \frac{\Phi}{1 - \Phi(n-1)\delta}, \quad \text{where } \Phi = \frac{R}{W} \times \frac{E_b}{N_0}. \tag{6.5}$$

Substituting for P/P_N in the equation for N_R yields:

$$N_R = \frac{1}{1 - \Phi(n-1)\delta} = \frac{1}{(1-\eta)}, \quad \text{where } \eta = \frac{R}{W} \times \frac{E_b}{N_0} \times (n-1)\delta. \tag{6.6}$$

It is useful to explore the way in which N_R varies with η. When the number of active mobiles, n, is one, $N_R = 1$, i.e., there is no increase in the noise power at the BTS because there are no interfering mobiles! At the other extreme, in theory N_R goes to ∞ when $\eta = 1$. In practice, the noise rise can, of course, never become infinite, because of finite mobile power. However, as the number of mobiles increases, each mobile must operate at a slightly higher power such that the Equation 6.3 is still met. Eventually, mobiles near the edge of the cell are unable to increase their power further and their communication link is dropped.

Solving for n with $\eta = 1$ gives rise to the theoretical 'pole' capacity of the system. For the reasons just discussed, systems are never designed to operate at this point or even close to this pole capacity. Figure 6.17 uses Equation 6.6 to show how the noise ratio varies as η moves from 0 to 1. Once η moves much above 0.7 (or 70% of the pole capacity), the noise starts to rise very quickly with each additional admitted mobile and the system starts to become less stable in terms of its coverage and capacity. In practical

Table 6.8. Uplink capacity as a function of propagation

	12.2 kbits/s AMR – capacity @ 75% of pole		
Channel model [43]	Static	Case 1	Case 2
Target E_b/N_0 (dB)	5.1	11.9	9
Number of users (n)	73	16	30

Common parameters Other cell interference = 0.5;
Spreading factor = 3840 / 12.2 = 315;
Voice activity factor = 0.67 (includes control overhead)

deployments it is unusual to operate much above 75% of pole, which equates to a noise ratio of about 6 dB.

Equation 6.6 is also instructive in illustrating the tools at the systems designer's disposal to increase capacity. If the user spreading factor, W/R, can be increased – perhaps by using lower rate codecs – more users can be permitted. Secondly, if it is possible to reduce E_b/N_0 by some means, the number of users can again be increased. The system designer should consider at least three means to reduce this figure:

- Use of the most powerful coding schemes possible (the use of turbo codes rather than convolutional codes can provide major benefits here for some applications),
- The use of interleaving and coding over several frames to provide resilience against bursty interference,
- The introduction of hybrid ARQ schemes that no longer attempt to *guarantee* error-free reception but instead accept a slightly higher probability of errors and the attendant need to resend some radio frames – in return for lower operating E_b/N_0 figures.

The final opportunity is to find a means of reducing δ, the activity factor. This represents a very considerable opportunity for uncompressed bursty traffic and indeed can be exploited for voice services by ensuring that only modern codecs with efficient voice activity detection (VAD) are utilised.

To complete this discussion of uplink pole capacity, it is appropriate to note that Equation 6.6 is only accurate for a single, isolated cell; no consideration has been given to interference, received in the serving cell, from traffic in surrounding cells. However, if the assumption is made that the surrounding cells contain similar applications and traffic levels, this effect can be included straightforwardly. Let a factor q be defined such that

$$q = \frac{\text{other cell interference power}}{\text{serving cell interference power}}.$$

Then an amended load factor η_{ul} for use in Equation 6.6 can be defined, such that:

$$\eta_{\text{ul}} = (1 + q) \times \frac{R}{W} \times \frac{E_b}{N_0} \times (n - 1)\,\delta. \qquad (6.7)$$

Table 6.8 shows the number of 12.2 k AMR voice users that can be supported in the uplink for three different channel conditions. Most of the parameters used to derive the results

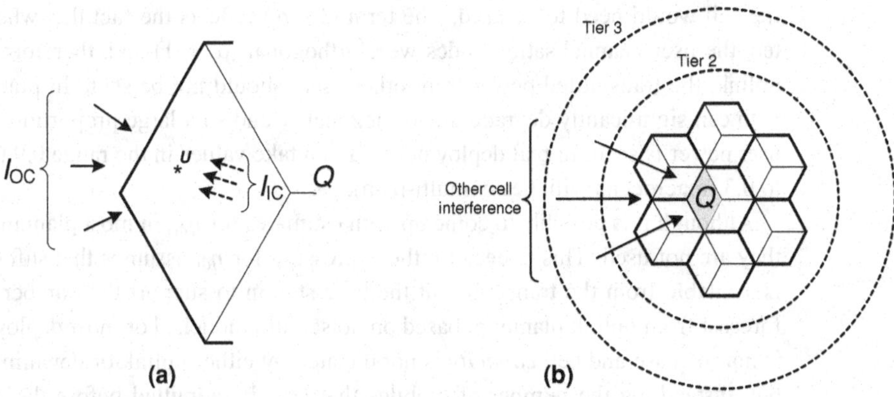

Figure 6.18 Downlink interference scenarios

of Table 6.8 are self-explanatory but it is useful to discuss briefly the value selected for other cell interference. The factors affecting this parameter are explored in some detail in [53] and can be summarised in order of impact as:

- The number of cells involved in soft handoff (most benefit captured with one other cell),
- The propagation exponent (large numbers here are good),
- Standard deviation of log–normal shadowing (small is good).

Depending on these parameters, values of q in the range 0.3 to 1.8 may be encountered.

6.4.2.2 Downlink 'pole' capacity

So far the discussion of coverage and capacity has been focused on the uplink. This is because the performance in the downlink is more complex and heavily dependent on the user location. However, it has already been seen that, because of the large disparity in output powers, maximum cell size will inevitably be set by the mobile. Therefore, all that remains to be addressed is downlink capacity. Figure 6.18(a) shows the interference scenario for the downlink. As for the uplink, it is possible to determine a pole capacity, as one of the key parameters to manage remains the noise ratio. Then, using the terminology of Figure 6.18(a), and for a single service deployed in the cell, η_{dl} can be expressed as:

$$\eta_{\mathrm{dl}} = \frac{R}{W} \times \frac{E_b}{N_0} \times (n - 1)\delta \times (\beta + (1 - \alpha)), \tag{6.8}$$

where β is a relative measure of the other cell interference (I_{oc}) and α is the orthogonality factor for in-cell interference, which is cell dependent and determines the proportion of I_{IC} that is seen by the mobile.

The first term is almost identical to the uplink but in this case represents interference power due to other downlink transmissions at the location of mobile u. The term β represents the other cell interference, but note that for the downlink β varies with the location of the mobile u. If this equation were to be used to estimate downlink capacity, a term representing an averaged value of β for mobiles randomly distributed throughout the cell would need to be used. The term $(1-\alpha)$ reflects the fact that when transmitted, the user channelisation codes were orthogonal ($\alpha = 1$) and, therefore, unlike the uplink, the transmitted power from other users should not be seen. In practice, multi-path can significantly degrade this orthogonality and so a large proportion of the other user power *is* seen. In real deployments, α can take values in the range 0.9 (microcells) to 0.3 (macrocells with heavy multi-path).

Although it is possible to come up with estimates for η_{dl}, in most planning situations they are not used. This is because the expression for η_{dl} assumes that sufficient power is available from the transmitter at the base station to support the number of mobiles forecast from uplink planning, based on noise ratio models. For most deployments, this is not the case and cell *capacity* is not dictated by either uplink or downlink noise rise but, instead, by the number of mobiles that may be admitted before the base-station transmitter power is exhausted. The assessment of uplink noise rise, however, still plays a key part in determining the WCDMA cell radius.

Instead, an estimate of capacity in the downlink is formed by considering the power that needs to be delivered to the mobile and determining how many instances of this transmit power can be supported by the base-station transmitter. The maximum number of mobiles, n, can be expressed as:

$$n = \frac{\text{available transmitter power}}{\text{average power required/mobile}} = \frac{I_{\text{OR}} \cdot W \cdot (1-\rho)}{\text{DPCH\_E}_{\text{C}} \cdot W \cdot \delta} = (1-\rho) \times \frac{1}{\frac{\text{DPCH\_E}_{\text{C}}}{I_{\text{OR}}} \cdot \delta} \tag{6.9}$$

where: ρ is the common channel overhead
DPCH_E_{C} is the average energy/DPCH chip
I_{OR} is the total transmit power spectral density

The term $\text{DPCH\_E}_{\text{C}}/\text{I}_{\text{OR}}$ is specified as a 'worst-case' test limit in mobile conformance testing. In practice, this value will vary depending on the location of the mobile u; the value will be large when the mobile is at the cell edge, suffering maximum interference and minimum wanted signal, and the converse will be true when the mobile is near the base station. Once again, an average value would need to be used to get a useful downlink capacity estimate. This topic is discussed further in [54] and [55].

6.4.2.3 Coverage and capacity for users with mixed service profiles

The foregoing discussion of uplink and downlink system performance and the development of approximate formulae for maximum cell range and capacities has provided useful insight into the mechanisms at work in WCDMA. However, in practice, only the estimates developed for maximum cell size at a given load are used. There are a number of reasons for this:

- All mobiles are assumed to operate with the same service,
- It is not possible to include the effects of local propagation variations,

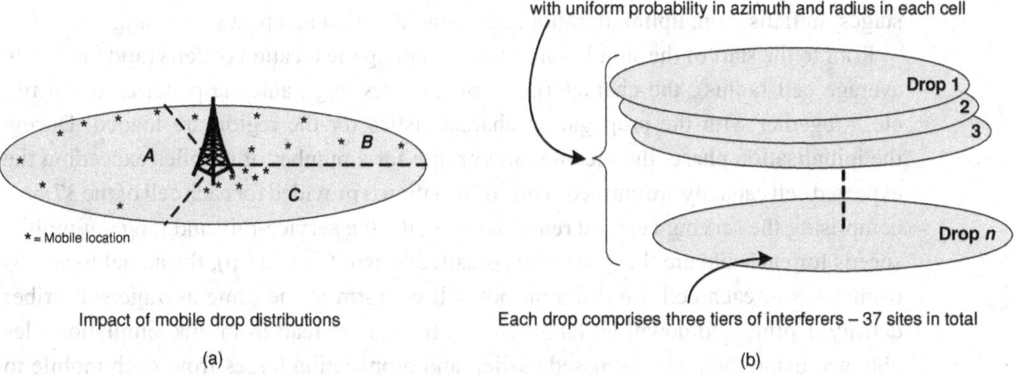

Figure 6.19 Modelling using static simulators

- It is not clear whether to use the uplink or downlink for evaluating system capacity,
- The effects of soft handoff are not considered,
- It is not possible to include distributions of mobile speeds.

In addition to these generic limitations, the accuracy of downlink estimates is severely constrained by the need to use the concept of 'mobiles at an average location' when forming estimates of interference and transmit power. For all these reasons, once a rough idea of the number of cell sites has been formed using mobile power and expected cell loading, detailed planning is almost always carried out using commercial tools, embodying a technique known as 'static simulation'. A summary of the key principles of static simulation is provided here, as an understanding of such tools is essential if accurate estimates of coverage and capacity are to be obtained for WCDMA deployments.

Because every cell in every site in a WCDMA network usually shares the same carrier frequency, the impact of interference from surrounding cells on the serving cell capacity is much higher than in GSM. Static simulators, therefore, need to consider many cells in assessing the level of 'other-cell' interference in the serving cell; typically three tiers of interfering cells are considered, as illustrated in Figure 6.18(b), and the cells in the outer tier are 'wrapped around' so that they are subject to interference from cells on the opposite boundary. The simulation process is carried out for a series of random mobile 'drops'. Figure 6.19(a) shows mobiles dropped into one of the 37 sites used in a typical static simulation. In addition to the physical location of each mobile, the mobile itself is further defined by:

- User speed,
- Supported service,
- Uplink data rate,
- Downlink data rate,
- Activity factor.

This information (the *mobile profile*), together with the multi-path environment in the deployment region, enables an appropriate target E_b/N_0 to be selected for each mobile.

The operation of static simulators may vary in detail but can usually be split into four stages: initialisation, uplink iteration, downlink iteration and post processing.

Prior to the start of the simulation, files containing the location of cells (and implicitly average cell radius), the characteristics of the sites (e.g., antenna patterns, down tilt, etc.,) together with the propagation characteristics for the region are loaded. During the initialisation phase, the location and profile for a number of mobiles exceeding the expected cell capacity are entered. This information is provided for each cell of the 37 sites comprising the serving cell and reuse tiers. Whilst the service-mix and range of mobile speeds for each site are the same (the so-called Monte Carlo drop), the actual locations of mobiles in each cell *are* different but still conform to the same average subscriber density. Uplink and downlink target E_b/N_0 figures are read from link simulation files obtained using methods discussed earlier, and propagation losses from each mobile to the surrounding cell sites are calculated.

In the next step, evaluation of the mobile powers for every mobile in the drop is carried out. The figures obtained in the previous step for the base station E_b/N_0 and propagation loss are used to estimate the transmit power required from each mobile; this is then adjusted to reflect soft handoff gain, together with a reduction in sensitivity in consideration of the noise rise at the base station receiver due to the anticipated load factor. Any mobiles requiring transmit powers above the maximum for the device are recognised as out of coverage and removed from the parent cell and a series of iterations is carried out until the received power from each mobile of each parent site is within a threshold value. If during any iteration the specified noise rise in any cell is exceeded, mobiles are removed to, if possible, be served by an adjacent cell. Iteration then resumes.

The third stage in the simulation process is evaluation of downlink transmit powers. Using the previously referenced propagation loss and a *mobile* E_b/N_0 target for the mobile under consideration, the required base-station transmit power will be evaluated in relation to the total interference power at the mobile location. As previously discussed, this will comprise thermal noise, in-cell noise (reduced by the orthogonality factor) and other cell noise. Note that the last two terms are also location dependent. This transmit power evaluation will be carried out for each base station in the soft handoff active set and for all mobiles in all cells of the drop. Because the total base-station transmit power (and thus interference) is dependent on the number of mobiles served, iteration will be required as mobiles are removed to ensure that individual cell sites remain within their total power allocation and to converge within the specified mobile E_b/N_0 threshold.

The last step in the simulation is post-processing. Typically, 50 to 100 drops of the type described in the previous three stages will be completed, depending upon the number of mobiles in the cell, and can be used to develop statistics that are used to optimise cell site planning including:

- Best server (uplink and downlink),
- Uplink load,
- Downlink load,
- Soft handoff areas,

- Soft handoff overhead,
- Capacity per cell (at specified outage),
- Coverage per cell (at specified outage).

This can readily identify cells that are not attracting traffic (perhaps wrong pilot-power allocation), overloaded cells, whether or not coverage is acceptable for all services at the planned cell radius, etc. Modern commercial tools incorporate great flexibility and can load three-dimensional cell site information, a variety of propagation models (including ray tracing, often useful in built-up areas). The modelling of WCDMA coverage and capacity is an extensive area in its own right; the interested reader can find further information in [56, 57, 58, 59, 60, 61].

6.5 HSDPA and HSUPA

UMTS was designed to support rich multimedia services and significantly lower the cost of voice services. To some extent it has delivered on both of these requirements. Circuit data rates of up to 384 kbits/s can be provided away from the cell edge and the use of a 5-MHz channel means that for small cells in the region of 500 m where the capacity is usually needed, more than 40 (12.2 kbits/s) voice users can be supported under ideal conditions of regular cell placement [62]. This significantly reduces the cost per voice user compared with GSM. However, in many deployments, where it is not always possible to site the cells at the ideal location, this efficiency can fall by a factor of two [63]. Even working with a figure of 50 voice users, this equates to a cell capacity of about 600 kbits/s in 5 MHz, corresponding to a spectral efficiency of 0.12 (bits/s)/Hz, or around 0.06 (bits/s)/Hz for more practical deployments. This represents only a marginal improvement when compared with 0.04 (bits/s)/Hz of the original hard-blocking GSM systems introduced earlier. To some extent, this reflects the inefficiency of CDMA systems in general for low-rate services, where power-control loops comparable in bandwidth to the traffic rate are needed. Clearly, full advantage is not being taken of the benefits of the wideband channel.

High speed downlink packet access (HSDPA) is responsible for a significant increase in overall spectral efficiency and capacity. It originated from a proposal originally made for 3GPP2 (1XTREME) [64], and was modified and adopted for UMTS. HSDPA introduces a new high-speed downlink shared channel (HS-DSCH) to support streaming, interactive and non-real-time services. This new channel is designed to integrate with and work alongside the existing Release 99 structure. The key changes and areas of the equipment affected are shown in Figure 6.20 and further discussed in [65, 66, 67, 68].

The key change in philosophy that makes an HS-DSCH more spectrally efficient than its Release 99 counterpart is abandoning the principle of servicing each mobile user in every TTI. To illustrate the inefficiency of this, consider the amount of power that would have to be allocated to a high-rate user at the cell edge in a deep Rayleigh fade! In contrast, for a Release 5 HSDPA system, the same user would, in principle, not be allocated power until it had moved out of the deep fade and was once more in good

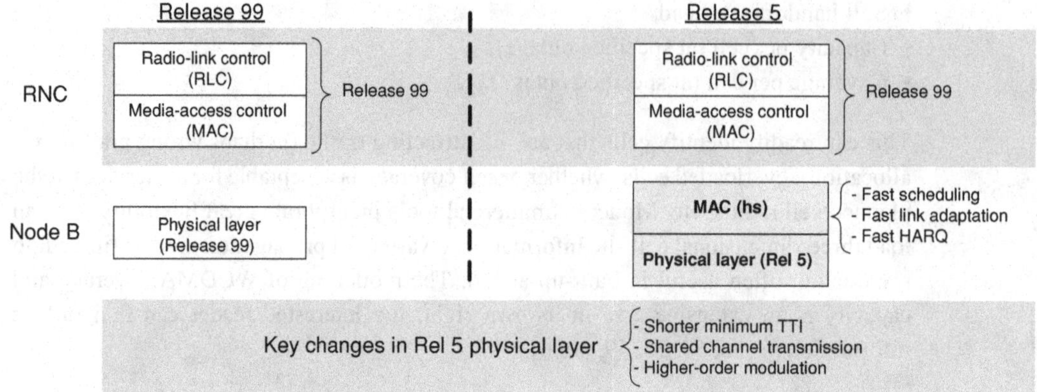

Figure 6.20 Key principles and changes to implement HSDPA

propagation conditions. (In practice this ideal situation may be diluted by the need for the scheduler to ensure that the committed information rates are met.) In this way, every scheduled user is at least in reasonable propagation conditions and the average power allocated is significantly reduced – creating capacity in a system that is usually downlink power limited.

The HS-DSCH is a fixed SF = 16 'fat' downlink pipe, which can be shared in a time-multiplexed fashion with many users. A new shorter radio frame of 2 ms is introduced and the minimum TTI is also 2 ms; this means that decisions on whether or not to schedule power to a particular user can be made sufficiently fast that the scheduler can usually track the Rayleigh fading and *not* allocate powers to users in deep fades. The other key change is that the HS-DSCH is not subject to fast power control. In fast power control, an average E_b/N_0 target is set, which guarantees an acceptable BLER in almost all cases. The HS-DSCH instead sets a lower E_b/N_0 target and uses fast hybrid ARQ to retransmit those blocks that occasionally get corrupted. This approach saves further power [69] and the shorter minimum TTI makes retransmission acceptable to most services. The final new feature introduced in Release 5 is the ability to configure the HS-DSCH to operate with Release 99 QPSK modulation or a new 16QAM mode.

Central to securing most of the benefits discussed above is the ability to make decisions quickly. In Release 99, measurements reported by the mobile have to be transmitted to the RNC where scheduling decisions are made and then the corresponding commands need to be delivered to the node B for implementation. Realisation of scheduling in this way means that real-time fade tracking is not possible. For this reason, a lot of the HS-DSCH media-access control (MAC) functionality is moved down to the node B. This approach does not disturb the Release 99 functionality, where fast power control continues to manage the performance of other channels. HSDPA represents a significant increase in UMTS performance; by using multi-code transmission peak rates of up to 14.4 Mbits/s can be delivered. The corresponding cell capacities are in the region of 2 Mbits/s to 3 Mbits/s, depending upon the traffic mix in Release 5 systems [70, 71].

A technique to complement HSDPA, high-speed uplink packet access (HSUPA), has also been developed and was introduced for Release 6; it is described in detail in [72, 73,

Table 6.9. Physical channels for HSDPA and HSUPA

Transport channel	Physical channel	Up or down	SF	Comments
HS-DSCH	HS-PDSCH	D	16	Used to carry the high-speed downlink shared channel (HS-DSCH)
	HS-SCCH	D	128	HARQ and TFI information associated with the HS-DSCH
	DPCH	D	Var.	Downlink DPCH carrying TPC for the uplink DPCH supporting the HS-DPCCH
	F-DPCH	D	256	Can be used as a more efficient alternative to the downlink DPCH
	HS-DPCCH	U	256	Supports HARQ ACK/NACK and CQI information for the HS-DSCH
	DPCCH	U	256	This uplink DPCH must be present to carry the HS-DPCCH information
E-DCH	E-DPDCH	U	Var.	This uplink DPCH is used to carry the E-DCH transport channel
	E-DPCCH	U	256	This uplink DPCH transmits control information associated with the E-DCH
	E-RGCH	D	128	Downlink DPCH carrying the uplink E-DCH relative grants
	E-HICH	D	128	Downlink DPCH carrying the uplink E-DCH HARQ acknowledgement indicator
	E-AGCH	D	256	Downlink physical channel carrying the uplink E-DCH absolute grant

74, 75. 76, 77, 78]. It, too, uses a short 2 ms radio frame and TTI, as well as fast scheduling and fast HARQ. However in this case, the resource being managed is the rise above thermal noise (RoT) at the node B receiver and it is the node B that enables individual mobile transmissions using fast scheduling to manage this process. Additionally, in WCDMA, as the cell range is typically determined by the mobile transmit power for a given service, higher-order modulation schemes with their attendant need for increased target E_b/N_0 figures were not considered for this release. As transmission is point-to-point, a dedicated channel format is used, but power control is implemented by means of downlink absolute grant channels (AGCH) and relative grant channels (RGCH). As in HSDPA, it is necessary to move the scheduling and HARQ functionality to the node B (in the form of a MAC-e) and soft combining is provided at the physical layer. Peak data rates of up to 5.76 Mbits/s are possible with cell capacities around 1.5 Mbits/s – an improvement of about 80% on the Release 99 uplink [79, 80].

Table 6.9 presents the new physical channels used on the uplink and downlink for both HSDPA and HSUPA. These are in addition to the corresponding Release 99 UMTS channels (shown in Table 6.5 and discussed previously) as UMTS Release 99 *and* HSDPA/HSUPA are intended to coexist on one or more 5 MHz RF carriers. Note that both solutions require the establishment of one or more dedicated physical channels (DPCH) with the associated 3–4 second set-up delay.

6.6 Access protocols and latency

6.6.1 WCDMA Release 99 call set-up

In comparing the attractiveness of WCDMA-based UMTS with GSM, a key consideration is the time taken to set up circuit-switched voice calls. A typical figure for a GSM mobile originated call is 2 seconds; the corresponding figure for UMTS is about 3.1 seconds in well-designed real networks and significantly worse in some deployments [81]. Thus, from a voice perspective, typical call set-up times are worse although higher voice capacity per carrier does provide operators with significantly reduced costs.

In respect of its primary goal 'to provide better support for the expected demand for rich multimedia services' discussed earlier in this chapter, the situation is no better: the set-up time for packet-switched calls is the same as for circuit calls. Consequently, with peak data rates up to 2 Mbits/s, the waiting period to establish the packet call will usually far outweigh the time taken to transmit and receive data from the server. This leads to operator decisions to set RNC timers (used to determine how long to maintain a bearer in the active 'Cell_DCH' state after the last packet activity) to large values or, indeed, to decide to support interactive packet traffic over circuit bearers – a regressive step. Whichever route is followed, up to 10 times the cell capacity intrinsically needed for support packet services may end up being dedicated to such applications, with the corresponding impact on service cost. The more bursty the application, the more capacity will be wasted, thus the efficiency will be especially low for VoIP and HTTP services. In addition, as the data rates of UMTS systems are increased by new high-speed downlink and uplink channels, such as HSDPA and E-DCH, the delay caused by RB set-up and channel allocation will contribute an even larger portion of total data transfer delay. Thus, the subscriber experience will contrast starkly with the total 100 ms to 200 ms response time from ADSL or even Wi-Fi.

So, what is the origin of the 3.1 seconds call set-up time? Figure 6.21 shows a high-level view of the transactions necessary to establish a WCDMA call in UMTS. There are four phases of transactions starting with establishing a connection between the mobile and the node B (RRC connection) and ending with completion of the link between the mobile and the called party. Although there are some slight differences between the allocation of delays between phases in the theoretical response, estimated by 3GPP [82], and the measured responses for each phase obtained in a real network [83], they broadly align. However, measurements in a practical network are larger for the two longest activities, core network signalling and RAB assignment. The major reason for the difference between the 'theoretical' and 'practical' columns lies in the time for real mobiles to execute their part of the transactions.

But why is even the theoretical estimate so large? The key to this is that in Release 99 WCDMA, more than 35 messages need to be exchanged between one mobile and three network elements. To maximise flexibility, large amounts of information are exchanged during call set-up (up to 7 blocks × 168 bits) while the signalling radio bearer rate (SRB) set-up between the UE and other network elements is typically about 15 kbits/s, but interleaved over four blocks to minimise the E_b/N_0 target and maximise coverage. Thus,

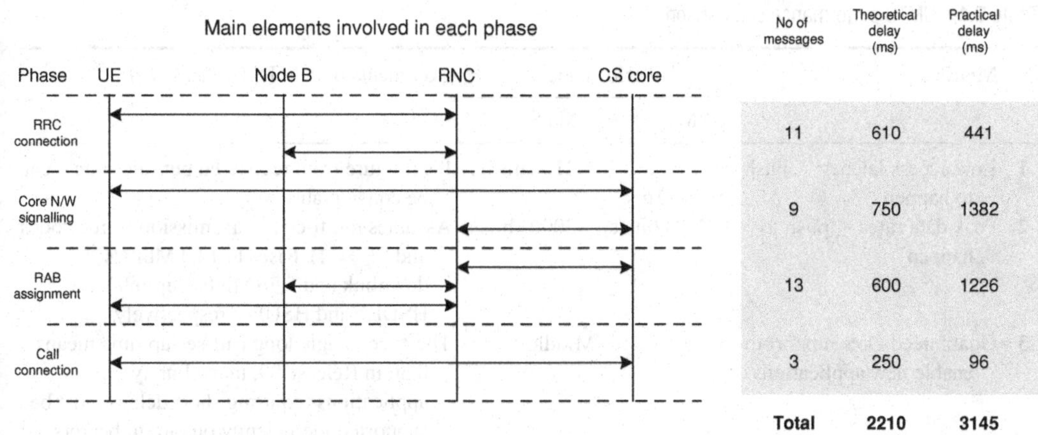

No of messages	Theoretical delay (ms)	Practical delay (ms)
11	610	441
9	750	1382
13	600	1226
3	250	96
Total	**2210**	**3145**

Figure 6.21 Release 99 call set-up – transactions and delays

about 300 ms are needed just to transport such messages. In practice, for badly designed networks, the situation can be much worse. In areas of the network where C/I is poor, unacknowledged-mode messages used during the RRC phase can get corrupted causing *the complete message set* to be retransmitted at layer 3 one or more times!

Since 2005, a work item has been underway in 3GPP to look at ways of reducing call set-up time. The outcome of this study was published at the end of 2006 and proposes changes as follows:

- Default SRBs and bearers for packet and circuit call set-up,
- A method for reducing the time taken to reconfigure bearers between the call-set-up and call-connection phases.

The impact of these changes in real networks remains to be seen.

6.6.2 HSDPA/HSUPA and user plane delay

As discussed earlier, neither HSDPA nor HSUPA is free from the Release 99 call-set-up delay discussed above; even if the downlink shared channel already exists, dedicated physical channels need to be established to support control information before the new subscriber is able to share the common downlink resource. However, once the user is sharing the channel, 'resume' latency – the time taken to access the shared channel on demand – is largely a function of the loading on the system and this is something that can be managed. On the uplink, the situation is similar.

For VoIP, and assuming an upper bound of about 285 ms for round-trip delay, an HSDPA/HSUPA combination can support about 20% more users than Release 99-based circuit-switched AMR [84]. This assumes that the 2 ms TTI is used in both up- and downlink and that delay accumulated across the node B/GSN/Internet transport does not exceed a round-trip total of about 75 ms, which is shown to be reasonable.

Table 6.10. UMTS performance comparison

	Metric	Performance		Comments on UMTS (Release 99)
		GSM	UMTS	
1	Low access latency – 'push to happen'	700 ms to 2000 ms	~3100 ms	Ideal figure from registered but idle state – can be considerably worse.
2	Peak data rates – 'push to happen'	~200 kbits/s	~2000 kbits/s	Assumes multi-code transmission (three codes and SF = 4). Rises to 14.4 Mbits/s downlink and 5.76 Mbits/s uplink with HSDPA and HSUPA, respectively.
3	Guaranteed QoS support to enable new applications	Worst	Middle	The exceedingly long call-set-up time means that, in Release 99, many bursty applications requiring short delays must be supported inefficiently on circuit bearers. VoIP over HSDPA / HSUPA can be efficient.
4	Enriched HLR functionality – support of new features	Middle	Middle	This is enabled by IMS and its successive releases. May, in practice, be limited by (3).
5	Low-cost configurations – developing market access	Best	Worst	Because of the need to allocate power continuously for signalling as well as traffic channels, the range of UMTS (Release 99 and 5/6) will always be inferior to GSM and LTE for deployment of comparable features. Coverage is further reduced if there is significant traffic.
6	Self-configuring network	Worst	Middle	Some improvement relative to GSM but much remains to be done.
7	Self-optimising network	Worst	Worst	Relies heavily on operator staff skills and specialist third-party software.
8	Common core network for all applications	Worst	Middle	Although possible using IMS, the call-set-up times remain a big disincentive, unless the default bearers planned for Release 7 do reduce these delays in practice.
9	Convection air-cooled BTS	Middle	Middle	Pico, micro or mini cells only. Macrocells invariably require fans or other forced-air approaches.
10	Common application support platform	Worst	Middle	See (8).
11	Reduced back-haul costs	Worst	Middle	The most expensive part of back-haul is the node 'B'–RNC link. Packet back-haul is possible with upgrades and is now being actively considered for volume deployment of HSDPA and HSUPA in the context of VoIP.
12	Higher cell capacities	Middle	Middle	UMTS is more cost-efficient than GSM for circuit voice when 5 MHz carriers are fully loaded. For this reason, and because of limited demand for higher rate services outside of cities, UMTS is largely only deployed in cities where capacity is still an issue.

(cont.)

Table 6.10. (*cont.*)

	Metric	Performance		Comments on UMTS (Release 99)
		GSM	UMTS	
13	Reuse of currently deployed equipment	N/A	Best	With the exception of the node 'B' Release 99 reuses all GSM elements, although many require significant upgrades.
14	Ability to reuse existing spectrum	Best	Worst	UMTS is severely limited by its 5 MHz channel bandwidth, with most operators only able to deploy two carriers. This also limits the possibility for re-mining GSM spectrum in due course.
15	Broadband data handover	N/A	N/A	
16	Revised standards IPR policy	N/A	N/A	IPR policy unchanged from GSM, and this has resulted in much higher IPR costs for UMTS

For applications, increasingly supported by IMS, the situation is not so positive. The issue is, once more, the time taken to establish the 'call'. Time needs to be added to the three-second packet call establishment delay to enable SIP (the IETF session initiation protocol) to complete the 'call' through to the application servers. Session initiation protocol is a text-based language, which means that the time to market for application development is short, but messages are long. Many of the most frequent messages sizes are in the range 500 to 800 bytes. These take a long time to transmit over the SRBs and total application set-up time can be around 7 seconds from initiation to application start. Message compression can reduce this to about 5 seconds in the worst case [81, 85].

The *wireless network metrics* established at the end of Chapter 2 have been completed for UMTS and added to the ratings for GSM. The resulting comparison is included in Table 6.10.

6.7 UMTS worked example

As has been discussed, the UMTS air interface is effectively made up of two distinct 'drops' of functionality: the Release 99 configuration – using fast power control and centralised decision making in the RNC – and HSDPA – which has adopted fast scheduling in the node B and HARQ to provide higher spectral efficiency. This worked example will explore the challenges of simultaneously providing voice and mobile broadband coverage and will detail the planning process for Release 99 (used to support voice) and HSDPA (used to provide broadband coverage).

6.7.1 Overview of UMTS air interface features

An overview of the UMTS air interface features is given in Table 6.11.

Table 6.11. UMTS air interface features

Feature	Main benefit		
	Capacity	Coverage	Quality
Inter-system handover	✓	✓	✓
Pilot power allocation	✓	✓	✓
Transmission diversity	✓		
Level of sectorisation	✓		
Soft handover region planning	✓	✓	✓
Secondary scrambling codes	✓		
Mast head amplifier		✓	✓
Code and power balance between HSDPA and Release 99	✓	✓	✓
Multiple carriers	✓		✓

6.7.2 Discussion of specific air interface features

As discussed in Chapter 3, the planning of any mobile network is a significant task. The detail provided in Section 3.3.6 under the 'Outline Planning Process' emphasises that the required output from that work is the achievable site density in different scenarios. Each of these will be based on a number of assumptions. When detailed deployment planning gets underway some of these assumptions will not reflect reality. Indeed, in a number of cases, 'reality' may not yield to intervention by the operator (e.g., the path-loss characteristics of a particular environment type). Thus, in practice, the planner must respond with decisions which *are* under his control – perhaps by looking for different sites, altering the sectorisation scheme or the power allocation to HSDPA. Hence, detailed planning must include the most appropriate design choices for different environments at different stages of the network rollout, alongside knowledge of the associated site densities that will be required.

In this design task, the network planner will invariably be assisted by commercial planning tools. These will initially be used in an iterative process to find the maximum cell radius at which coverage, capacity and other constraints are met by the proposed network configuration. Once a baseline solution has been selected, a further phase to refine the design will usually follow. The aim of this phase will be to improve the cell range by detailed optimisation of the network configuration and parameters.

In practice, the expertise of the system engineer will be called upon to varying degrees in each phase to identify the factors that limit performance and then take appropriate actions to address these limitations. For UMTS, the main tools at the planner's disposal can be summarised as:

- Add or remove carriers in response to capacity constraints,
- Add or remove secondary scrambling codes in response to code blocking,
- Alter pilot power allocations in response to coverage constraints,
- Introduce mast head amplifiers to meet uplink coverage constraints,
- Introduce transmission diversity to meet downlink capacity constraints,

- Alter the power balance (and code allocation) between R99 UMTS channels and R5 HSDPA channels in response to constrained coverage or capacity,
- Adjust the mapping of services to R99 or HSDPA bearers to optimise load balance.

6.7.2.1 GSM and GPRS legacy

An operator with an existing GSM network has a major head start over a new entrant deploying UMTS on a green-field site. This is because in UMTS, more so than other technologies, the network operator's existing footprint is very important. When rolling out any network, establishing coverage is fundamental, and the wide availability of dual mode GSM–UMTS handsets means that UMTS coverage of low-traffic areas can often be deferred, with traffic carried by the GSM network and a corresponding benefit to profitability. However, where capacity is already an issue on the GSM network – or where the operator has no GSM footprint and might have opted to negotiate access to a competitor's GSM network – any relaxation in coverage requirements brings with it an associated additional 2G cost. Reliance on an existing GSM network also places more importance on inter-system handover and the need to ensure that reselection works seamlessly at coverage borders or in response to congestion on one system.

An operator with an existing GSM network may also be able to gain insight from GSM usage statistics to construct the 3G traffic mix and geographical distribution. However, it is not always straightforward to translate usage of GSM and GPRS services into expected usage of UMTS services because the two customer bases and the attractiveness of services and tariffs may differ greatly. Further, there are inevitably reasons why only a proportion of GSM sites can be used to form the UMTS network. A mature GSM network will probably have passed through several iterations of capacity improvement measures, which may have tailored the site density and sectorisation to a level beyond that required in the early stages of rollout of a 3G network. Conversely, the propagation and penetration losses experienced at the lower GSM frequencies may compare favourably with those expected at the UMTS frequencies, giving rise to larger cell sizes in less populous areas. Further, the low-bit-rate voice-centred GSM network will usually give rise to larger cells than the higher-rate services that a UMTS operator may wish to deploy. Finally, it is not always physically possible to add additional equipment or mount additional antennas at existing sites.

6.7.2.2 Coverage

Whilst high-level dimensioning provides valuable information to confirm the basic network configuration, it is a generic solution that provides a site density that assumes an homogenous plan. In reality, it is recognised that the final site density may need to exceed the initial plan by around 10%. Some of the reasons for this include:

- The difficulties of site acquisition mean that a perfect grid-like layout cannot be maintained,
- Site locations need to reflect the local variation in subscriber densities,

- It is sometimes desirable to modify the site density figures to accommodate non-uniform radio propagation characteristics within environment types, which dimensioning cannot fully recognise.

With respect to this last point, the introduction of elevation and clutter data – and in some cases building data – enables a more accurate RF prediction model to be utilised. It also allows finer granularity in some of the other assumptions necessary for RF prediction or system simulation, for instance allowing penetration losses, subscriber speed distributions and fast fading margins to reflect the scenario detail.

Basic network coverage will typically be assessed with reference to minimum thresholds for both pilot received-signal code power (RSCP) and E_c/I_0. The distribution of the best server's RSCP and E_c/I_0 may also be analysed to show the ease with which users can be expected to access the network. Radio frequency prediction can be combined with the link budgets for different bearers to produce coverage statistics for services of interest. Coverage at this stage is assessed by comparing the predicted path loss from the best server with the maximum permitted path loss according to link budgets; this can also be adjusted for differing penetration losses to examine in-building coverage.

There are obviously simplifications in this analysis. For instance, soft handover is not adequately addressed, variable interference is represented solely as a constant design margin and the required E_b/N_0 for each service is considered to be constant over all subscriber speeds and RF conditions. However, despite these limitations, it is possible for the system engineer to identify certain coverage-related issues including:

- Regions with no apparent coverage,
- Large differences over sectors offering coverage with the best path loss,
- Sectors offering unintended coverage in distant regions,
- Larger regions than planned being served by multiple sectors.

Actions can then be taken to adjust coverage, e.g., by changing antenna azimuth, tilt, height or pattern or to reposition, remove or add sites. When the most significant problems have been identified and resolved, system simulation can be used to explore network coverage in more detail.

The problems above should be largely eliminated at this stage before moving to full-system simulation to explore capacity issues.

6.7.2.3 Load planning

Before potentially expensive system-load simulations are run, it is important to establish where the subscribers are likely to be! In this respect, clutter data – gathered for more accurate characterisation of propagation – can also be used to refine earlier assumptions of uniform (or cell-edge) traffic distribution, for instance by recognising the presence of high-rise flats. Such buildings will raise local subscriber density quite significantly. This same information can also be used to weight traffic density of individual services according to the presence of specific clutter types, whilst maintaining the overall desired density.

System simulations enable performance statistics to be categorised based on specific attributes, such as the service used, the mobility of the user, assigned penetration loss, handset properties (e.g., HSDPA capabilities) and the selected serving-cell. Selected serving-cell statistics allow the system engineer to identify quickly the sectors with the most problems and take appropriate remedial action at a local level. However, there will typically also be system-wide alterations whose applicability is assessed via simulation.

The performance of individual UEs should also be analysed by simulation. This will normally be carried out separately for each service type. Depending on the model's complexity, in a given cell there may be users who have been rejected via admission or congestion control procedures. If such behaviour is unintended, they will need to be resolved before proceeding further. Analysis of the performance of admitted users also provides valuable information. Downlink and uplink power requirements associated with a particular mobile, along with the available power headroom are important to monitor. For the uplink, headroom is likely to be defined in relation to the mobile maximum transit power, whereas in the downlink it is likely to be in relation to software-configured limits on the transceiver specific to each service. If in certain situations the headroom approaches zero, user performance can be adversely affected, for example becoming vulnerable to transient increases in propagation loss. Comparison of the actual FER or BLER with the design targets for each service can also provide useful insight; discrepancies in these statistics can often be traced back to limited headroom.

In addition to the points raised above for Release 99 networks, analysis for HSDPA-based services analysis must look at:

- Modulation scheme usage,
- Utilisation of HSDPA traffic channel,
- Number of simultaneous users,
- Packet delay (introduced via scheduling and retransmission),
- Throughput per user,
- Total throughput per sector.

Measures that may be taken to address the problems discussed above can be categorised by scope:
Local (cell by cell) adaptation:

- Sector reconfigurations (e.g., antenna azimuth, tilt, height or pattern),
- Site additions or removals,
- Addition of features at troublesome sites, e.g.:
  - Mast head amplifiers (MHA),
  - Tx diversity,
  - Rx diversity.
- Carrier additions or removals.

Network-wide adaptation:

- Alteration of pilot-power allocation,
- Alteration of traffic-power constraints,
- Alteration of HSDPA/R99 power balance (or code allocation).

Global alteration:

- Alteration of mapping of services to HSDPA/R99 bearers.

6.7.2.4 Cell isolation

Simulations will also output best server plots. These may vary from those based on path loss alone if a non-uniform pilot plan has been adopted. Care must always be taken when significantly adjusting an individual sector's pilot powers, owing to the potential for introducing imbalance on the uplink path loss to the serving sector. Such plots can also highlight sectors whose best server coverage area varies greatly from surrounding sectors, although it is important to remember that the ultimate aim is to balance load and not geographical service area.

Additional plots are normally produced to show the number of sectors where E_c/I_0 (or RSCP) lies within the soft handover and pilot pollution thresholds of the best server (see also Appendix 6.1). The ultimate intention is to maintain dominant sectors at all locations, whilst restricting the size of the soft handover regions to reasonable proportions, and eliminating areas with multiple non-dominant pilots (i.e., pilot pollution). However, it must be ensured that the desire to reduce cell overlap does not come at the expense of introducing coverage holes. In addition to geographic plots, statistics generated to identify the sites and sectors with the highest average soft handover factor, or pilot pollution factor, are often useful – as is the comparison of the network-wide soft handover factor against plan.

6.7.2.5 Load balancing

Balancing the load in the system across sectors is a key design aim. Load is normally assessed in the uplink via the noise rise at each sector and its variation across sectors. Because uplink noise rise normally acts as a constraint within uplink admission control, it also provides indications of which sectors are approaching their capacity limit. In the downlink, the load is often assessed via the proportion of transmission power being utilised. However, because transmission power can be pooled in different ways, it is necessary to examine this at both a sector level and a site (or carrier) level. Even if the pooled transmission power resource remains within its limits, it is still undesirable for individual sectors to exceed their share of the overall transmission power resource. Such imbalanced usage is better utilised as a transient ability to meet instantaneous traffic demands rather than as accepted design practice.

Prior to the introduction of HSDPA, transmission power availability normally acted as more of a constraint in the downlink than the availability of spreading code resources. If code blocking was an issue, then secondary scrambling codes could be used to introduce an additional code tree, albeit with a resultant loss in orthogonality. However, HSDPA will increase the pressures on codes for Rel. 99 traffic.

Vendor-specific hardware or software capabilities may also introduce limitations on the traffic that can be supported on a given sector, site or carrier by imposing limits on the number of active users, active bearers of specified rates or the cumulative bearer rate.

6.7.2.6 HSDPA–Release 99 balance

For a single-carrier solution, codes for HSDPA come from the same resource as Release 99 so care must be taken in deciding upon the split between the two solutions. A fairly standard five-code allocation will block 5/16 of the code tree. Ten or 15 codes – which are necessary to achieve the highest throughputs for HSDPA mobile categories that support such configurations – would block 10/16 or 15/16 of the code tree and the carrier could become dedicated to HSDPA only.

Power sharing between Release 99 and HSDPA also needs to be considered. In early deployments, this was normally statically configured, but more complex dynamic power sharing algorithms are now used, whereby HSDPA can move closer to utilising all of the unused power remaining from the Release 99 allocation at any given time. Such schemes allow a static power reservation for HSDPA, which can be exceeded (possibly up to a configured limit) if the additional power is not required for Release 99 at that instant. This is similar to the GPRS switchable time slot mode. Clearly, such schemes intrinsically have more complex admission control; additionally, at present, Release 99 traffic tends to be prioritised above HSDPA in such scenarios.

6.7.2.7 Neighbour list construction

System simulations often provide the means to populate initial neighbour lists for each sector automatically. For example, simple algorithms may be used that examine the non-dominant sectors present in the regions where a specific sector is the best server, and the E_c/I_0 of these sectors in relation to the best server. This can apply to both intra-frequency and inter-frequency neighbours. In an integrated planning tool, this may extend further still to the examination of the RSSI of GSM sectors, allowing inter-system neighbour lists to be generated between GSM and UMTS.

To facilitate either inter-system or inter-frequency handover, the network may require communication with a given mobile to enter compressed mode. This mode introduces time gaps in the uplink or downlink during which the UE can measure potential neighbours. Compressed mode can operate via either interaction with scheduling, the use of decreased spreading codes or increased puncturing. Unnecessary utilisation of compressed mode can have a detrimental effect on system capacity and individual user performance – hence, care must be exercised in choosing the thresholds that control its use. Even dual-receiver UEs may require the use of compressed mode when measuring neighbours operating on frequencies close to their current uplink transmissions.

6.7.2.8 Scrambling code planning

Some planning tools are able to select scrambling code allocations that minimise the likelihood of encountering code ambiguity. Inter-sector path loss information is used to restrict sectors using the same scrambling code to those having the highest mutual path loss. This problem is not normally significant, however, as the set of scrambling codes is relatively large, at 512.

6.7.3 Scenario subset selection

For the worked example, the 'Chapter 6' requirement of Table 4.2 will be addressed. Whilst commercial planning tools would be used in practice, system dimensioning will be carried out here using a spreadsheet approach to illustrate the process both for Release 99 and for HSDPA deployments. The primary requirement is to provide a 500 kbits/s broadband data service for an urban population density of 1000/km^2 with a contention ratio of 20:1. The scenario is assumed to take place at a time when there is a very low take-up of HSDPA services (less than 20% of the subscriber base). Voice coverage using UMTS Release 99 is also to be provided.

6.7.4 Questions to answer

For the UMTS example, the following design considerations will be addressed:

- What is the cell radius if the system is designed for indoor UMTS voice coverage?
- What is the minimum data rate that must be provided over the HSDPA air interface to meet the service that has been defined?
- What is the cell radius over which HSDPA can provide this data rate?
- How much power should be allocated to HSDPA?
- How many codes should be allocated to HSDPA?

6.7.5 Propagation model selection

3GPP has defined many bands for UMTS operation, but the most widely deployed solutions are around the 2 GHz region. At this frequency, the COST 231 Hata model is often selected as the basis for link budget and other relatively simple planning calculations even though the UMTS downlink frequency exceeds the 0.5 GHz to 2 GHz range over which the model applies. However, a more serious limitation of this model is that it is only accurate for mobile-to-base station separations exceeding 1 km, which may well not be the case for HSDPA deployments offering the highest data rates. Instead, the path loss model proposed by 3GPP in [44] will be used:

$$\text{Path loss [dB]} = 37.6 \times \log_{10}(\text{distance [km]}) + 128.$$

6.7.6 Initial deployment assumptions

The 'class' of HSDPA devices to be deployed in the network forms an important input to the design process. A device's HSDPA class or category constrains the modulation scheme that may be utilised, along with the largest transport block size and the number of HS-DSCH spreading codes. The HSDPA category can also impose more complex constraints on the scheduler, for example, imposing a minimum number of transmission-time intervals between subsequent transmissions to the same UE and limits on the number of soft-channel bits that the UE can support.

Based on the HSDPA devices that are currently available, this example will assume that the devices are HSDPA categories 6, 8 and 12. All of these categories support a minimum inter-TTI of 1, which means that the maximum theoretical sustained data rate for these devices does not differ from their maximum instantaneous data rate. One significant difference is that category 8 devices can utilise up to ten codes simultaneously whilst categories 6 and 12 are constrained to five simultaneous codes. Category 5 devices suffer the additional constraint of only supporting QPSK modulation.

Because the scenario focuses on urban coverage, it is assumed that initial dimensioning has selected a sectorisation level of 3 as the most appropriate network for the design.

6.7.7 Estimation of maximum cell size

A basic link budget for UMTS voice is presented in Table 6.12. It follows the normal process of calculating the EIRP of the signal after considering transmitter power limits, antenna gains and cabling losses. Whilst the 21 dBm limit of the traffic channel in the uplink is a hardware constraint of the mobile, the 33 dBm limit chosen for the downlink is a software parameter in the transceiver and this design decision could be revisited were it to result in a solution that was downlink limited.

The interference and noise powers are then accumulated, recognising that the mobile receiver noise figure will be higher than that for the base station receiver. An interference margin is introduced. The margin to include for downlink interference is debatable as the actual interference present will vary within the coverage area and can only sensibly be modelled by simulation; nevertheless, a constant 8 dB figure is used. However, for a relatively low data rate service such as voice, the *uplink* noise rise will normally prove to be the limiting factor and is the important figure. For a lightly to moderately loaded network, an uplink margin of 3 dB represents a sensible design goal. At this point, the sum of interference and noise powers is reduced to reflect the benefits of the integration of signal and noise powers during the de-spreading process. The figure of 25 dB reflects the ratio of the channel bandwidth to the service information bandwidth.

Finally, to determine base station and mobile sensitivities, it is important to select a single E_b/N_0 value for a given service. This must adequately represent E_b/N_0 requirements for the variety of mobility patterns, propagation profiles, etc., that individual UEs will encounter.

The resulting receiver sensitivity then forms the basis for exploring two high-level scenarios – outdoor coverage, in which there is some body loss, and indoor coverage with a moderate penetration loss. Both include similar margins to maintain some power headroom in order to combat fast fading and conversely benefit from a soft handover (or macro-diversity) gain. Handover gain will not always be applicable in the case of high data rate data services because it is sometimes disabled as a result of the high overhead involved. Note, also, that the log–normal fading margin included is actually a function of both the standard deviation of the fading and the required coverage probability at either the cell edge or over the entire coverage area.

In this example, the coverage limit is 1.4 km, determined by the uplink link budget for indoor voice. This cell range is not untypical for an urban deployment.

Table 6.12. Link budget for 12.2 kbits/s speech with Release 99 UMTS

Parameter	Units	Downlink Value	Downlink Subtotal	Uplink Value	Uplink Subtotal	Comments
Allocated Tx power	dBm	33		21		
Antenna gain	dB	17		0		
Cable and connector losses	dB	−3		0		
Antenna gain	dB	0		17		Receive aperture
Cable and connector losses	dB	0		−3		
Sub total (A)	dBm		**47**		**35**	Potential receive power
Thermal noise floor	dBm/Hz	−174		−174		Note 1
Noise bandwidth	dB	65.8		65.8		3.84 MHz
Receiver noise figure	dB	7		5		
Interference margin	dB	8		3		Allowable 'noise' rise
Processing gain	dB	−25		−25		$10 \log_{10}(3.84 \times 10^6/12.2 \times 10^3)$
Required E_b/N_0	dB	7		5.5		Note 2
Receiver sensitivity (B)	dBm		**−111.2**		**−119.7**	
Soft handover gain **(C)**	dB	3		3		
Link budget	dB		**161.2**		**157.7**	A − B + C

Parameter	Units	Downlink Indoor	Downlink Outdoor	Uplink Indoor	Uplink Outdoor	Comments
Link budget	**dB**	**161.2**	**161.2**	**157.7**	**157.7**	= (A − B + C) from above
Less margins and losses						
Log-normal fade margin	dB	8.7	8.7	8.7	8.7	Note 3
Fast fading margin	dB	4	4	4	4	
Additional losses	dB	2	2	2	2	Body losses, etc.
Penetration loss	dB	9.5	0	9.5	0	Note 4
Maximum propagation loss	**dB**	**137**	**146.5**	**133.5**	**143**	Note 5

Outdoor cell size: 143 = 128 + 37.6 log$_{10}$(R); R = 2.51 km **Indoor cell size: 133.5 = 128 + 37.6 log$_{10}$(R); R = 1.4 km**	See note 6

Note 1: Spectral density = kT = $(1.38 \times 10^{-23} \times 288)$
Note 2: The higher mobile E_b/N_0 figure reflects less rake fingers
Note 3: 8dB standard deviation, 95% coverage probability
Note 4: Normally in the range 5 dB to 20 dB
Note 5: Uplink propagation loss used to define cell size
Note 6: 3GPP model used [44]: Loss (dB) = $(128 + 37.6 \log_{10}$ (distance [km]))

Table 6.13. HSDPA data rate requirements

	Reference	Units	Value	Subtotals	Comments
Cell radius		km	1.4		Established by Release 99 speech
Sector area		km^2	1.7		Cells on a hexagonal grid
Population density		/km^2	1000		
Subscriber penetration		%	100		
Market share		%	20		
HSDPA penetration		%	20		
Number of HSDPA users	(a)			**68**	
Scaling factor for HARQ		–	1.111		1/0.90009
Scaling for RLC/MAC O/H		–	1.087		1/0.92
User peak data rate		kbits/s	500		
Bearer rate (inc O/H)	(b)	kbits/s		**603.8**	
Contention ratio		–	20:1		
Average contending users	(c)	–		3.4	= (a)/20
Required sector bandwidth		kbits/s		**2053**	= (c) × (b)

6.7.8 Required data rate determination

As set out in the requirements of Chapter 4, the service specified for this example is a 500 kbits/s 'data pipe'. If the system is being deployed to meet the basic goal of indoor UMTS voice coverage, the sites can be considered to lie on a hexagonal grid with an edge width of 1.4 km, corresponding to an average coverage area of 1.7 km^2 for each sector. (Note that with this definition the actual site separation is slightly less than the cell radius arrived at via the dimensioning exercise above.) This coverage area will be used to determine the number of HSDPA users demanding coverage.

Theoretical data rates in HSDPA are normally quoted for a particular transport block size, i.e., the size of the MAC PDU exchanged at every transmission time interval between the transport and physical layers at the transmitter. Each MAC PDU contains a header and is typically formed from the concatenation of multiple RLC PDUs. Each of these contains an RLC header as well as fragments of PDUs from higher layers, such as TCP/IP, which will also introduce their own overheads. For this reason, an 8% margin will be included to account for the overheads introduced at the RLC and MAC layers.

As discussed in the following section, signal-to-interference ratios form a key input to the HSDPA link budget and will be based upon a figure that permits a 10% probability of error in the first transmission of a transport block set (10% BLER). It follows that an additional overhead should be included to reflect the required HARQ retransmissions. A pessimistic assumption is made that the incremental redundancy or chase combining used does not significantly improve the probability of subsequent transmissions being successfully decoded, along with the slightly optimistic assumption that all transmissions will succeed after four attempts. The resultant net throughput rate due to HARQ retransmissions can easily be shown to be:

$$1/(0.1^0 + 0.1^1 + 0.1^2 + 0.1^3) = 0.90009.$$

Table 6.13 summarises the level of data traffic that each sector must serve.

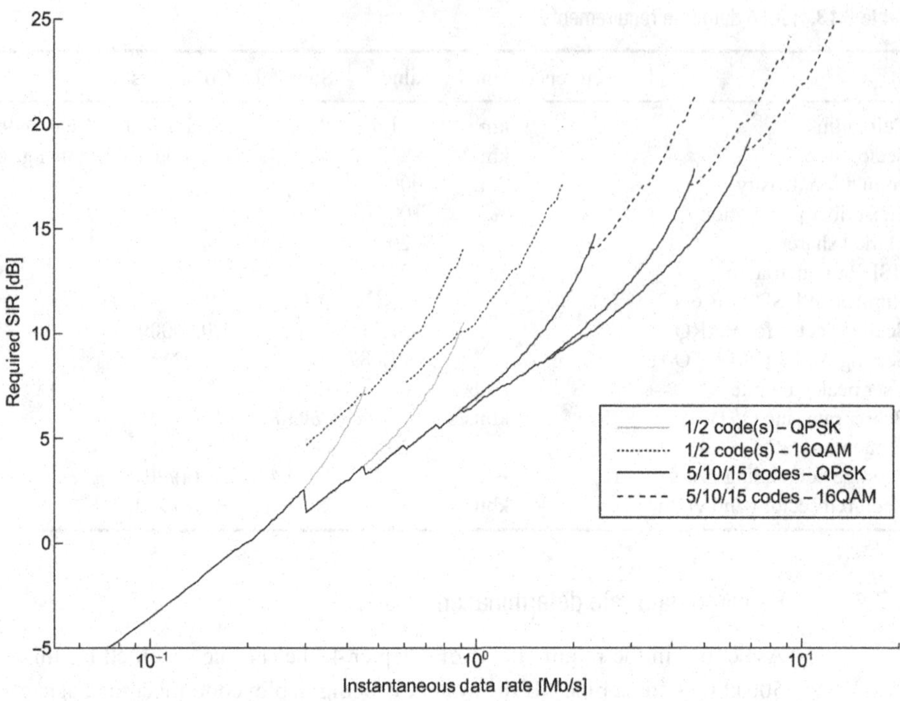

Figure 6.22 Data rate vs. HSDPA signal to interference ratio (10% BLER)

The table indicates that individual user throughputs of at least 604 kb/s at any location are required, in order to allow them to burst instantaneously at their full quoted data rate. Further, the aggregate throughput over the sector must be at least 2.1 Mb/s in order to meet the desired contention ratio.

6.7.9 Estimation of HSDPA coverage

Coverage estimation for HSDPA is more complicated than for Release 99 UMTS, largely because of the numerous possible transmit permutations. HSDPA allows the same bearer rate to be delivered in many ways through a combination of various transport block sizes, different levels of error-correcting coding, two different modulation schemes and the utilisation of between 1 and 15 HS-DSCH channels (or spreading codes). This flexibility means that the S/N requirement for a particular service at a given location will be dependent on the transmit format adopted and the capability (category) of the handset being supported. This reflects two main factors.

Firstly, there are differences in performance when decoding QPSK and 16QAM depending upon the levels of interference – with QPSK operating less efficiently than 16QAM at higher C/I levels. The second factor is the level of redundancy applied within the coding. Very-high-rate codes (those with little coding protection) operate less efficiently – even at high C/I – than more moderate coding rates achieved by spreading the information over a larger number of HS-DSCH channels. Figure 6.22 indicates the average signal-to-noise ratio required to decode the wanted signal, at 10% BLER, for the more efficient combinations of block size, levels of error-correcting coding, modulation scheme and numbers of HS-DSCH channels.

The effect of being limited to 5, 10 or 15 codes or to QPSK modulation is also illustrated by this figure. Some of the less efficient combinations that might be chosen because of the sector configuration or mobile categories in use are also included. Vertical offsets between lines in the figure equate to a different SIR requirement to deliver the same data rate, whilst horizontal offsets between lines equate to a different data rate achievable under the same SIR conditions.

A number of link budgets achieving an HSDPA data rate that satisfies the previously determined 604 kb/s requirement are presented in Table 6.14. A common choice of modulation and coding scheme is employed; the differences in the budget are to do with transmit power, interference levels and shadowing margin. Before embarking on the link budget discussion, it is useful to understand the rationale behind the chosen modulation and coding scheme.

The choice of QPSK and a spreading factor of 16 yields a data rate capability for each HS-DSCH code of:

$$2\,\text{bits} \times \frac{3.84 \times 10^6}{16} = 0.48\,\text{Mbits/s over the air interface}$$

–where 3.84 Mchips/s is the UMTS chip rate.

A 4-code mode has been selected, thus the total air interface rate is:

$$4 \times 0.48 = 1.92\,\text{Mbits/s}.$$

This is consistent with that required, taking account of the adopted coding rate of 0.32:

$$\frac{0.604}{0.32} = 1.888\,\text{Mbits/s}.$$

Each populated column of the link budget represents a slightly different scenario with the first three cases representing HS-DSCH power of 3, 5 or 7 watts. The link budget has a similar structure to the UMTS case but the data service to be carried has, of course, to be defined differently and is now represented by a transport block size, modulation scheme, number of codes and coding rate. The required signal-to-interference ratio is essentially a function of these parameters.

The accumulation of the EIRP for HSDPA traffic channels is done in a similar manner to before; in addition the dimensioned load of the network and maximum transmission power of a sector must also be provided to aid the estimation of interference. Figure 6.22 provides the signal-to-interference requirements under static conditions in the absence of fading. Hence adjustments have to be made to obtain the signal-to-interference requirements in more realistic situations where the signal fades during the 2 ms HSDPA sub-frame. Because these requirements are not defined in terms of E_b/N_0, but rather in terms of the signal-to-interference ratio accumulated over all HS-DSCH codes, the applied processing gain is defined solely by the spreading factor and so remains constant over all HSDPA data rates.

Table 6.14. HSDPA link budget

Parameter	Ref.	Unit	HSDPA link budget scenarios								Comments
			Scenario 1		Scenario 2		Scenario 3		Scenario 4		
			Value	Sub total	Value	Sub total	Value	Sub total	Value	Sub total	
Blocks	1	/s	500		500		500		500		2 ms frames
Spreading factor	2	–	16		16		16		16		
Modulation scheme		–	QPSK		QPSK		QPSK		QPSK		
Coding rate	3	–	0.32		0.32		0.32		0.32		Variable – 0.21, 0.28, 0.32, 0.35, etc.
Number of codes	4	–	4		4		4		4		1 to 15 (mobile dependent)
Transport block size	5	bits	1217		1217		1217		1217		Dependent on parameters 2 to 4
Bearer rate	6	**kbits/s**		**608.5**		**608.5**		**608.5**		**608.5**	= (1) × (5)
HS-DSCH power allocation	7	dBm	34.8		37		38.5		38.5		
Transmit antenna gain		dB	17		17		17		17		
Cable, connector losses		dB	−3		−3		−3		−3		
Receive antenna gain		dB	0		0		0		0		
Rx cable connector losses		dB	0		0		0		0		
Sub total	8	**dBm**		**48.8**		**51**		**52.5**		**52.5**	Potential received power

Parameter	Row	Unit									Notes
Total sector power		dBm	43		43		43		43		Drives other cell interference
Dimensioned sector load		%	60		60		60		60		
Thermal noise density		dBm/Hz	−174		−174		−174		−174		
Noise bandwidth		Hz	65.8		65.8		65.8		65.8		
Receiver noise figure		–	7		7		7		7		
Thermal noise power	9	dBm	−101.2		−101.2		−101.2		−101.2		
Serving cell interference	10	dBm	−90.9		−92.6		−93.5		−91.4		Note 1
Other cell interference	11	dBm	−85.4		−84.5		−83.8		−81.4		Note 2
Noise + interference	12	dBm		**−84.2**		**−83.8**		**−83.3**		**−81.0**	Note 3
Processing gain (SF = 16)	13	dB	−12.0		−12.0		−12.0		−12.0		$10\log_{10}[16]$
Required S/(I + N)	14	db	4.7		4.7		4.7		4.7		From Figure 6.22
Receiver sensitivity	15	dBm		**−91.6**		**−91.1**		**−90.7**		**−88.3**	Note 4
Link budget	16	dB		**140.4**		**142.1**		**143.2**		**140.8**	Equals (8) − (15)
Shadowing margin		dB	8.3		8.3		8.3		3.5		Note 5
Scheduler margin		dB	2		2		2		2		Non-ideal scheduling
Other losses		dB	2		2		2		2		
Propagation loss	17	dB		**128.1**		**129.8**		**130.9**		**133.3**	
Cell range	18	km		**1.01**		**1.12**		**1.19**		**1.39**	Note 6
Actual propagation loss	19	dB		128.12		129.8		130.79		133.32	Note 7

Note 1: From geometric model including orthogonality, shadowing, path loss
Note 2: From geometric model of surrounding cells, including shadowing, path loss
Note 3: Linear addition of powers of (9), (10) and (11) converted to dBm
Note 4: The sensitivity varies(!!); this arises from the way the powers in (9), (10) and (11) vary with distance
Note 5: All scenarios assume log-normal shadowing, scenario 4 at 67% coverage probability, the others at 85%
Note 6: Although this is set out as a conventional link budget, the link budget equation has to be solved for range by iteration
Note 7: Row (17) vs. (19) shows the result of the converged iteration

The 3 dB soft handover gain that boosted coverage in the UMTS link budget cannot be included in the present link budget. However, the 4 dB fast-fading margin that was previously included is replaced with a 2 dB scheduling margin in HSDPA, which recovers some of this loss. The justification for this is that a full fast-fading margin is not required when there are multiple HSDPA users, as the scheduler will attempt to maximise throughput to some degree by not scheduling users when they are in deep fades. The fast-fading margin cannot be entirely removed, as the delay between reporting and scheduling means that the level of fading reported will not correlate exactly with the fading experienced when data are transmitted. Furthermore, there is also a contrasting pressure on the scheduler to maximise fairness to some (configurable) degree to ensure that devices with queued data, because local RF conditions are momentarily relatively poor, can still achieve acceptable throughputs over longer time periods.

Interference is treated in a more detailed manner than for the UMTS case. In HSDPA, it would be inefficient to set a constant interference margin, as this would have to cover the worst case. It would ignore the complex variation of interference with user location and thus the varying significance of this power in relation to omnipresent thermal noise. Interference is, therefore, modelled in a more realistic manner by making assumptions about the average transmission power in each sector, accounting for both control channels and the level of traffic expected. This assumption can be verified via simulation; several rings of sites in a hexagonal grid are modelled with the previously determined cell radius, path loss model and shadowing standard deviation to obtain a polynomial approximation for the interference experienced as a function of distance from the serving sector. This model is complicated and consists of interference from the other sectors at the same site (which decreases with distance from the serving sector) and interference from other sites (some of which increases and some of which decreases with distance from the serving sector). Hence, as the edge of the HSDPA coverage for a particular data rate is estimated by the link budget spreadsheet, the interference term can be automatically updated to reflect the shrinking of the data-rate-specific coverage area in relation to the voice coverage area. The link-budget spreadsheet starts from the assumption that coverage can be provided for the given data rate over the full UMTS voice cell range and iteratively reduces the reported HSDPA cell range until the link budget can be satisfied.

As already discussed, the degree of code orthogonality is reduced in practice by multipath propagation. Thus an orthogonality factor is included in the budget to account for same-sector interference, which can be updated as a function of user range and reflect this variation. This does, however, complicate the application of other margins, such as log–normal shadowing, which previously needed only to have been considered after receiver sensitivity had been calculated. These margins must now *additionally* be applied to *reduce* the same cell interference term, as it is clear that both the signal and the interference from the same cell will experience the same additional losses. It should be clear that the case that reflects the full shadowing margin will be the most constraining in terms of coverage, as the ratio between the signal and the same-cell interference will be unchanged from the case where there is no shadowing, but the other sector interference will have increased weight owing to the lower signal level.

The first three scenarios of Table 6.14 show that as the power allocation to HS-DSCH is increased from 3 W to 7 W, the region over which edge reliability for the selected data rate can be maintained at 85% is increased from 1 km to around 1.2 km. Although below the 1.4 km target range, this may represent the likely power limit that the operator is prepared to allocate to HSDPA until device and service uptake increases. The final scenario explores how much the shadowing margin would have to be reduced to meet the desired data rate with a 7 W power allocation, and indicates that reliability would fall to 67% at the 1.4 km cell edge. The relatively low data rate target means that the different device categories do not require separate link budgets to be prepared, as 16QAM does not offer a significant gain over QPSK in this case and only four spreading codes are necessary.

6.7.10 Sector throughput

Some impression could be gathered as to whether the overall throughput requirements can be met by preparing several link budgets in a similar way to the above. This would enable an estimate of capacity to be made by summing throughput over a series of annular rings where the link budget (and associated data rates) can be assumed to be constant. However, there are significant limitations with this approach.

First, as already hinted, many separate link budgets would have to be derived for the possible combinations of data rate, coding rate and other parameters that vary as a function of the mobile device category in use (for instance the modulation scheme and number of codes). Secondly, the non-uniform distribution of such devices makes it difficult to predict accurately what aggregate throughput would be achieved without the use of statistical approaches. Finally, and probably the greatest obstacle to the prediction of accurate throughput, is the need to reflect the dynamics of the particular HSDPA scheduler in the vendor equipment, which must strive to maintain fairness in resource allocation across all users.

For these reasons, sector capacity is best established using dynamic simulation, which embodies the specific scheduler algorithms embodied in the equipment. Sector throughput information for different mixes of mobile device category and offered service will, therefore, normally be provided by the manufacturer as the scheduler implementation is not standardised. However the 2.1 Mbits/s sector throughput required for this example is well within the 3 Mbits/s capacity obtained using such techniques and discussed in Section 6.5.

References

1 R. P. O. Huber, A decade of European collaboration in the development of integrated broadband communications, *IEE Second International Conference on Broadband Services, Systems and Networks* (1993) 113–123.

2 3GPP, *General Packet Radio Service (GPRS); Service Description; Stage 2 (Release 1999)*, 3GPP TS 23.060 version 3.16.0 (2003).

3 3GPP, *Quality of Service (QoS) Concept and Architecture (Release 1999)*, 3GPP TS 23.107 version 3.9.0 (2002).

4 3GPP, *Mobile Radio Interface Layer 3 Specification; Core Network Protocols; Stage 3 (Release 1999)*, 3GPP TS 24.008 version 3.20.0 (2005).

5 3GPP, *Physical Channels and Mapping of Transport Channels on to Physical Channels (TDD)*, 3GPP TS 25.221 version 3.1.0.

6 3GPP, *Multiplexing and Channel Coding (TDD)*, TS 25.222 version 3.1.0.

7 3GPP, *Spreading and Modulation (TDD)*, TS 25.223 version 3.1.0.

8 3GPP, *Physical Layer Procedures (TDD)*, TS 25.224 version 3.1.0.

9 3GPP, *UTRA (UE) TDD; Radio Transmission and Reception*, TS 25.102 version 3.1.0.

10 E. H. Dinan and B. Jabbari, Spreading codes for direct sequence CDMA and wideband CDMA cellular networks, *IEEE Communications Magazine*, **36** 9 (1998) 48–54.

11 I. Ahmed, *Scrambling Code Generation for WCDMA on the StarCore™ SC140/SC1400 Cores*, Freescale Semiconductor Application Note AN2254 Rev. 1 (2004).

12 B. Sklar, Rayleigh fading channels in mobile digital communication systems part I: characterization, *IEEE Communications Magazine*, **35** 7 (1997) 102–109.

13 G. T. Martin and M. Faulkner, 1.9 GHz measurement-based analysis of diversity power versus the number of rake receiver tines at various system bandwidths, *IEEE Symposium on Personal, Indoor and Mobile Radio Communications*, (1997) 1069–1073.

14 B. N. Vejlgaard, P. Mogensen and J. B. Knudsen, Grouped rake finger management principle for wideband CDMA, *IEEE Vehicular Technology Conference 2000* (2000) 87–91.

15 A. J. Viterbi, A. M. Viterbi, K. S. Gilhousen and E. Zehavi, Soft handoff extends CDMA cell capacity and increases reverse link capacity, *IEEE Journal on Selected Areas in Communications*, **12** 8 (1994) 1281–1288.

16 J. Reig, L. Rubio and N. Cardona, Comparison of capacity in downlink WCDMA systems using soft handover techniques with SIR-based power control and site selection diversity transmission, *IEEE Vehicular Technology Conference* **4** (2004) 1999–2002.

17 3GPP, *Physical Layer Procedures (FDD) Release (1999)*, TS 25.214 version 3.12.0 (2003).

18 C. Zhou *et al.*, Transfer power allocation of WCDMA synchronization channel, and common pilot channel, *Proceedings of the International Conference on Communications Technology, ICCT 2000*, **1** (2000) 355–358.

19 3GPP, *Physical Channels and Mapping of Transport Channels onto Physical Channels (FDD) (Release 1999)*, TS 25.211 version 3.12.0 (2002).

20 A. Gerdenitsch, S. Jakl, Y. Y. Chong and M. Toeltsch, A rule-based algorithm for common pilot channel and antenna tilt optimization in UMTS FDD networks, *ETRI Journal*, **26** 5 (2004) 437–442.

21 R. T. Love, K. A. Beshir, D. Schaeffer and R. S. Nikides, A pilot optimization technique for CDMA cellular systems, *IEEE Vehicular Technology Conference*, **4** (1999) 2238–2242.

22 K. Valkealahti, A. Hoglund, J. Parkkinen and A. Hamalainen, WCDMA common pilot power control for load and coverage balancing, *Proceedings of the 13th IEEE International Symposium on Personal, Indoor and Mobile Radio Communications*, **3** (2002) 1412–1416.

23 V. DaSilva and E. S. Sousa, Performance of orthogonal CDMA codes for quasi-synchronous communication systems, *International Conference on Universal Personal Communications, ICUPC 1993*, **2** 995–999.

24 V. M. DaSilva, E. S. Sousa and V. Jovanovic, Performance of the forward link of a CDMA cellular network, *International Symposium on Spread Spectrum Techniques and Applications, ISSSTA '94*, **1** 213–217.

25 N. B. Mehta, L. J. Greenstein, T. M. Willis and Z. Kostic, Analysis and results for the orthogonality factor in WCDMA downlinks, *IEEE Vehicular Technology Conference Spring 2002*, **1** (2002) 100–104.

26 B. J. Wysocki and T. A. Wysocki, Modified Walsh–Hadamard sequences for DS CDMA wireless systems, *International Journal of Adaptive Control and Signal Processing*, **16** (2002) 589–602.

27 K. I. Pedersen and P. E. Mogensen, The downlink orthogonality factors influence on WCDMA system performance, *IEEE Vehicular Technology Conference Fall 2002*, **4** (2002) 2061–2065.

28 3GPP, *Spreading and Modulation (FDD) (Release 1999)*, TS 25.213 version 3.9.0 (2003).

29 B. Olin, H. Nyberg and M. Lundevall, A novel approach to WCDMA radio network dimensioning, *IEEE Vehicular Technology Conference Fall 2004*, **5** (2004) 3443–3447.

30 3GPP, *BS Radio Transmission and Reception (FDD) (Release 1999)*, 3GPP TS 25.104 version 3.13.0 (2005).

31 3GPP, *Radio Resource Control (RRC) Protocol Specification*, 3GPP TS 25.331 version 3.21.0 (2004).

32 3GPP, *Radio Interface Protocol Architecture (Release 1999)*, 3GPP TS 25.301 version 3.11.0 (2002).

33 3GPP, *Services Provided by the Physical Layer (Release 1999)*, 3GPP TS 25.302 version 3.16.0 (2003).

34 3GPP, *Multiplexing and Channel Coding (FDD)(Release 1999)*, 3GPP TS 25.212 version 3.11.0 (2002).

35 S. Aftelak and D. Bhatoolaul, Rate matching attribute settings and error rate performance sensitivity for selected UMTS FDD services, *IEEE Vehicular Technology Conference 2003-Fall*, **3** (2003) 1558–1562.

36 D. Lee and C. Liu, TFC selection for MAC scheduling in WCDMA, *IEEE Vehicular Technology Conference 2003-Fall*, **4** (2003) 2328–2332.

37 ETSI, *Universal Mobile Telecommunications System (UMTS); Selection Procedures for the Choice of Radio Transmission Technologies of the UMTS*, Technical report TR 101 112 version 3.2.0 (1998), Appendix B.

38 R. M. Buehrer, S. Arunachalam, K. H. Wu and A. Tonello, Spatial Channel Model and Measurements for IMT-2000 Systems, *IEEE Vehicular Technology Conference Spring 2001*, **1** (2001) 342–346.

39 N. Andrade *et al.*, A comprehensive 3G link level simulator, *Proceedings of 35th Annual Simulation Symposium, 2002*, (2002) 381–388.

40 NTT DoCoMo, *Link Level Simulation Results for UL Performance Requirements (FDD)*, 3GPP TSG RAN WG4, Document TSGR4(99)698.

41 3GPP, *User Equipment (UE) Radio Transmission and Reception (FDD) (Release 1999)*, TS 25.101 version 3.19.0 (2006).

42 H. Holma and A. Toskala, *WCDMA for UMTS* (John Wiley & Sons, 2001) pp. 226–227.

43 3GPP, *BS Radio Transmission and Reception (FDD) (Release 1999)*, 3GPP TS 25.104 version 3.13.0 (2005).

44 3GPP, *RF System Scenarios (Release 1999)*, 3GPP TR 25.942 version 3.3.0 (2002).

45 3GPP, *RF System Scenarios (Release 1999)*, 3GPP TR 25.942 version 2.0.0 (1999).

46 K. Sipila, J. Laiho-Steffens, A. Wacker and M. Jasberg, Modeling the impact of the fast power control on the WCDMA uplink, *IEEE Vehicular Technology Conference Spring 1999*, **2** (1999) 1266–1270.

47 M. M. Zonoozi and P. Dassanayake, Shadow fading in mobile radio channel, *5th IEEE International Conference on Universal Personal Communications*, **2** (1996) 291–295.

48 V. Erceg, L. J. Greenstein, S. Y. Tjandra *et al.*, An empirically based path loss model for wireless channels in suburban environments, *IEEE Journal on Selected Areas in Communications*, **17** 7 (1999) 1205–1211.

49 IEEE Vehicular Technology Society Committee on Radio Propagation, Coverage prediction for mobile radio systems operating in the 800/900 MHz frequency range, *IEEE Transactions on Vehicular Technology*, **37** 1 (1988) 3–72.

50 K. Gilhousen, I. M. Jacobs, R. Padovani *et al.*, On the capacity of a cellular CDMA system, *IEEE Transactions on Vehicular Technology*, **40** 2 (1991) 303–312.

51 V. V. Veeravalli, A. Sendonaris and N. Jain, CDMA coverage, capacity and pole capacity, *IEEE Vehicular Technology Conference Spring 1997*, **3** (1997) 1450–1454.

52 P. Padovani, B. Butler and R. Boesel, CDMA digital cellular: field test results, *IEEE Vehicular Technology Conference 1994*, **1** (1994) 1–15.

53 S. S. Kolahi, A. G. Williamson and K. W. Sowerby, Other-cell interference in CDMA systems, *IEE Electronics Letters*, **40** 18 (2004) 1134–1135.

54 QUALCOMM Incorporated, *Air Interface Cell Capacity of WCDMA Systems, Engineering Services Group Note 80-W0989–1*, Revision A (August, 2006).

55 Spirent Communications Inc., *Impact of Multipath Propagation Effects on Channel Demodulation and Network Capacity* (Spirent Communications, 2005).

56 J. Laiho, A. Wacker, T. Novosad and A. Hamalainen, Verification of WCDMA radio network planning prediction methods with fully dynamic network simulator, *IEEE Vehicular Technology Conference 2001 Fall*, **1** (2001) 526–530.

57 A. Wacker, J. Laiho-Steffens, K. Sipila and M. Jasberg, Static simulator for studying WCDMA radio network planning issues, *IEEE Vehicular Technology Conference July 1999*, **3** (1999) 2436–2440.

58 J. Laiho-Steffens, A. Wacker and K. Sipila, Verification of 3G radio network dimensioning rules with static network simulations, *IEEE Vehicular Technology Conference 2000 Spring*, **1** (2000) 478–482.

59 D. Molkdar, S. Burley and J. Wallington, Comparison between simulation and analytical methods of UMTS air interface capacity dimensioning, *IEEE Vehicular Technology Conference 2002-Fall*, **3** (2000) 1596–1601.

60 A. Wacker, J. Laiho-Steffens, K. Sipila and K. Heiska, The impact of the base station sectorisation on WCDMA radio network performance, *IEEE Vehicular Technology Conference 1999*, **5** (1999) 2611–2615.

61 J. Laiho-Steffens, A. Wacker and P. Aikio, The impact of the radio network planning and site configuration on the WCDMA network capacity and quality of service, *IEEE Vehicular Technology Conference 2000-Spring*, **2** (2000) 1006–1010.

62 K. Sipila, K.-C. Honkasalo, J. Laiho-Steffens and A. Wacker, Estimation of capacity and required transmission power of WCDMA downlink based on a downlink pole equation, *IEEE Vehicular Technology Conference 2000-Spring*, **2** (2000) 1002–1005.

63 S. Dehghan, D. Lister, R. Owen and P. Jones, W-CDMA capacity and planning issues, *IEE Electronics and Communication Engineering Journal*, **12** 3 (2000) 101–118.

64 R. Love, A. Ghosh, R. Nikides *et al.*, High speed downlink packet access performance, *IEEE Vehicular Technology Conference 2001-Spring*, **3** (2001) 2234–2238.

65 3GPP, *Physical Layer Aspects of UTRA High Speed Downlink Packet Access (Release 2000)*, 3G TR25.848 version 0.5.0 (2000).

66 3GPP, *High Speed Downlink Packet Access (HSDPA); Overall Description; Stage 2 (Release 5)*, 3GPP TS 25.308 version 5.7.0 (2004).

67 S. Parkvall, E. Dahlman, P. Frenger, P. Beming and M. Persson, The evolution of WCDMA towards higher speed downlink packet data access, *IEEE Vehicular Technology Conference 2001-Spring*, **3** (2001) 2287–2291.

68 A. Furuskar, S. Parkvall, M. Persson and M. Samuelsson, Performance of WCDMA high speed packet data, *IEEE Vehicular Technology Conference 2002-Spring*, **3** (2002) 1116–1120.

69 R. Love, B. Classon, A. Ghosh and M. Cudak, Incremental redundancy for evolutions of 3G CDMA systems, *IEEE Vehicular Technology Conference 2002-Spring*, **1** (2002) 454–458.

70 R. Love, A. Ghosh, X. Weimin and R. Ratasuk, Performance of 3GPP high speed downlink packet access (HSDPA), *IEEE Vehicular Technology Conference 2004-Fall*, **5** (2004) 3359–3363.

71 R. Love, K. Stewart, R. Bachu and A. Ghosh, MMSE equalization for UMTS HSDPA, *IEEE Vehicular Technology Conference 2003-Fall*, **4** (2003) 2416–2420.

72 3GPP, *Feasibility Study for Enhanced Uplink for UTRA FDD (Release 6)*, 3GPP TR 25.896 version 6.0.0 (2004).

73 3GPP, *FDD Enhanced Uplink; Overall description; Stage 2 (Release 6)*, 3GPP TS 25.309 version 6.6.0 (2006).

74 3GPP, *Physical Channels and Mapping of Transport Channels onto Physical Channels (FDD) (Release 6)*, 3GPP TS 25.211 version 6.7.0 (2005).

75 3GPP, *Spreading and Modulation (FDD) (Release 6)*, 3GPP TS 25.213 version 6.0.0 (2003).

76 3GPP, *Radio Interface Protocol Architecture (Release 6)*, 3GPP TS 25.301 version 6.4.0 (2005).

77 A. Ghosh, R. Love, N. Whinnet *et al.*, Overview of enhanced uplink for 3GPP W-CDMA, *IEEE Vehicular Technology Conference 2004-Spring*, **4** (2004) 2261–2265.

78 S. Parkvall, J. Peisa, J. Torsner, M. Sagfors and P. Malm, WCDMA enhanced uplink – principles and basic operation, *IEEE Vehicular Technology Conference 2005-Spring*, **3** (2005) 1411–1415.

79 W. Xiao, R. Ratasuk, A. Ghosh and R. Love, Scheduling and resource allocation of enhanced uplink for 3GPP W-CDMA, *IEEE International Symposium on Personal, Indoor and Mobile Radio Communications 2005*, **3** (2005) 1905–1909.

80 K. W. Helmersson, E. Englund, M. Edvardsson *et al.*, System performance of WCDMA enhanced uplink, *IEEE Vehicular Technology Conference 2005-Spring*, **3** (2005) 1427–1431.

81 G. Foster, M. I. Pous, D. Pesch, A. Sesmun and V. Kenneally, Performance estimation of efficient UMTS packet voice call control, *IEEE Vehicular Technology Conference 2002-Fall*, **3** (2002) 1447–1451.

82 GPP, *Signalling Enhancements for Circuit-Switched (CS) and Packet-Switched (PS) Connections; Analyses and Recommendations (Release 7)*, 3GPP TR 25.815 version 7.0.0 (2006).

83 C. Johnson *et al.*, Evaluating and refining call setup delay, *Fifth IEE International Conference on 3G Mobile Communication Technologies*, (2004) 332–336.

84 K. Yong-Seok, VoIP Service on HSDPA in Mixed Traffic Scenarios, *The Sixth IEEE International Conference on Computer and Information Technology* (2006) 79–84.

85 M. Melnyk and A. Jukan, On signaling efficiency for call setup in all-IP wireless networks, *IEEE International Conference on Communications, 2006, ICC '06*, **5** (2006) 1939–1945.

Appendix 6.1 Definition of E_b/N_0 and cell size calculation

In the discussion of Section 6.3, the richness and flexibility afforded through the combination of a variety of physical and transport channel multiplexing schemes will have become apparent; however, with this flexibility comes the need to make choices. How much coding should be added to the payload data? Over what period should data be interleaved for a given mobile speed and propagation profile? Unless other steps are taken, this very flexibility brings with it the possibility of making non-optimum physical and transport channel choices; i.e., there may be other schemes that meet the QoS requirements for the particular application but that use fewer system resources and could therefore be implemented at lower cost (or higher profit margin). This opacity is compounded by the very nature of CDMA, where the 'spreading gain' needs to be considered in determining equipment sensitivity. But the spreading gain, in fact, relates to the coded and interleaved symbol rate – *not* the user bit rate. How can the relative value of different schemes to the user be assessed?

Definitions of E_b/N_0 and related figures of merit

As a result of trial measurements, the power P_R needed at the terminals of a receiver (point **A** in Figure 6.23) to deliver a service under particular propagation conditions, mobile speed, etc., and with a specified BLER, is known. The service has a user rate R bits/second and is delivered over a physical bearer of rate C symbols/second. The channel symbols are spread by a code with a chip rate of W chips/second. The receiver input bandwidth is assumed to be matched to the chip rate, W, and has a noise figure NF, this being the factor by which the receiver's noise contribution is greater than that of a perfect receiver, one that introduces no additional noise. (In such a receiver, the noise is due only to the inevitable agitation of electrons in its conductive components and its density equals kT, where k is the Boltzmann constant and T is the temperature in degrees Kelvin.)

From the discussion of Section 6.2.1 and consideration of Figure 6.23, it might seem reasonable that P_R should be set using $E_{channel}/N_0$ as a figure of merit at point **C**. This figure, which represents the ratio of the energy accumulated over the symbol period to the noise spectral density (for a given symbol error rate), reflects implementation losses, such as the receiver noise figure, and limitations arising in the rake receiver, such as non-perfect channel estimation and rake finger placement. This relationship can easily be seen from Figure 6.23 and is:

$$\frac{E_{channel}}{N_0} = \frac{G \cdot P_R}{C} \times \frac{1}{G \cdot kT \cdot NF} = \frac{P_R}{kT \cdot C \cdot NF}.$$

Then P_R can be related to this channel figure of merit:

$$P_R = \underset{(1)}{\frac{E_{channel}}{N_0}} \times \underset{(2)}{kTW} \times \underset{(3)}{NF} \times \underset{(4)}{\frac{C}{W}}. \tag{6.10}$$

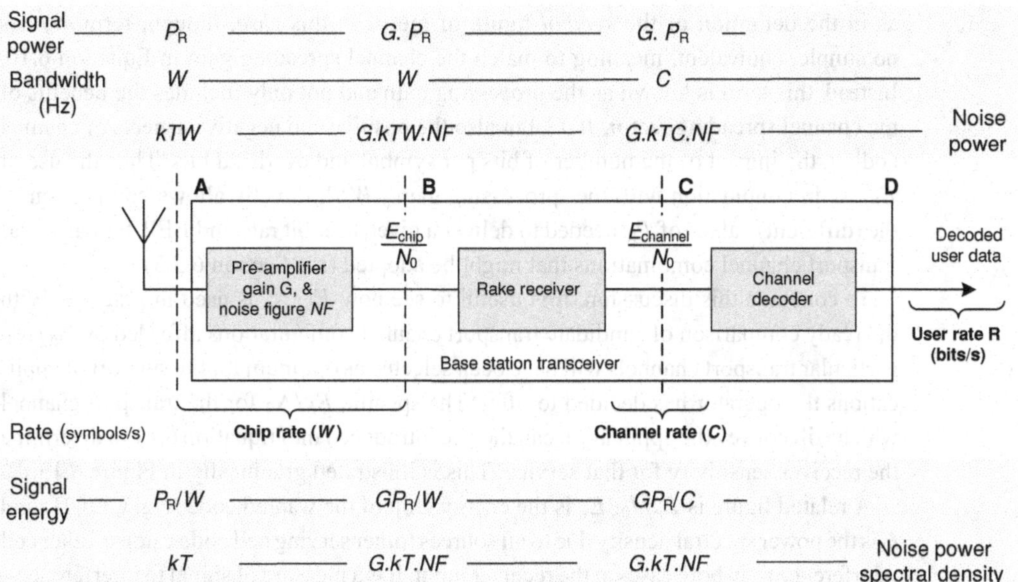

Figure 6.23 Reference configuration for E_b/N_0 definition

It is useful to understand the meaning of each of these terms. The first term is the proposed figure of merit referenced to point **C** necessary to deliver the required symbol error rate. However, observe that this may also be written as $P_R/kTCNF$, the signal-to-noise ratio at the output of the rake receiver. The second and third terms together represent the thermal noise power at point **B**, the input to the rake receiver. The final term is the reciprocal of the channel spreading factor, W/C; terms (2), (3) and (4) taken together thus represent the thermal noise power present at the output of the rake receiver and, when multiplied by the channel figure of merit, explicitly represent the required signal power P_R at point **A**, needed to sustain the desired symbol error rate.

However, whilst the symbol error rate after the rake receiver can easily be measured, it is not a good metric as far as the user is concerned. The symbol rate is only loosely related to the end-user bit rate through factors such as the number of bits per symbol, the interleaving period, the coding rate and other overheads. Instead, a more useful term would address the power P_R required to deliver a specified *user* bit rate at a desired BER. Then, by analogy, a new figure of merit E_b/N_0 can be defined at point **C**:

$$\frac{E_b}{N_0} = \frac{G \cdot P_R}{R} \times \frac{1}{G \cdot kT \cdot NF} = \frac{P_R}{kT \cdot R \cdot NF}.$$

P_R can be related to E_b/N_0:

$$P_R = \underset{(1)}{\frac{E_b}{N_0}} \times \underset{(2)}{kTW} \times \underset{(3)}{NF} \times \underset{(4)}{\frac{R}{W}}. \tag{6.11}$$

The figure of merit has changed from $E_{channel}/N_0$ to E_b/N_0, with the noise term now referenced to the energy per bit; terms (2) and (3), however, have the same meaning as in the definition of the *symbol* figure of merit. In this case, though, term (4) has no simple, equivalent, meaning to match the channel spreading gain in Equation 6.10. Instead, this term is known as the processing gain and not only includes the benefits of the channel spreading factor, W/C, but also the positive and negative aspects of channel coding, the impact of the number of bits per symbol and overhead bits. Thus the use of E_b/N_0 in conjunction with the 'processing gain', W/R, directly allows comparison of the (different) values of P_R needed to deliver a given user bit rate and BER for particular transport channel combinations that might be adopted (see Section 6.3.3).

To conclude this discussion, it is useful to see how E_b/N_0 is used in practice. With the ready comparison of candidate transport channel configurations afforded by E_b/N_0, particular transport channels will have been selected as optimum for the support of applications the operator has decided to offer. The specific E_b/N_0 for the transport channel which will convey the application can then be introduced into Equation 6.11 to determine the receiver sensitivity for that service. This is illustrated graphically in Figure 6.13.

A related figure is E_c/I_0. E_c is the energy/chip of the wanted code (e.g. CPICH) and I_0 is the power spectral density due to all sources (other serving cell codes, noise, other cell interference) – in both cases at the receiver input. It is a measure of signal to interference – but for the wanted code only and is often used by the mobile to select the next serving cell.

Cell size evaluation

This section provides the detail behind the calculation of relative cell sizes in Section 6.4.2 addressing WCDMA coverage and capacity. Table 6.15 below summarises the key input parameters used for the absolute cell size calculation.

The general discussion of propagation in Chapter 3 and the review of the table parameters provided in the main text means that this need not be addressed further here. However, it is useful to take the general form of the equation for propagation loss in Table 6.15 and see how this can be used to illuminate the way maximum cell size can be expected to vary with choice of service.

The reference equation for macro-cell propagation is taken from *RF System Scenarios* [1] and is intended to be representative of the environment within which the UMTS equipment will operate. More generically, the equation can be written:

$$L_B\,(\mathrm{dB}) = A + 37.6 \cdot \log_{10}(R),$$

where L_B is the link budget representing the difference between the system gains and losses.

For the purposes of comparing maximum cell sizes, and provided the higher-rate services are still delivered to the mobile phone (rather than, for instance, to a data card), it is reasonable that the only parameter to alter as a result of the service change should be the receiver sensitivity. For comparison purposes, cell ranges for other services will be normalised to that for 12.2-kbit speech, since this will be the largest cell range. Then, noting that the link budgets, L, are in dB:

$$L_{12.2\mathrm{kbits/s}} - L_{\mathrm{new\ service}} = 37.6\,(\log_{10}[R/R_{\mathrm{new}}])\,.$$

Table 6.15. Input parameters for cell size calculation

Gains		Losses	
Parameter	Value	Parameter	Value
Soft handover gain (a)	3 dB	Shadow fading (p)	8 dB
BTS antenna gain (b)	18 dB	Cable loss (q)	2 dB
Mobile antenna gain (c)	1 dB	Body loss (r)	2 dB
Mobile power (d)	21 dBm	Fast fading margin (s)	5 dB
Noise rise (e)	2 dB	Building penetration loss (t)	0
Receiver sensitivity (f)	−123 dBm		
$G = (d + c + b + a) - (e + f)$	**164 dB**	$L = (p + q + r + s + t)$	**17 dB**

$G - L = 147 = 128 + 37.6\log_{10}(R)$ where R is in km
$R = 3.2$ km (for macro-cell with pedestrian 12.2 kbits/s users only)

Inspection of the link-budget terms and recognising that only one term will change with each new service, this expression can be simplified further. Let Rx_{ref} = the receiver sensitivity for the 12.2 kbits/s speech and Rx_{new} be the receiver sensitivity for some new service. Then:

$$R_{\text{new}} = R/\left(10^{+[\Delta Rx/37.6]}\right) \text{ where } \Delta Rx = (Rx_{\text{new}} - Rx_{\text{ref}}).$$

The resulting cell sizes relative to the 12.2 kbits/s voice service are shown in Table 6.16.

Table 6.16. Cell sizes relative to 12.2 kbit voice

Service (kbits/s)	12.2	64	144	384
Thermal noise + NF (dBm)	−103	−103	−103	−103
E_b/N_0 (dB) [a]	5	1.7	0.9	1.0
Processing gain (dB)	25.0	17.8	14.3	10.0
Rx (dBm)	−123	−119.1	−116.4	−112.0
ΔRx	0	3.9	6.6	11
Range (% of 12.2 kbit cell)	100	79	67	51

[a] E_b/N_0 figures from the reference sensitivities in [2].

References

1 3GPP, *RF System Scenarios (Release 1999)*, 3GPP TR 25.942 version 3.3.0 (2002).
2 3GPP, *BS Radio Transmission and Reception (FDD) (Release 1999)*, 3GPP TS 25.104 version 3.13.0 (2005).

7 Cellular OFDM RAN planning and design

At the beginning of 2003, results from the first commercial deployments of UMTS were coming in and the inefficiency of Release 99 both spectrally and in terms of its long 'call' set-up times were becoming apparent. Coincident with this, Wi-Fi networks were becoming omnipresent in businesses and cities and broadband was achieving significant penetration in homes through much of the developed world. The user expectation was shifting from 'dial-up' latencies of tens of seconds to delays of less than 200 ms! These events together made it clear that there was a need for change if operators using 3GPP-based networks were to remain competitive. The requirements for a 'long-term evolution' of UMTS can thus be traced to this time when a study, which eventually led to the publication of a document *Evolution of 3GPP System* [1], commenced. The results from this study made it clear that, in future systems, support of all services should be via a single 'all IP network' (AIPN), a fact already recognised in the 3GPP study '*IP Multimedia Services*' with an initial functionality release in Release 5 (June 2002). What was different, however, was that it also identified that both the core network and access systems needed to be updated or replaced in order to provide a better user experience. Even at this initial stage, control and user plane latencies were itemised as major issues with latencies of <100 ms targeted alongside peak user rates of 100 Mbits/s. The primary findings from this and subsequent studies are shown in Table 7.1. This initial report was followed by AIPN feasibility studies [2] and subsequently publication of *Service Requirements for the All-IP Network (AIPN)* [3].

With the development of the proposed requirements for *Evolution of 3GPP System* during the course of 2003, and the attendant expectations for dramatic reductions in latency and corresponding increases in peak user rates, much thought was given to whether or not these requirements could be met by upgrades to the current WCDMA radio access network. This debate was crystallised in a special meeting held in Toronto in November 2004, which resulted in a decision (in December 2004) to commence a *Study Item on Evolved UTRA and UTRAN* [4]. This work, together with the subsequent feasibility study [5] and the development of requirements for E-UTRAN [6] starting in June 2005 meant that work on evolved 3G was underway. Before discussing the architecture for the evolved UMTS system, it is appropriate to note the extensive work carried out to determine whether evolution or replacement of the UMTS WCDMA was required. Evaluation of six concepts was started in April 2005 and concluded in September 2006 with the publication of the report *Physical Layer Aspects for Evolved Universal Terrestrial Radio Access (UTRA)* [7]. The decision (in December 2005) to go

Table 7.1. Key findings from the study *Evolution of 3GPP System* commenced in January 2003

	General requirements [1]	Subsequent further definition [2, 3, 4, 5, 6, 8]
1	A seamless integrated network comprising a variety of access systems connected to a common IP based network	• Service independent common IP network • Provide a high level of basic system performance including low communication delay, low connection set-up time and high communication quality • E-UTRAN + evolved core + internet < 70ms round trip • E-UTRAN + evolved core < 25 ms delay jitter
2	Focus on the inter-working between 3GPP and other networks	• Inter-work with a variety of wireless broadband networks based on IP technologies including those not specified by 3GPP • The network shall be able to accommodate fixed access systems and to inter-work with fixed networks • The AIPN shall provide appropriate mechanisms to support independent operation of services, networks or access systems
3	A similarity of services and applications across the different systems	• The network shall support service provision and provide mobility functionality within and across the different access systems • Flexible charging model triggered by QoS, transport, content or events • More efficient support of MBMS
4	Shorter radio access latencies (both call set-up and round-trip time)	• Control plane latency: <100 ms idle to Cell_DCH state • User plane latency: <5 ms
5	Long-term target peak data rates of up to 100 Mbps in full mobility	• Device capability: two receive and one transmit antennas • Peak data rate of 100 Mb/s within a 20 MHz downlink • Peak data rate of 50 Mb/s within a 20 MHz uplink • Peak data rates to scale linearly with channel bandwidth
6	Spectral efficiency target, for best effort packet communication is 2–3 bps/Hz	• Conditions as (5) • Downlink spectrum efficiency – 5 bps/Hz • Uplink spectrum efficiency – 2.5 bps/Hz

for a new OFDM-based air interface scheme was driven by:

- Simpler UE processing,
- Easier to meet latency requirements,
- Scaleable to a smaller minimum channel bandwidth,
- Reduced complexity for wider channel bandwidths and MIMO.

Concurrent with the above events, the IEEE standards body Task Group 802.16e (Physical and medium access control layers for combined fixed and mobile operation in licensed bands) was developing the lower layers for a new mobile solution; it too utilised OFDM technology. The industry in Korea and America decided to take the IEEE 802.16e standard published in 2006 and add to it higher-layer protocols to enable it as a mobile cellular solution. The industry body driving this work is the WiMAX Forum and it is

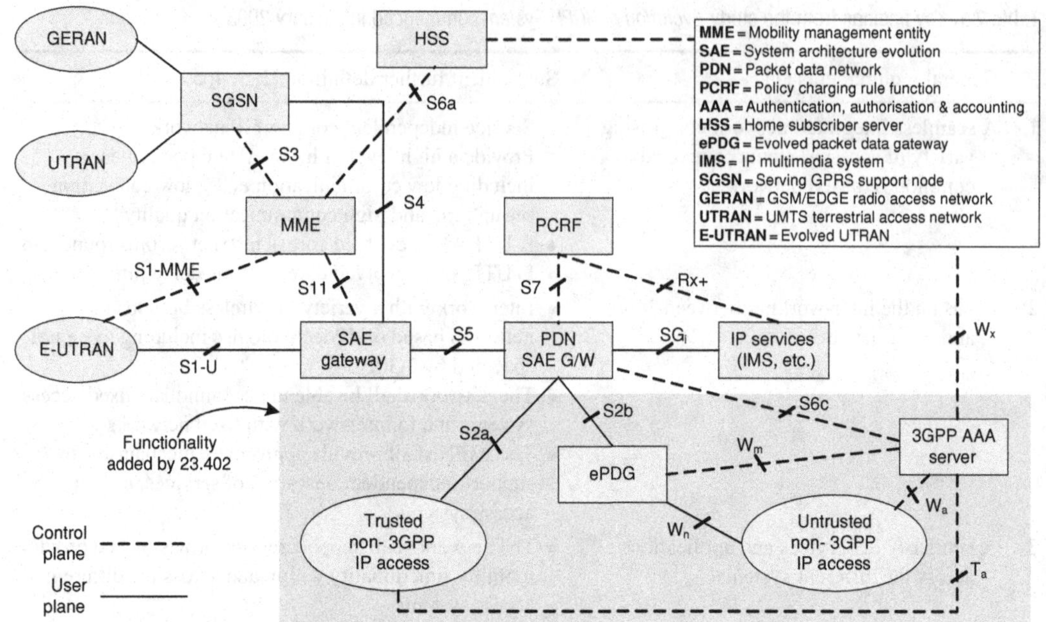

Figure 7.1 3GPP system architecture evolution according to TS 23.402

currently forecasting initial deployment of its mobile solution as early as mid 2008, some two years in advance of its 3GPP counterpart.

In writing this chapter on OFDM based solutions, it would have been possible to have documented WiMAX as the air interface specification; the initial release is complete and equipment is being deployed to this standard. Nevertheless, in keeping with the rest of the book, it is the 3GPP cellular standards solution that will be discussed, not least because the mobile aspects of 802.16e are not nearly as well documented at present as its PHY and MAC layer counterparts. WiMAX will, however, be used to illustrate the planning process.

Finally, note that at the time of writing, the 3GPP specifications for the evolved packet core (EPC) and evolved UTRAN (E-UTRAN) are very much under development and potentially subject to change. However, the physical layer for the E-UTRAN is fairly stable and the foundations of the total system architecture are substantially agreed, albeit with some operator–vendor-selectable options. The expected key characteristics of 3GPP UMTS long-term evolution (LTE) can, therefore, be established, documented and put alongside those for GSM and UMTS for consideration by new system designers.

7.1 E-UTRAN and evolved packet core architecture

7.1.1 Evolved packet core

The architecture for the EPC is shown in Figure 7.1 and illustrates the new network elements along with interfaces, generally following an '*Sn/α*' convention where *n* takes values in the range 1 to 11 and α takes values in the range *a* through *c*. Perhaps the most acute reminder that the architecture for 'Evolved UMTS' is work in progress is the fact that it is necessary to refer to two specifications for the system architecture evolution (SAE). These two documents, TS 23.401 [9] and TS 23.402 [10], reflect strongly held differences about the best way forward; these differences cut across vendors and operators alike. Document TS 23.401 is focused on ensuring efficient operation between E-UTRAN and the existing GERAN and UMTS networks. Development risk is reduced by not only retaining GTP – the GPRS tunnelling protocol introduced in Chapter 2 – for the S1-U, S3 and S4 interfaces, but also for the S5 interface and its roaming equivalent, the S8a (providing the interface to the PDN SAE gateway, PCRF and IMS elements in the home network).

In contrast, TS 23.402 seeks to meet the full requirements set out in Table 7.1 by providing interfaces to both 3GPP and non-3GPP networks (see shaded section in Figure 7.1). In this architecture, GTP is retained for the S1-U, S3 and S4 interfaces but for the S5 and its roaming equivalent the S8b interface, an internet engineering task force (IETF) protocol, *proxy mobile IP* (PMIP), is proposed.

PMIP, discussed more fully in [11], would site the PMIP *home agent* (HA) in the PDN SAE gateway (in PMIP terminology – the local mobility anchor (LMA)) and the PMIP *proxy agent* (PA) at the SAE gateway or its equivalent in the non-3GPP mobile networks. The SAE gateway or equivalent is known in PMIP parlance as the *mobile access gateway* (MAG). When a mobile first powers up, for instance in the E-UTRAN, and establishes a GTP tunnel through to the SAE gateway, the PMIP proxy agent signals the home agent and establishes a tunnel between the MAG IP address and the LMA IP address. If the mobile eventually moves out of coverage of E-UTRAN cells served by the current SAE gateway, the PA in the new serving SAE gateway signals the HA and establishes a fresh tunnel. The PMIP mobility management process is unchanged whether mobility is within SAE gateways, across SAE gateways, between a home agent in the mobile home network and a roaming SAE gateway or indeed the MAG equivalent in non-3GPP networks. They are all peer-to-peer transactions and hold the prospect of near seamless inter-network application support.

The final building blocks for mobility across heterogeneous access networks are network discovery, network selection and network negotiation prior to handover. The draft IEEE standard 802.21 [12] is being considered for this purpose and these functions would potentially be located at the PDN SAE gateway. The TS 23.402 architecture thus introduces additional risks to the programme, but holds the prospect of a much richer solution that meets the original requirements set out in *Evolution of 3GPP System*. Where relevant, the TS 23.402 architecture will be assumed for all future discussion in Chapter 7.

A common goal for network operators is to extend the life of existing network assets for as long as possible, and to avoid disruption to the existing network during upgrades; this is reflected in the architecture for the evolved 3G system. The legacy GERAN and UTRAN networks remain supported by the SGSN and, as the S4 interface is based on the G_n reference point (the open interface between the SGSN and GGSN in GPRS), can work into the legacy GGSN until near the point of network upgrade.

The serving SAE GW is the gateway that terminates the interface towards EUTRAN. For each UE associated with the SAE system, at a given point of time, there is a single serving SAE GW. Serving SAE GW functions include:

- The local mobility anchor point for inter-e node B handover,
- Mobility anchoring for inter-3GPP mobility (terminating S4 and relaying the traffic between 2G/3G system and PDN SAE GW),
- Lawful interception.

The mobility management entity (MME) is a control-plane element responsible for providing mobility support, both within E-UTRAN and between legacy networks and E-UTRAN. Key functionality supported by this element includes:

- Inter-core network node signalling for mobility between 3GPP access networks,
- Idle-mode UE tracking and reachability (e.g., paging),
- Roaming (S6a towards the home HSS),
- Authentication,
- Non-access stratum signalling (NAS – the functional layer running between the UE and the core network).

The PDN SAE GW is the SAE gateway, which terminates the SGi interface towards the packet data network (PDN). If a UE is accessing multiple PDNs, there may be more than one PDN SAE GW for that UE. PDN SAE GW functions include:

- Policy enforcement,
- Per-user based packet filtering,
- Charging support,
- Lawful interception.

Policy enforcement in the PDN SAE gateway is carried out in conjunction with the policy charging rule function via the S7 interface. The PCRF is linked to the specific application and its requirements via the Rx+ interface, and is responsible for implementing the flexible charging model of Clause (3), Table 7.1. Specifically, the PDN SAE gateway can optionally enforce the following charging models:

- Volume-based charging,
- Time-based charging,
- Volume-based and time-based charging,
- Event-based charging.

Central to the satisfactory operation of arbitrary applications over an ALL IP network is the guaranteed delivery of the required QoS associated with that arbitrary application. The QoS requirements can be established using tools such as *ARC* discussed in Chapter 3, but how are such QoS requirements to be delivered? This is still the subject of discussion in standards but it looks likely that QoS will be enforced at the level of an SAE bearer, where an SAE bearer is defined to be a logical aggregate of one or more service data flows (SDFs), running between a UE and a PDN SAE-GW. A service data flow (SDF) is the packet stream associated with the correct operation of a specific application. Service data flows mapped to the same SAE bearer receive the same bearer-level packet forwarding treatment (e.g., scheduling policy, queue-management policy, rate-shaping policy, RLC configuration). Providing different QoS to two SDFs thus requires that a separate SAE bearer is established for each SDF.

All other elements in the evolved packet core (e.g., HSS, IMS) are based on the elements introduced in earlier releases.

7.1.2 E-UTRAN

The call set-up performance of UMTS, whether Release 99, 5 or 6, is so poor that one of the six major goals for evolved UMTS is dedicated to improving this; this ignores a second goal focused on peak data rates, which is also designed to improve the user application experience dramatically. It is not surprising, therefore, that the new radio access network looks quite different from its GSM or UMTS counterparts. Analysis of the origins of the UMTS set-up delays in Section 6.6 highlighted two major components to this delay:

- Extensive message exchange between the mobile and three network elements,
- Low bandwidth control plane links relative to the size of messages exchanged.

The evolved UMTS system addresses the set-up delay issue by making two key architectural decisions:

- Offer newly authenticated users a default bearer, simultaneously with authentication,
- Move the radio and S1 bearer resource management down to the e-node B.

This is a major architectural rethink.

The default IP access service provides the basic 'always-on' IP packet-bearer service. It is expected to be used for applications that do not require any service-specific policies or charging rules and might be covered by flat rate or bundled charging. The default IP access service is described by a predetermined context in the network and established for a UE immediately after the subscriber has been authenticated and authorised by the network. The service provides the UE with IPv6 or IPv4 connectivity to operator services, other UEs, private IP networks or the Internet. The default IP access service supports mobility of the terminal. Optionally, the establishment of one or more dedicated SAE bearers may be triggered by the completion of the attach procedure. The establishment of these additional bearers is determined by operator policy and may be based on subscription information. This new architecture avoids bearer negotiation entirely for many services. This is discussed further in [7].

Figure 7.2(a) shows the distribution of the RAN and core functionality for Release 99 and Release 8 (evolved UMTS). To minimise call set-up delay, all decisions regarding radio resource management, admission control, radio bearer control and connection management are made within the e-node B platform, which will have high bandwidth messaging and data paths. All scheduling of traffic onto the physical layer is also now executed within the new node B, bringing potential capacity benefits, along with selection of the MME at UE attachment and routing of user-plane data towards the SAE gateway.

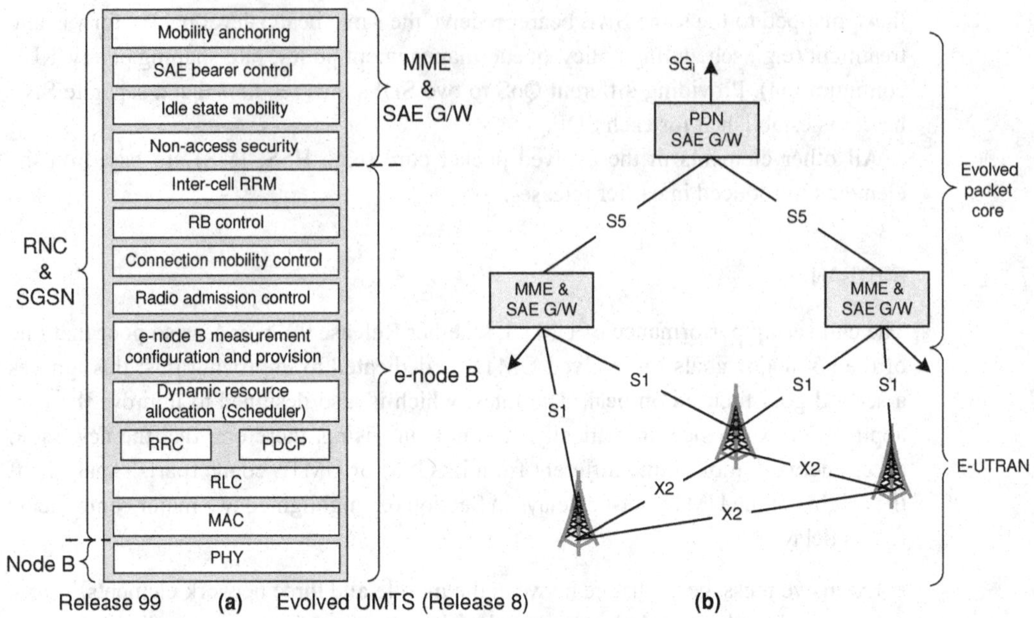

Figure 7.2 Revised functionality distribution and network architecture for the E-UTRAN

MME, the control plane entity and the SAE and PDN SAE gateways, were discussed as part of the evolved packet core architecture. However, it is worth pointing out that, as a result of the movement of so much functionality to the e-node Bs, there is very little air interface specific functionality in either of the gateway elements and thus user-plane latency for sensibly dimensioned commercial routers should easily be less than 1 ms [13]. The only significant downside of the new architecture is the need to introduce additional communication links between the node Bs. This is necessary both to ensure that individual node Bs have the visibility of adjacent-cell information required to make handover decisions previously made centrally in the RNC (loading, transport resources, etc.) and, sometimes, to execute handover. The resulting architecture is shown in Figure 7.2(b) and is discussed further in [14].

7.2 The OFDM air interface

7.2.1 Principles and benefits of OFDM

Points (1) to (4) of the key requirements listed in Table 7.1 have largely been addressed by architectural decisions in either the EPC or E-UTRAN in the previous section. The last two points (5) and (6) – peak rates of up to 100 Mbits/s and spectral efficiencies of up to 5 (bits/s)/Hz, – are entirely dependent on the radio air interface. In the introduction to this chapter on evolved UMTS, reference was made to the decision in December 2005 to go with the OFDM air interface, rather than try to evolve the existing WCDMA solution. This should not come as a great surprise. In the discussion of the UMTS

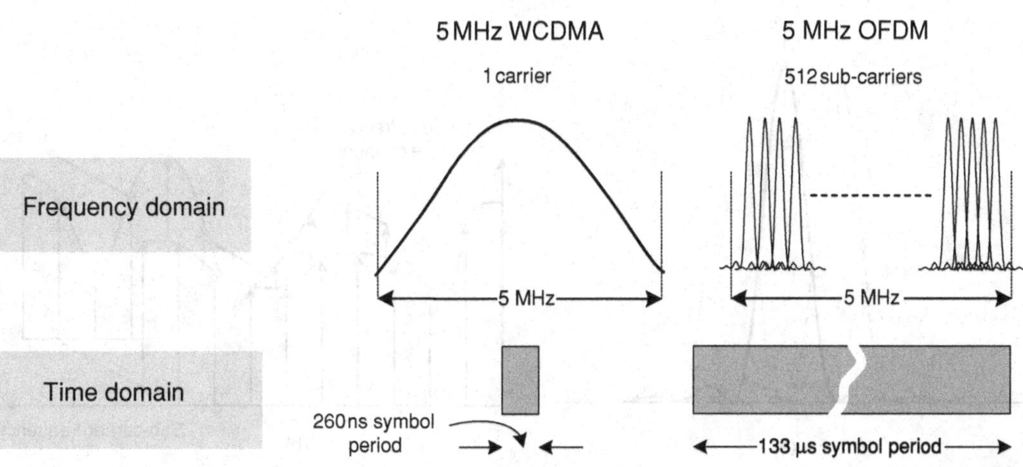

Figure 7.3 Principles of the OFDM air interface

rake receiver, with its chip rate of 3.84 Mchip/s, it was pointed out that for higher user data rates, spreading factors as low as SF = 4 have to be configured and that under these circumstances the rake receiver can no longer fully equalise the channel multi-path. This limitation is captured by Love *et al.* [15], where the performances of rake receiver and MLSE equaliser-based implementations of HSDPA handsets are compared. In scenarios where there is significant multi-path, such as 3GPP pedestrian B, the MLSE-based solution offers cell-capacity improvements of more than 30% and this is reflected in commercial implementations of HSDPA handsets where equalisers are normally used. However, this is still a long way short of the spectral efficiency of 5 bps/Hz targeted for 3GPP evolution. The performance limitation of CDMA systems is captured even more graphically in [16] where Baum *et al.* compare system-level performance of CDMA and OFDM in a 5-MHz channel. In a GSM typical urban channel, the rake receiver-based CDMA solution is never able to support more than 1 bit/symbol, even for C/I ratios of 25 dB. The corresponding figure for the OFDM solution rises to 4.5 bits/symbol and shows no signs of saturation; this disparity in performance would only get worse if higher CDMA chip rates were to be used to spread information over 10 MHz or 20 MHz channels.

The key change that makes OFDM a much better solution for high data rate wide-area mobile networks is the move from one carrier per channel to hundreds of carriers per channel. Figure 7.3 illustrates the way in which the total data transmitted on *one* carrier for WCDMA can now be shared amongst hundreds of carriers in the proposed 5 MHz E-UTRAN implementation. Thus, even if it were possible to deploy a zero spreading factor WCDMA system (i.e., the WCDMA chip period of 260 ns was taken as the data symbol duration), in the same 5 MHz channel, OFDM would utilise a symbol length of $512/2 \times 260\,\text{ns} = 66.7\,\mu\text{s}$ if half the sub-carriers were used for data. This means that the transmitted symbols are hardly affected by inter-symbol interference, substantially eliminating the BER floor seen in some WCDMA formats. There are, in fact, a number

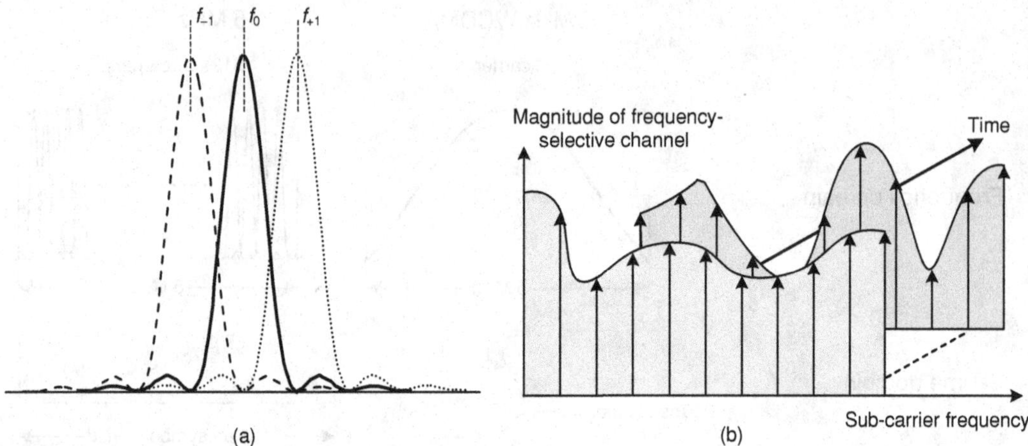

Figure 7.4 Orthogonal sub-carrier multiplexing

of other benefits accruing to OFDM, which together create higher cell capacities. These can usefully be grouped into several areas.

Perhaps the primary advantage of OFDM accrues from the orthogonality in the frequency domain, which gives the modulation its name. The concept of frequency division multiplexing has been around since the beginning of radio transmission and is central to the operation of narrow-band cellular systems, such as GSM. GSM RF channel-time slot combinations are the means of multiplexing many users within a limited spectrum resource, so what is so special about OFDM? Figure 7.4(a) illustrates the central concept of OFDM. The separation of the hundreds of sub-carriers in each OFDM channel is carefully chosen such that the transmission peak in one channel intrinsically aligns with the nulls of the adjacent channel frequency side lobes. In principle this means that no power is seen from adjacent sub-carriers; in practice, only limited adjacent sub-carrier power *is* seen. This represents a very high information-packing density whilst incurring minimal interference.

Figure 7.4(b) illustrates another key benefit of OFDM. It shows that the broadband channel exhibits frequency selective fading, which changes with time. To manage these limitations, OFDM modulation schemes not only allow the scheduler to choose when to transmit in time (as is done for instance in HSDPA) but also to choose only those *frequency* sub-carriers, at that point in time, which exhibit good propagation characteristics. In this way, the average C/I required for OFDM systems is reduced relative to a single broadband carrier. As ever, this benefit does not come for free and it is necessary to dedicate a significant number of resource elements ($=$ 1 sub-carrier $\times$ 1 symbol) to ensure that sufficiently accurate estimates of the propagation channel are available. Despite this overhead, OFDM systems deliver significant increases in spectral efficiency. In the usual way, coding is introduced to the user data prior to transmission, to afford further protection.

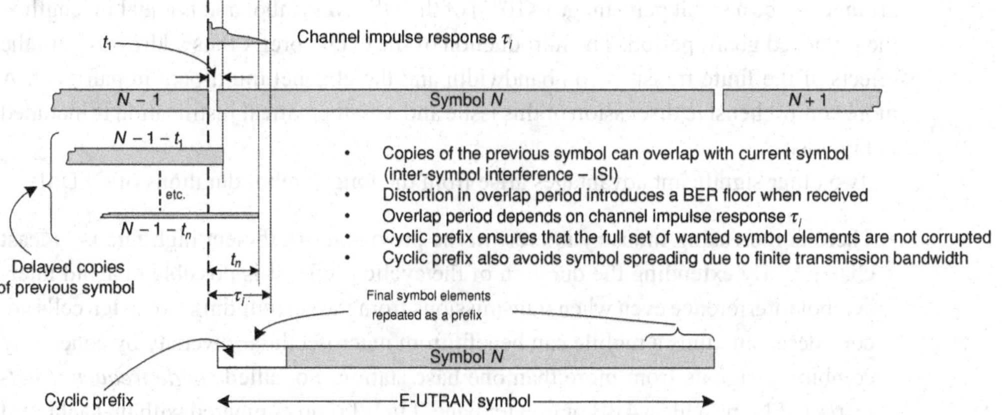

Figure 7.5 OFDM and the cyclic prefix

Other benefits that arise from the use of OFDM sub-carriers include:

- Non-frequency selective (flat) fading in narrow 15 kHz sub-carrier channels, which enables trivial and channel or data rate independent equalisation.
- The ability to schedule by sub-carrier as well as time slot means that the probability of inter-cell interference is reduced. For this reason transmit power control is optional.
- Consideration is being given to co-ordination between cells to minimise interference, e.g., avoid scheduling users that are physically close, with the same time slot and sub-carrier set.

The second group of benefits stems directly from the longer symbol durations, an artefact of modulation schemes with large numbers of sub-carriers. As illustrated in Figure 7.5, the multi-path delay spreads encountered in cellular environments will, at worst, cause only a limited level of inter-symbol interference. In practice this, too, can be avoided if a guard period is inserted between successive symbols; this addresses the residual impairments that originate in the propagation channel. However, if no other steps are taken, there will still be symbol spreading in the time domain, but this time intrinsically originating in the modulation process. Consider for a moment the way in which payload traffic is introduced to the sub-carriers. Each sub-carrier is multiplied by an arbitrary sequence of rectangular symbols representing the information to be conveyed. Rectangular symbols have an infinite bandwidth requirement, whereas they will in practice be transmitted through a band-limited channel. Consideration of the Fourier frequency–time duality makes it clear that the impact of band-limiting in the frequency domain will be 'smearing' of the time-domain symbol, in principle over an infinite period, introducing inter-symbol interference once more. Fortunately, this problem can be eliminated by repeating some of the last elements of the time-domain symbol, as a prefix to form a new composite symbol, which then appears to be periodic. This 'periodic' waveform can then be accommodated within the transmit channel. Because the channel only has a finite memory (the impulse response time), the length of the cyclic prefix is usually arranged to be a small percentage (<10%) of the OFDM symbol and is equal in length to the proposed guard period. The introduction of the cyclic prefix thus addresses both the effects of the finite transmission bandwidth and the channel multi-path impairment. A more comprehensive discussion of this issue and a mathematical justification is included in [17, 18].

Two other significant advantages arise from the long symbol durations of OFDM:

- There is increasing industry interest in the provision of efficient high rate broadcast channels. By extending the duration of the cyclic prefix, it is possible to avoid inter-symbol interference even when transmissions from the surrounding five to ten cells are considered and thus a mobile can benefit from macro-cellular diversity by coherently combining signals from more than one base station. So-called *single frequency networks* (SFN) provide a 4 dB or greater benefit in S/I ratio compared with that achieved by a mobile served by a single site [19]. This translates directly into increased cell capacity even after accounting for the loss of capacity introduced by the longer cyclic prefix.
- The long symbol duration means that it is possible to time align uplink as well as downlink transmissions so that any overlap is guaranteed to be addressed by the cyclic prefix period. Orthogonality is thus preserved fully for both the uplink and downlink, substantially eliminating other user interference, with the corresponding capacity gains.

The remaining advantages of OFDM arise from the simplicity of implementation. In GSM and UMTS (HSDPA), equalisation must be included to remove the effects of channel multi-path with delay spreads much greater than the symbol duration. The complexity of these equalisers is dependent on the multi-path environment and effectively on the data rate, as the symbol period eventually has to reduce to realise the higher throughput. For OFDM, the approach is more straightforward. More sub-carriers are added to exploit the additional channel bandwidth and the symbol rate per sub-carrier does not change. Additionally, because propagation in the OFDM sub-carriers is narrowband, as already discussed, only a single tap equaliser is needed to correct for the flat phase and amplitude error determined using the training sequence. Equalisation occurs in the frequency domain, after the FFT in the receiver.

The ability to produce straightforward variants of the product for different channel bandwidths simply by altering the number of OFDM sub-carriers has a second system level benefit; the basic architecture remains unchanged as the channel bandwidth is varied over the planned range of 1.4 MHz to 20 MHz. These benefits, combined with the availability of efficient implementations of the FFT and IFFT processes, make OFDM attractive to realise in both handsets and infrastructure.

There is no solution yet invented that does not have at least some intrinsic disadvantages and the OFDM air interface is no exception. The primary issues to be addressed are:

- Frequency synchronisation with BTS / mobile,
- Time synchronisation with BTS / mobile,
- Channel estimates to enable link adaptation,
- More linear power amplifiers.

These problems are clearly not unique to OFDM but the level of precision required, particularly for frequency synchronisation if full orthogonality is to be maintained, is significantly greater.

Frequency synchronisation is a much more important issue for OFDM than for prior air-interface solutions. Even small frequency shifts (due to oscillator drift or Doppler shift) will cause the frequency domain orthogonality to be severely degraded with the corresponding adverse impacts on performance. This is addressed by the transmission of synchronisation information and the choice of 15 kHz sub-carrier spacing; in combination, this is sufficient to avoid significant degradation at speeds up to 250 km/hour and at frequencies up to 2.5 GHz.

In the case of time synchronisation, the much larger symbol length generally makes this problem easier, although the system now needs to send time-alignment information to each mobile to ensure that signals are orthogonal in the uplink as well as the downlink. A further issue is a problem unique to broadband OFDM systems: the need to obtain accurate C/I estimates for hundreds of sub-carrier channels. It turns out that FER curves of instantaneous channel conditions are nearly parallel, so that if one point of the instantaneous FER curve is found, then the entire FER curve can be obtained by shifting the static FER curve to match the required point [20]. Reference *resource elements* are included for channel estimation purposes in more than one symbol per time slot and across several sub-carriers. Finally, as will be discussed in the next section, the requirement for more linear power amplifiers to cope with the intrinsically higher peak-to-mean ratio of OFDM systems needs to be addressed.

7.2.2 OFDM implementation

Figure 7.6 shows a typical implementation of an OFDM transmitter chain. It is assumed that the user information bit stream shown on the left of this diagram has been subject to coding, interleaving and rate matching, in a process similar to that already described for UMTS. Sets of x bits are collected and mapped to the appropriate constellation for the x-ary modulation scheme. For the E-UTRAN, QPSK, 16-QAM and 64QAM are the supported schemes and one of these will be selected based on the C/I in the scheduled sub-carriers. Note that the modulation scheme will vary dynamically, tracking changes in C/I level.

In principle, blocks of N, x-bit symbols are accumulated to form inputs S_1 to S_N of the inverse fast Fourier transform (IFFT). Note that, in accumulating the N samples, the period of the modulating waveform has now increased to N times that of the x-bit input symbol; this reflects the reduced data rate on each of N sub-carriers. For every input block of N symbols, the IFFT will deliver N time samples of the output OFDM symbol. The composite OFDM symbol is formed by copying the last Q samples of each block and inserting these prior to the first sample from the IFFT to form the cyclic prefix. Following digital to analogue conversion and frequency shifting to the appropriate carrier frequency, the composite symbol stream is available for transmission. Demodulation in the receiver essentially implements the inverse of the transmit processes but, additionally, a single tap equaliser stage follows the FFT.

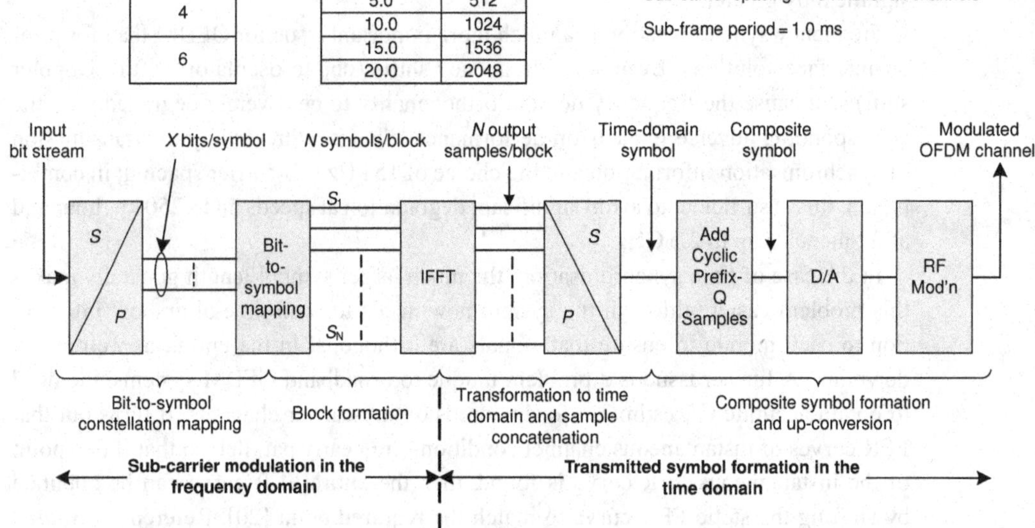

Figure 7.6 Expected E-UTRAN OFDM downlink implementation

In practice, as in other sampled systems, care must be taken in OFDM to ensure that the bandwidth of the sampled signal does not give rise to overlapping spectral components, which cannot be filtered out, thus destroying signal fidelity. Therefore, the outermost inputs to the IFFT are not used. It turns out that only about 65% of the IFFT inputs are used, the others being permanently set to zero. To be clear, this should be viewed as an oversizing of the FFT and does not reflect on the efficiency of the air interface.

Figure 7.7 illustrates some of the ways in which users share the downlink resource. The node B scheduler dynamically controls which time or frequency resources are allocated to a certain user at a given time. Downlink control signalling informs UEs what resources and respective transmission formats have been allocated. The scheduler can instantaneously choose the best multiplexing strategy from the available methods; e.g., frequency localised (adjacent sub-carriers) or frequency-distributed transmission. The flexibility in selecting resource blocks and multiplexing users will influence the available scheduling performance. Scheduling is tightly integrated with link adaptation and HARQ, which employs similar principles to the approach introduced in HSDPA. The decision on which user transmissions to multiplex within a given sub-frame will be influenced by:

- Required QoS performance,
- The number of blocks buffered in the node B, ready for scheduling,
- Pending retransmissions,
- Channel quality indicator reports from the UEs,
- UE capabilities.

Table 7.2. E-UTRAN – key system parameters vs. channel bandwidth

		Channel bandwidth (MHz)					
		1.4	3	5	10	15	20
Sub-frame duration		1 ms					
Sub-carrier spacing		15 kHz					
Sampling frequency (MHz)		1.92	3.84	7.68	15.36	23.04	30.72
IFFT size (N)		128	256	512	1024	1536	2048
Occupied sub-carriers (DL)		73	181	301	601	901	1201
Number of OFDM symbols per sub-frame		7/6 for short or long cyclic prefix respectively					
Cyclic prefix length (μs)	Short	4.69 μs					
	Long	16.67 μs					

The basic *resource block* comprises a number of sub-carriers and OFDM symbols. In principle (and indeed as shown in Figure 7.7) each of these sub-carriers is a contiguous sequence. In practice, particularly for slow-moving mobiles, there is advantage in dispersing these sub-carriers across the whole sub-carrier set to gain the benefits of frequency diversity. When this configuration is chosen, the resource block is known as a *virtual resource block.*

The exact number of sub-carriers depends on the precise transmit format in use and hence the number of resource blocks available to be shared with one or more users is also variable. Some of the configurations for the IFFT size and number of active sub-carriers are shown in Table 7.2 [5]. A better understanding of the relationship between these parameters can perhaps be obtained by the way in which the overall sub-frame of 1 ms is established below for the 5 MHz configuration:

- The sampling period is 1/7.68 MHz = 130 ns,
- The IFFT set is accumulated over 512 samples = 66.7 μs,

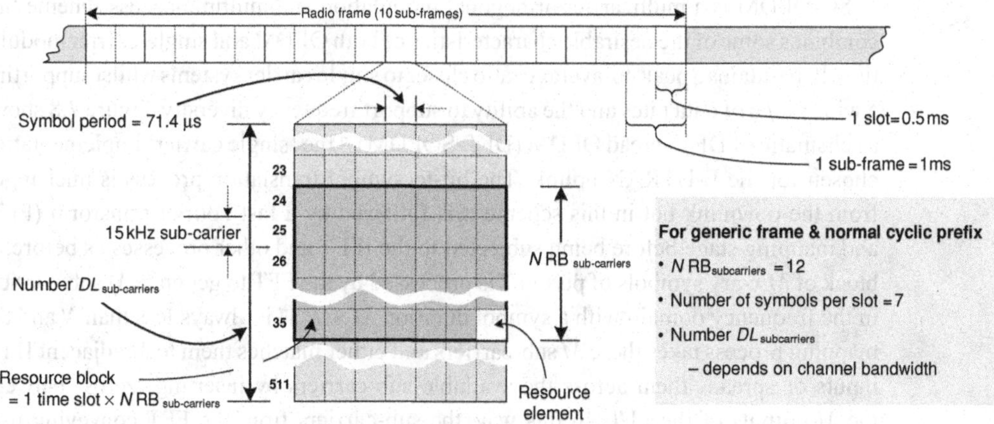

Figure 7.7 Downlink transmission format

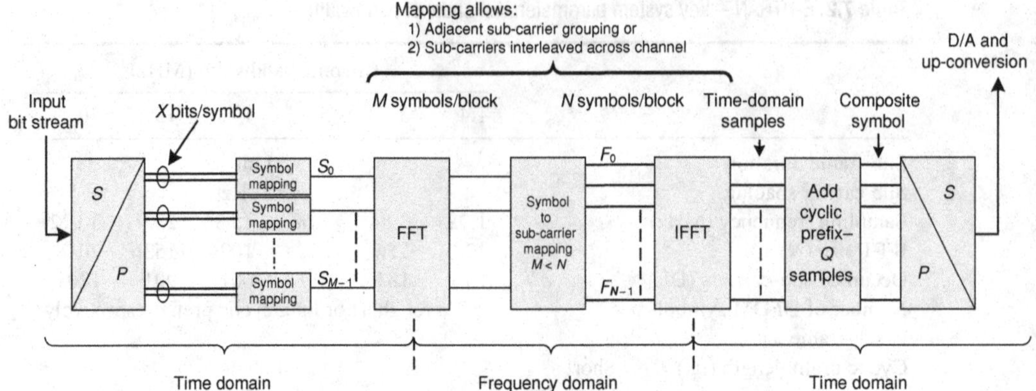

Figure 7.8 A representative DFT SOFDM implementation

- Adding the short cyclic prefix of 4.69 μs gives a composite symbol period of 71.4 μs,
- For deployments using the short cyclic prefix a slot comprises seven symbols = 500 μs,
- Two slots give a sub-frame period of 1 ms.

On the uplink, an OFDM-based transmission scheme is again used and supports QPSK and 16QAM modulation schemes. However, one further characteristic of conventional OFDM becomes a significant disadvantage for mobile handsets: the high peak-to-average power ratio (PAPR) compared with single carrier solutions. The peak-to-mean ratio for unmodified OFDM is more than 8 dB if the ratio is to be exceeded less than 0.1% of the time [21]. If normal OFDM were to be used for the mobile terminal, for a given amplifier peak power capability, the average output power from the mobile would be significantly reduced, with corresponding adverse impacts on cell range. For this reason, a variant of OFDM that has come to be known as 'single carrier' OFDM (SC-OFDM), which reduces the PAPR by almost 3 dB (relative to OFDM) for the QPSK case, has been adopted [22, 23].

SC-OFDM is a multicarrier orthogonal modulation and multiple-access scheme that combines some of the desirable characteristics of both OFDM and single-carrier modulation. It maintains a peak-to-average ratio closer to single carrier systems whilst supporting a wide range of data rates and the ability to support frequency diversity. Figure 7.8 shows a schematic of DFT spread OFDM (DFT-SOFDM) – the 'single carrier' implementation chosen for the E-UTRAN uplink. The bit-to-symbol translation process is unchanged from the downlink but in this scheme it is followed by a fast Fourier transform (FFT) and mapping stage before being subjected to the IFFT and other processes as before. A block of M x-ary symbols of period T is processed by the FFT to generate M sub-carriers in the frequency domain with a symbol duration $M \times T$. M is always less than N and the mapping process takes these M sub-carriers and either matches them to M adjacent IFFT inputs or spreads them across the available sub-carriers by inserting 'zeros' between the M outputs of the FFT. In this way, the sub-carriers from the FFT conveying user data are either 'bunched together' in one part of the channel or interleaved across that spectrum, which allows the benefit of frequency diversity to be gained, if this is desirable. Depending upon the service to be supported, M can take values up to the number of sub-carriers available for user data and N is equal to the IFFT size indicated in Table 7.2, and depends upon the operating channel bandwidth. To complete this very brief overview of DFT-SOFDM, it is perhaps interesting to consider what happens if M *is* set equal to N and the mapping process is removed. By definition, the output of the IFFT *must* be the original x-ary symbol stream present at the input to the FFT; i.e., this symbol stream would directly modulate the single carrier – hence the term 'single-carrier' OFDM. The FFT and symbol mapping can thus be viewed as a phase and amplitude 'pre-coding' which reduces the peak-to-mean ratio.

It is appropriate at this point to highlight that support of MIMO (multiple-input–multiple-output) antenna techniques is an integral part of the E-UTRAN solution. On the downlink, E-UTRAN supports transmission of user data streams on up to four separate antennas. A separate identical resource grid of the type shown in Figure 7.7 is available for each antenna supporting user data transmission. The system is specified for up to four antennas at the e-node B and either one, two or four receive antennas at the mobile, depending on the mobile category. However, current mobiles only utilise one transmit channel regardless of category [14, 24]. Under suitable propagation conditions, it is the ability to transmit multiple data streams of user information that make possible some of the improvements in spectral efficiency and cell capacity.

7.2.3 E-UTRAN – expected performance

In estimating the coverage and capacity of any broadband system it is necessary first to consider whether the modulation scheme proposed is capable of supporting the required peak data rates at the link level. Discussion in [16] highlighted that, even in GSM typical urban propagation conditions, OFDM is capable of supporting well in excess of 4.5 bits/symbol. In fact, even after consideration of amplifier phase and amplitude distortions, it was decided that the E-UTRAN would be capable of supporting 6 bits/symbol through a 64QAM modulation scheme in the downlink. In the uplink, largely driven by realisation considerations and the maximum cell range possible with the small and finite mobile power, 16QAM was selected as the highest-rate modulation scheme. Then for the downlink, it is possible to estimate the maximum supported (peak) data rate as follows:

$$\text{peak rate} = \text{symbol rate} \times \text{bits/symbol} \times \text{number of active sub-carriers} \times (1-\gamma),$$

where γ is an overhead factor for synchronisation, channel estimation, etc.,

$$= 14 \times 10^3 \times 6 \times 1200 \times (1-\gamma)$$

$$= \text{maximum rates in the range 60 Mbits/s to 70 Mbits/s for a 30\% to 40\% overhead.}$$

Similar calculations for the uplink give peak rates in the range 40 Mbits/s to 50 Mbits/s. Given that the modulation schemes selected can support the link data rates above, a simulation framework needs to be established that captures all of the benefits and limitations of the new system.

Table 7.3. Simulated cell capacities and user rates for E-UTRAN

		Mbits/s		Spectrum efficiency
		HSPA (5 MHz)	E-UTRAN (20 MHz)	Multiplier for E-UTRAN
Downlink	Capacity	4	40	2.5
	Average rate	0.4	3.8	2.4
	Peak rate	14.4	60[a]	–
Uplink	Capacity	1.5	14	2.3
	Average rate	0.15	1.4	2.3
	Peak rate	5.74	40[b]	–

[a] Instrumented rate = 14 ksymbols/second × 6 bits/symbol × 1200 sub-carriers × 0.6 (to reflect overhead) ~ 60 Mbits/s
[b] Instrumented rate as for downlink but using 4 bits/symbol gives ~ 40 Mbits/s

The cell throughput will depend upon the C/I encountered by mobiles at their locations and the mapping of FER for particular modulation schemes to C/I. This latter activity presents particular challenges for OFDM in a broadband frequency-selective channel because of the large number of sub-carriers for which the C/I needs to be established prior to the mapping process. Blankenship *et al.* [20] proposed a method of predicting the FER as a function of C/I, for a given modulation rate, coding scheme and propagation channel. A link-level simulation based on this approach will thus enable the benefits of flat fading in sub-carriers and fast frame-by-frame scheduling across those sub-carriers to be captured. Classon *et al.* [25] and Sun *et al.* [26] quantify the benefits of time–frequency scheduling to be in the range 20% to 30%, compared with time scheduling alone and show that only low rates of feedback are necessary to provide performance close to that with ideal feedback.

The other benefits of OFDM, such as uplink and downlink time and frequency orthogonality, inter-cell interference co-ordination, etc., will only be captured by system-level simulations. Similar techniques to those already discussed for UMTS are used but with link-level information, which reflects the frame-by-frame time–frequency scheduling already discussed. Classon *et al.* [27] have carried out a comprehensive system-level performance simulation using the current E-UTRAN parameter working assumptions. Results in all cases are for the full buffer case but reflect different cell sizes and user speeds. A representative view of the uplink and downlink performance detailed in this paper is provided in Table 7.3.

Both the peak user rate and spectral efficiency numbers for the downlink and the uplink remain less than the goals targeted for E-UTRAN. The final piece of the solution is to consider the benefits of MIMO in conjunction with the OFDM air interface. The performance improvements resulting from the use of MIMO systems can arise from array gain, diversity gain, spatial multiplexing gain, interference reduction or a combination of these factors. In a typical cellular environment, where there is significant scattering, the signal propagates from the transmitter to the receiver along a number of different paths, collectively referenced as multi-path, and spatial multiplexing can be exploited. If

Table 7.4. E-UTRAN performance comparison

	Metric	Performance			Comments on E-UTRAN
		GSM	UMTS	E-UTRAN	
1	Low access latency (push to happen)	700 ms to 2000 ms	~3100 ms	<100 ms	From registered but idle state. Dramatic reduction through (1) consolidation of most functions in the e-node B and (2) allocation of default bearer on authentication
2	Peak data rates (push to happen)	~200 kbits/s	~2000 kbits/s	>100 Mbits/s	Assumes deployment of 2×2 MIMO
3	Guaranteed QoS support to enable new applications	Worst	Middle	Best	The combination of 100 ms access time and lower user plane delay mean that true QoS can be efficiently delivered for packet services
4	Enriched HLR functionality – support of new features	Middle	Middle	Best	This is enabled by IMS and its successive releases. The PCRF and Gx^+ interfaces enable a much more flexible charging regime. The low-latency wireless access makes IP based services truly viable
5	Low-cost configurations – developing market access	Best	Worst	Middle	It is likely that the E-UTRAN will remain at a disadvantage to GSM until product volumes become comparable. However, in due course, the flexible OFDM air interface will allow the efficient support of voice users at large cell radii
6	Self-configuring network	Worst	Middle	Best	A self-configuration process is defined in 3GPP TS 36.300. The reality remains to be seen
7	Self-optimising network	Worst	Worst	Best	A self-optimisation process is defined in 3GPP TS 36.300. The reality remains to be seen
8	Common core network for all applications	Worst	Middle	Best	The 3GPP 23.402 specification defines an evolved packet core which can be common to wired and wireless interfaces and provide one interface to IMS based common services
9	Convection air-cooled BTS	Middle	Middle	Middle	Pico, micro, mini cells only. Macrocells invariably require fans or other forced-air approaches
10	Common application support platform	Worst	Middle	Best	See (3) and (8)

(*cont.*)

Table 7.4. *(cont.)*

	Metric	Performance			Comments on E-UTRAN
		GSM	UMTS	E-UTRAN	
11	Reduced back-haul costs	Worst	Middle	Middle	Efficient packet back-haul is supported in E-UTRAN. It is not yet clear whether the reduced 'S1' transport costs will offset the need to support logical 'X2' links
12	Higher cell capacities	Middle	Middle	Best	The intrinsic costs of a base station are increasingly centred on the RF power amplifier and other high-power elements. As the output powers are essentially common for all three solutions the cost per user drops with cell capacity, being the lowest for LTE
13	Reuse of currently deployed equipment	N/A	Best	Best	All current GSM/UMTS equipment is retained (with the exception of the GGSN), and IMS equipment is also reused
14	Ability to reuse existing spectrum	Best	Worst	Middle	E-UTRAN is able to operate at channel bandwidths down to 1.4 MHz, making it deployable in the narrowest GSM spectrum allocations. Second only to GSM in this respect
15	Broadband data handover	N/A	N/A	Best	The 3GPP TS 23.402 specification defines an EPC which is designed to support seamless transfer of services across arbitrary access solutions
16	Revised standards IPR policy	N/A	N/A	Best	The GSM IPR policy has been reviewed within ETSI, driven by an operator interest group (NGMN). The updates leave the IPR policy essentially unchanged but make IPR ownership more visible

there are *N* transmit–receive antenna pairs, under these conditions, MIMO is normally configured to take advantage of the multi-path by:

(a) Converting the input data stream into *N* parallel transmitted streams, or
(b) Transmitting duplicate (possibly time-coded) versions of the data across each antenna pair.

Configuration (a) clearly has the potential to increase the peak data rate and system capacity, whilst configuration (b) provides redundancy to reduce the BER or increase the cell size for the same BER. The current E-UTRAN specifications provide support for up to two transmit–receive antennas on the mobile and up to four transmit–receive antennas at the e-node B. The locations where cell capacity and peak data rates are important tend to be in city areas, where cell sizes are small. Under these conditions, the link budget should allow configuration (a) to be deployed, which in turn allows peak rate improvements of up to '× 2' (for a 2 × 2 configuration) with an attendant increase in cell capacity dependent on mobile locations. Measured data in suburban areas for this configuration [28, 29] indicate that the average capacity multiplier is around 1.7 times. Further improvement should be possible with the larger e-node B array.

The *wireless network metrics* established at the end of Chapter 2 have been completed for the E-UTRAN and added to the ratings for GSM and UMTS. The comparison is included as Table 7.4.

7.3 OFDM RAN planning example

As mentioned earlier in this chapter, it was decided to describe the OFDM approach in terms of the 3GPP solution, currently under development, to be consistent with the description of other cellular solutions in this book, but nevertheless to provide a worked example based on the WiMAX solution, as this is already commercially deployed. The term WiMAX is commonly used as a name for the family of IEEE standards produced by the 802.16 working group, from late 1999 onwards (for a detailed genealogy of the 802.16 family of standards, consult [30]). For the purposes of RAN design, the discussion here will be focused on the 802.16e standard, which was finalised in late 2005. Earlier versions, such as 802.16a and 802.16d, were primarily aimed at static wireless broadband applications, whereas 802.16e extended the scope of WiMAX to address mobility, whilst still embracing the needs of static users. Note that the emphasis in this example is on how radio resources can best be applied to convey a mix of applications.

7.3.1 WiMAX air interface features

This sub-section aims to provide the reader with sufficient background in WiMAX to obtain a full appreciation of the worked example, which is the subject of this section.

Table 7.5 makes it clear that 802.16e allows for a very wide variety of design choices which provides flexibility in meeting the requirements of most deployment scenarios. The most important parameters of the WiMAX RAN solution are described in the next sub-sections.

7.3.1.1 Operating frequencies

The 802.16e standard addresses frequencies between 2 and 11 GHz. Frequencies in this range may be licensed or unlicensed. Licensed spectrum may restrict access only to those operators who have paid a license fee (and may in some cases require the use of specific radio-access technologies). Unlicensed spectrum allows any operator to use the frequency, subject to locally defined maximum transmission guidelines, etc. The matrix

Table 7.5. Summary of WiMAX air interface features

	Benefits			
Feature	Capacity	Coverage	QoS	Deployment flexibility
Operating frequency choices	✓	✓		✔
Flexible channel bandwidth	✓	✓		✔
TDD or FDD duplexing				✔
OFDM sub-channelisation	✔	✓		✓
Adaptive modulation and coding	✔			
Variable cyclic prefix		✓		✔
Fractional reuse	✔	✓	✓	✓
ARQ and hybrid ARQ	✔	✔		✓
Integral MIMO	✔	✔		✓
Quality of service support			✔	

(✔) Primary benefit (✓) Secondary benefit

of frequencies available in particular countries and the particular conditions that apply are complicated and subject to change. The current deployments of WiMAX infrastructure are largely in a small number of frequency bands, centred around 2.3, 2.5 and 3.5 GHz.

7.3.1.2 Channel bandwidths

In addition to the carrier frequency, 802.16e also defines a range of bandwidths for the radio communications channel, for instance 5 MHz, 7 MHz and 10 MHz. Generally, a higher channel bandwidth will provide higher peak rates and capacities.

7.3.1.3 Time- or frequency-duplexing

Time-division duplexing (TDD) is the duplex mode for current WiMAX deployments but the FDD (frequency-division duplexing) mode is intended to be standardised in the future to give a choice for later deployments. The TDD mode allows the access point (AP) and subscriber station (SS) to share the same frequency for both link directions by allocating different time periods to uplink and downlink users within one radio frame. Various uplink–downlink allocations are defined. The FDD mode requires separate frequencies to be used in each direction but makes full-duplex communications possible. The TDD mode offers benefits in terms of reduced equipment complexity, and the ability to support the asymmetric uplink-to-downlink bearer rates required in some of the newer applications. The FDD mode, on the other hand, is well matched to voice and other largely symmetric applications but has the disadvantage that paired uplink and downlink frequency channels must be found.

7.3.1.4 OFDM/OFDMA sub-channelisation

The general arrangement of the WiMAX frame structure is shown in Figure 7.9. The vertical axis is frequency and comprises a number of *sub-channels*, which in turn are each made up of groups of OFDM *sub-carriers*. The horizontal axis is time. The major

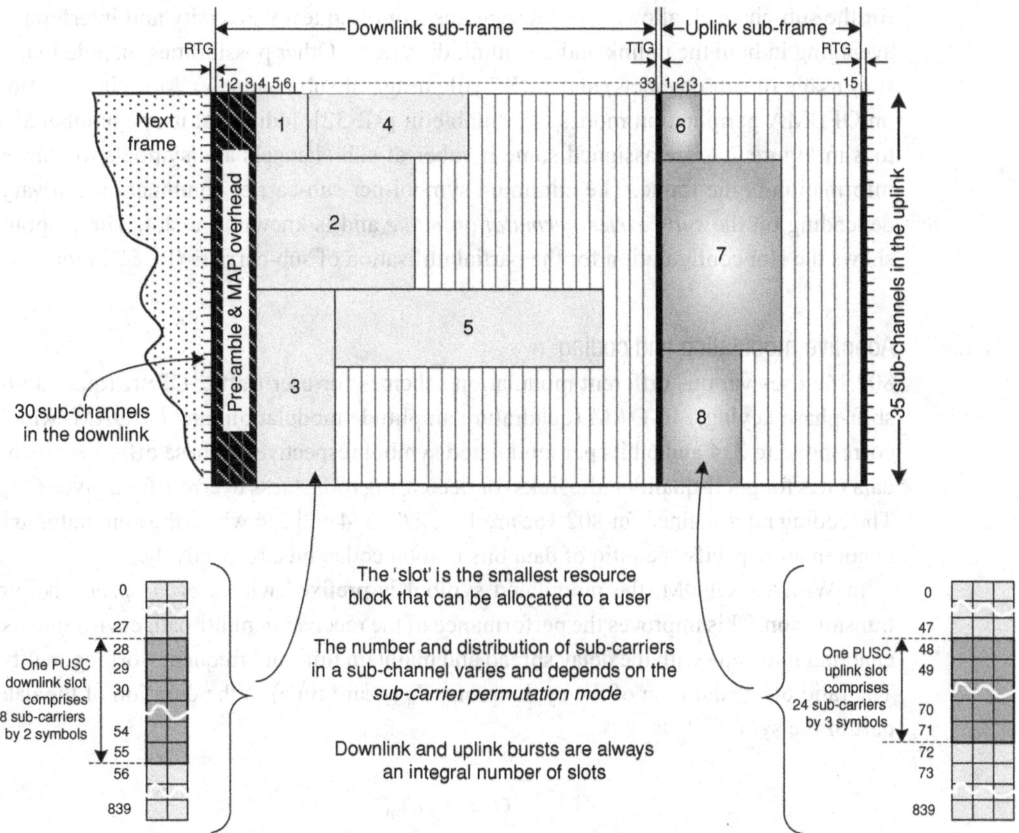

Figure 7.9 Example WiMAX frame structure

granularity here is the *radio frame*, which comprises 48 symbols shared between *downlink* and *uplink sub-frames*; for all initial WiMAX versions the radio frame period is 5 ms. Note that there are short gaps (a fraction of a symbol) between the end of transmission and the start of the reception period and also between the end of reception prior to the start of the next transmission. These gaps, known as the transmit and receive transition gaps (TTG and RTG), respectively, allow the sensitive receiver to be protected from the high-power transmissions (which are of course at the same frequency).

The number of orthogonal sub-carriers in the OFDM signal varies depending on channel bandwidth, and examples include 512 sub-carriers for 5 MHz bandwidth and 1024 sub-carriers for 7 or 10 MHz bandwidth. Note that not all of these sub-carriers are available for user data, as some are needed as guard and pilot sub-carriers.

For multiple-access OFDM systems, users are multiplexed onto the air interface using a technique called OFDMA (orthogonal frequency division multiple access). Different users transmit and receive on distinct groups of sub-carriers, which are grouped together to form *sub-channels*. Various allocation schemes have been defined in 802.16e, which determine the permutations of sub-carrier allocation to sub-channels. Partial utilisation of sub-channels (PUSC) is one such method, which pseudorandomly selects sub-carriers for the sub-channel, allowing capacity gains from frequency diversity and interference averaging in both the uplink and downlink directions. Other possibilities include FUSC (full usage of sub-channels) and TUSC (tile usage of sub-channels). More information on OFDMA permutation modes is available in [31, 32]. Individual users, numbered 1 to 8 in Figure 7.9, are assigned some number of sub-channels and symbols to convey information in the frame. The minimum symbol-per-sub-carrier combination can vary, depending on the *sub-carrier permutation mode* and is known as a slot. The diagram shows the slot configuration for the partial utilisation of sub-carriers (PUSC) mode.

7.3.1.5 Adaptive modulation and coding

802.16e uses various different modulation schemes for user traffic: QPSK (quadrature shift-phase keying), 16-QAM (quadrature amplitude modulation) and 64-QAM, which correspond to 2, 4 and 6 bits per modulated symbol respectively. These offer increasing data rates for good-quality radio links (or decreasing robustness to errors for a given C/I). The coding rates defined for 802.16e are 1/2, 2/3, 3/4 and 5/6 where the numerator and denominator specify the ratio of data bits to total coded bits, respectively.

In WiMAX OFDM, the modulated symbol is prefixed with a *cyclic prefix* before transmission. This improves the performance of the receiver in multi-path environments, enabling it to cope with the delay spread and maintain time and frequency orthogonality. The ratio of the duration of the cyclic prefix T_g (guard time) to the duration of the data part of the symbol T_d is:

$$G = T_g/T_d.$$

In radio environments with extensive multi-path, the value of G needs to be higher than is the case for deployments with a limited delay spread. In contrast to LTE, the 802.16 PHY layer makes provision for four different values of G (1/4, 1/8, 1/16 and 1/32) – although only 1/8 is mandatory in the first release. 802.16e also specifies support for convolutional code (CC) and convolutional turbo-code (CTC) coding schemes, for error correction. These options allow a wide variety of physical data rates to be provided for optimum support in different radio environments. Example data rates for various modulations, code rate, coding schemes and channel bandwidths (for the case of PUSC sub-channelisation) are given in Table 7.6.

Adaptive modulation and coding (AMC) is the mechanism in 802.16e whereby the channel modulation and coding are adjusted to deliver acceptable bit-error rates according to radio-channel conditions (this is distinct from the AMC permutation mode). The user terminals provide channel quality indicator (CQI) information, such as carrier-to-interference (CINR) values over a dedicated signalling channel (CQICH) to the access point, which can then use this information to perform rate adaptation in real time.

For system design purposes it is necessary to estimate the minimum C/I or C/N values that are required for the different modulation schemes to be used. This will vary in practical deployments depending on the physical environment and the nature of interferers, and will also be heavily dependent on equipment sensitivities and the use of

Table 7.6. Example WiMAX uplink and downlink data rates for a 60:40 DL-to-UL ratio

Modulation	Coding	Normalised C/N (dB)	5 MHz channel bandwidth DL rate (Mbps)	5 MHz channel bandwidth UL rate (Mbps)	10 MHz channel bandwidth DL rate (Mbps)	10 MHz channel bandwidth UL rate (Mbps)
QPSK	½ CTC	6	3.17	2.29	6.33	4.71
	2/3 CTC	7.5	4.22	3.05	8.44	6.28
	¾ CTC	9	4.75	3.43	9.50	7.06
16-QAM	½ CTC	12	6.34	4.57	12.67	9.41
	2/3 CTC	14.5	8.45	6.09	16.89	12.55
	¾ CTC	15	9.50	6.85	19.01	14.11
	5/6 CTC	17.5	10.56	7.61	21.12	15.68
64-QAM	½ CTC	18	9.50	6.85	19.01	14.11
	2/3 CTC	20	12.67	9.14	25.34	18.82
	¾ CTC	21	14.26	10.28	28.51	21.17
	5/6 CTC	23	15.84	11.42	31.68	23.52

advanced features. However, some guidelines have been published in 802.16e (in support of the requirement for power control procedures) and these are also shown in Table 7.6.

7.3.1.6 Sectorisation and fractional reuse

802.16e, in principle, supports a frequency reuse of one, whereby the same 5 or 10 MHz channel is used at all sites, on all sectors. This promises greater spectral efficiency, enabling more users to be supported on the system, but has potential problems at the boundaries of cells, where co-channel interference effects can sometimes degrade user performance at these locations. Sub-channelisation mitigates this problem, for example when using PUSC, as not all sub-carriers are used by the sub-channels, and the pseudo-random allocation of sub-carriers to sub-channels means that the probability of frequency clashes is reduced. Furthermore, the number of sub-channels used at the periphery of cells can be reduced by performing sub-channel segmentation: a segment is defined as a sub-set of the available sub-channels. Systems planned in this way are said to employ fractional frequency reuse, since a smaller percentage of the sub-channels is in use at cell boundaries to minimise interference, whilst users near the base site can use all available channels with consequent throughput and capacity gains.

7.3.1.7 ARQ and hybrid ARQ

Automatic repeat request (ARQ) is a mechanism used by the MAC to request retransmission of an erroneously received block. Whenever a block is correctly received, the receiver sends an ACK (acknowledgement) message back to the transmitter, providing a robust data-link layer. In common with other packet-based protocols, a sliding window scheme can be employed to improve throughput, whereby a number of blocks can be sent without waiting for an ACK, and then the receiver can send a 'cumulative ACK' to acknowledge all the blocks so far sent. Selective ACK allows the receiver to indicate which blocks in a sequence have been correctly received, and request retransmission of those which were in error. Acknowledgement messages can be sent over a management connection, or piggybacked on an existing connection between the receiver and transmitter. Note that a connection can choose to use ARQ or not, by negotiation at session creation; once established, the connection must continue to carry either ARQ or non-ARQ traffic but not both.

Hybrid ARQ (HARQ) is a mechanism implemented in the PHY for retransmission of blocks that have been incorrectly received. In the simplest case of HARQ, *chase combining*, the PHY retransmits the errored block, and the receiver uses the newly received block and also the stored errored-block, in an attempt to decode the block correctly. A more sophisticated form of HARQ, *incremental redundancy* (IR), requires the transmitter to re-encode the block received in error and transmit up to four versions to the receiver for subsequent decoding. Incremental redundancy is significantly more complicated and costly to implement in the radio network and hence chase combining is more commonly used. HARQ can improve the resilience of the PHY against interference but its use will introduce latency into the transmission path over the air interface, because of the need to occasionally retransmit blocks. This ability to operate at reduced C/I may be exploited either to increase cell capacity or cell radius. Typical implementations will allow HARQ to be disabled for applications with low latency requirements, although the link budget will have to be enhanced accordingly.

7.3.1.8 Advanced antenna technologies

The 802.16e standard includes techniques for improving signal quality or enhancing capacity through the use of *multiple-input–multiple-output* (MIMO) antenna technology. The benefits are as briefly outlined in Section 7.2.3.

7.3.1.9 Quality of service (QoS) support

One distinct difference between LTE and WiMAX is that WiMAX 802.16e defines five distinct categories for QoS management, most of which have been derived from DOCSIS standards for cable broadband [33]. These service classes are as follows:

- *Unsolicited grant service (UGS)* This is designed for constant bit-rate applications where there is a need for a fixed bandwidth allocation to deliver fixed-size packets. The most common example of this type of service is VoIP without silence suppression (e.g., G711).
- *Extended real-time polling service (ertPS)* This service class has been introduced in 802.16e to support real-time applications with variable data rates, where the minimum data rate and maximum delay requirements need to be specified. This class is intended to be used for applications such as VoIP with silence suppression.
- *Real-time polling service (rtPS)* Intended for applications with variable-sized packets with a maximum latency requirement. Example applications of this sort include streaming audio and video.

Table 7.7. Service flow QoS parameters

	Maximum sustained rate	Minimum reserved rate	Jitter tolerance	Maximum latency tolerance	Traffic priority
Unsolicited grant service (UGS)[a]	✓		✓	✓	
Extended real-time polling service (ertPS)	✓	✓	✓	✓	✓
Real-time polling service (rtPS)	✓	✓		✓	✓
Non-real-time polling service (nrtPS)	✓	✓			✓
Best effort service (BE)	✓				✓

[a] Minimum reserved rate and traffic priority are not applicable to UGS as the rate is constant and traffic delivery is always guaranteed

- *Non-real-time polling service (nrtPS)* Similar to rtPS without the latency requirement, but where some minimum sustained rate can be specified. This class can be used for bulk data transfer applications, such as FTP.
- *Best effort (BE)* This class allows the maximum sustained rate and traffic priority to be specified, but without a minimum service guarantee. The BE class is intended for background applications with no maximum latency requirement.

The service classes allow different combinations of QoS parameters to be specified, as summarised in Table 7.7. Once service classes have been established for the air interface, the scheduler has to ensure that data are transmitted at the required rate within the jitter and latency tolerances. The system designer, on the other hand, has to ensure that the network is appropriately dimensioned to support the applications of interest to the subscriber – the scheduler does not have detailed knowledge of the application traffic characteristics. Buffers may be introduced at the application or network layer to remove some of the 'burstiness' of application traffic, provided that unacceptable delays are not introduced; this will enable the radio resources to be more efficiently utilised by the scheduler. In the following worked example, the two-step process introduced in Section 3.3.4 will be used, whereby the buffer size and minimum bearer rates are dimensioned for bursty applications; this enables the available bandwidth on the PHY to be partitioned logically into virtual circuits for the purposes of system dimensioning.

7.3.2 Planning – preparation

7.3.2.1 Scenario sub-set selection

For this worked example, system dimensioning will be driven from the Chapter 7 Scenario outlined in Table 4.2. At the highest level, the requirement is to provide a city with population density of 5000/km^2 with wireless broadband access.

Table 7.8. Characteristics for the worked example applications

	Burst size characteristics	Mean data rate	Latency	Jitter	BER	Asymmetry UL/DL
VoIP (G729, non-VAD)	Fixed	27 kbps	<100 ms	10 ms	10^{-3}–10^{-4}	1
Web browsing	Pareto	64 kbps–2 Mbps	2–5 s	1–2 s	10^{-4}–10^{-6}	0.0001

The two applications to be supported are:

- *Web browsing*: Highly 'bursty' in nature with occasionally heavy throughput requirement, representing 10% of overall service activations,
- *VoIP*: A constant bit-rate application with a demand for very low delay, representing 90% of service activations; in this example it is assumed that the G729 codec is being used, without silence suppression.

The specific characteristics of the particular applications in this example are shown in Table 7.8.

To calculate individual application bandwidth requirements, a 'session activity rate' of one session activation per hour will be used. It will be assumed that the majority of the users are static or nomadic (i.e., they have limited mobility) and coverage is only required outdoors. If indoor coverage is required, it will be provided by customer premises equipment (CPE) with externally mounted antennas. Coverage for both applications is required over the entire cell area. In each case, in the interests of space, only the downlink will be considered, although planning for the uplink would be performed in a similar manner.

7.3.2.2 Key questions to address

For the WiMAX system design, the following key questions need to be addressed:

- What propagation models are appropriate?
- What service classes should be used for the applications?
- What are the parameters of these service classes?
- What is the maximum possible cell size?
- What is the system capacity, and can the subscriber density goals be met?

Each of these questions will now be considered in turn in the following sections.

7.3.2.3 Propagation model selection

802.16e is specified to operate in the frequency range 2–11 GHz, and current WiMAX deployments typically operate at 2.5 or 3.5 GHz, which is beyond the normal range for popular cellular models, such as Okumura–Hata and Walfisch–Ikegami. If the deployment is to be undertaken in a dense urban environment, it may be necessary to consider modelling the propagation with a ray-tracing or deterministic model. For empirical modelling, the Erceg-B or SUI models have been successfully employed. In this example, propagation and coverage analysis has been performed using a combination of Erceg-B and deterministic models, using a radio planning tool as referenced in Section 4.2. More information on these propagation models is available in [34].

7.3.3 System budget allocation

7.3.3.1 Service class selection

VoIP traffic (without VAD) is characterised by a constant stream of fixed-size packets, typically every 20 ms, with stringent demands on latency. Real-time applications with a constant bit rate are best allocated to the unsolicited grant service (UGS) class of WiMAX. After service flow creation, the SS is given a regular periodic allocation without the need for explicit bandwidth allocation requests and their associated delays.

Web-browsing traffic is very bursty and uneven; long periods of inactivity can be followed by a sudden need to transmit large volumes of data in response to a user-generated request for a web page. Historically, such applications would have been allocated to the best effort (BE) class, as no attempt would have been made to guarantee a response within a specified time after the 'mouse click'. However, users now increasingly demand that applications respond within a reasonable time interval. In the scenario for this example, a maximum delay of 6 seconds has been specified.

As has been discussed in Chapter 3, use of the buffer pipe model to make more efficient use of transport does require complete 'ownership' of the bearer until the session has been completed. Thus it would seem appropriate to also allocate the web-browsing traffic to a UGS bearer. In fact, even though the bearer must be available to the application 100% of the time (just in case a large burst should arrive) there are in fact significant gaps (see Section 3.3.4.2), which can be exploited by the extended real-time polling service (ertPS). This service enables the quiet periods associated with 'reading time' to be utilised by other applications, such as those tolerating best effort (BE), improving the overall system throughput and efficiency.

Having selected the service classes, the specific parameters of the service flow request can now be established for each application – this is referred to as the *provisioned QoS parameter set* and represents the QoS characteristics that will be requested from the network.

7.3.3.2 Required QoS parameter set for VoIP

For VoIP, which will be using the UGS service flow designation, it is necessary to define:

- Maximum latency tolerance,
- Maximum sustained rate,
- Jitter tolerance.

These will now be studied in more detail.

Figure 7.10 shows the system budget components for VoIP. When estimating the maximum latency tolerance, it is necessary to consider the contributions to the delay from the various network elements along the transmission path between the end-user devices. The overall delay experienced by a packet traversing this path is simply the sum

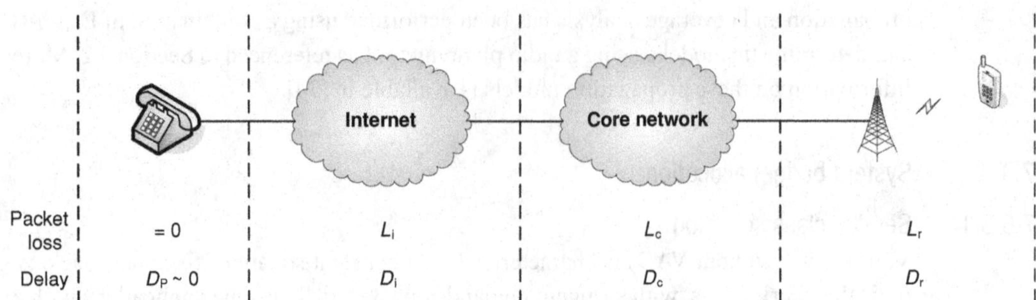

Figure 7.10 System budget components for VoIP

of the delays:

$$D_{tot} = D_P + D_i + D_c + D_r + D_{buffer}.$$

The delay D_i due to transmission over the Internet is highly variable, and depends on the distance to the server, number of hops traversed, volumes of traffic, etc., and should be measured in practical deployments (for example using the 'ping' command) – however for the purposes of this example let the mean $D_i = 70$ ms. D_P, the delay through the PSTN network, is taken to be zero in this example. If, however, international connections have to be expected, it will be necessary to replace the Internet connection with an operator 'intranet' to manage this delay down to 10 ms or so.

The delay D_c for traversal of the core network is normally very low, as this is on a dedicated private IP network; in a correctly dimensioned network this will be <10 ms and so to be conservative 10 ms will be used.

The delay D_r for transmission over the air interface is the parameter of interest that will be specified in the service class parameters; this delay has two components:

- $D_{retrans}$: delay from retransmissions of incorrectly transmitted blocks, for example, due to HARQ,
- $D_{scheduler}$: scheduling delay whilst radio resources are obtained for data-block transmission.

The propagation delay associated with transmission of the radio signal can be assumed to be negligible compared to the other delays in this example.

The system designer must decide if HARQ is to be enabled or not, by establishing whether potential improvements to the link budget and overall system throughput are at the expense of an unacceptable increase in latency. $D_{retrans}$ can be estimated by determining the probability that a block will be delivered in error one or more times. In WiMAX, the PHY manages transmissions and retransmissions in terms of blocks – if a block is wrongly received, it may be retransmitted by HARQ (either type 1 or type 2 HARQ). The probability that a block will need to be retransmitted decreases as the quality of the radio link improves. This is illustrated in Figure 7.11, which shows the results of simulation for a profile with QPSK modulation, for different types of HARQ. At 6 dB SNR, the average number of transmissions required varies from 1.7 (with HARQ type 2) up to just under

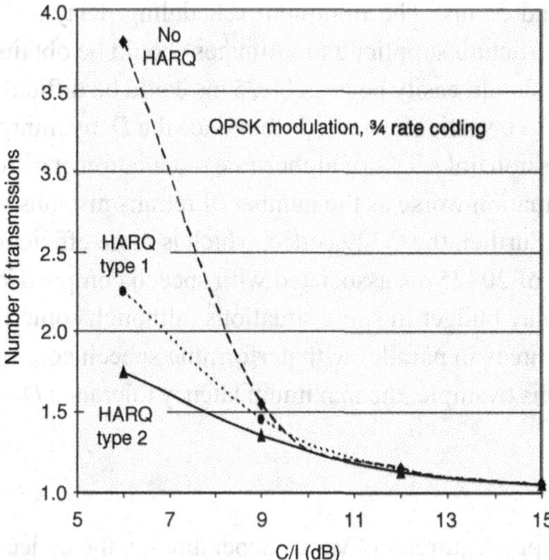

Figure 7.11 Average number of transmissions per block vs. signal-to-noise ratio [35]

4 without any HARQ. If our WiMAX system has been designed to provide ubiquitous QPSK coverage, it is safe to assume that the vast majority of blocks will be successfully received with no more than two transmissions required. If HARQ is invoked, the faulty block will be queued for retransmission in a subsequent frame. Obviously, the time taken to do this will depend on the scheduling algorithms implemented by the manufacturer; for the purposes of this example it is assumed that an incorrectly received block will be transmitted no more than three frames later.

Consequently, the delay introduced by retransmission is:

$$D_{\text{retrans}} = (N_{\text{tranmissions}} - 1) \times N_{\text{scheduler}} \times T_{\text{frame}},$$

where

$$N_{\text{transmissions}} = \text{number of transmissions required for successful block delivery}$$
$$N_{\text{scheduler}} = \text{number of frames before block can be rescheduled (worst case)}$$
$$T_{\text{frame}} = \text{duration of a WiMAX frame} = 5\,\text{ms},$$

so

$$D_{\text{retrans}} = (2 - 1) \times 3 \times 5 = 15\,\text{ms}.$$

Hence, the total delay budget for our VoIP application is now as follows, remembering that a buffer will not be used for constant-bit-rate traffic and, thus, $D_{\text{scheduler}}$ can be evaluated:

$$\begin{aligned} D_{\text{tot}} &= D_{\text{P}} + D_{\text{i}} + D_{\text{c}} + D_{\text{retrans}} + D_{\text{scheduler}} + D_{\text{buffer}} \\ &= 0 + 70 + 10 + 15 + D_{\text{scheduler}} + 0 \end{aligned}$$

The target maximum latency for one leg of the VoIP application is of the order of 100–150 ms, hence, depending upon the delay goal chosen, the maximum scheduling delay permissible is between 5 and 55 ms. The minimum scheduling delay achievable will depend on the chosen infrastructure supplier and estimates should be obtained from the manufacturer. Whilst 55 ms should easily be possible, 5 ms could be difficult and consequently it may be necessary to investigate methods to reduce the D_i by improved routing and faster dedicated transmission links. Use of higher-rate modulation or different coding schemes might make this situation worse as the number of retransmissions will increase because of the higher BER. Further, the G729 codec, which is more efficient in terms of bandwidth, also has a delay of 20–25 ms associated with speech compression; this may result in an unacceptable delay budget in some situations (although some implementations will request radio resources in parallel with performing speech compression, thus reducing this latency). For this example, the maximum latency tolerance D_r will be set at 20 ms.

Maximum sustained rate

For VoIP traffic, the throughput requirement varies depending on the codec being used; with a G.711 codec, the coded data rate is 64 kbps rising to about 83 kbps when the signalling overhead is included. By comparison, G.729 has a coded data rate of only 8 kbits/s but the total throughput requirement is up to 27 kbits/s, accounting for the signalling overhead. For UGS this is equal to the guaranteed minimum rate since the traffic is constant bit rate.

Jitter tolerance and packet loss ratio

As well as being intolerant to delay, VoIP is also notoriously susceptible to jitter (variability in delay), which can reduce voice quality. The service flow parameters should be designed to meet a network-level mean opinion score (MOS) for the voice quality. The e-model [36] shows how VoIP MOS varies as a function of jitter and delay – and can be used to determine the allowable jitter tolerance. For VoIP jitter will need to be 5 ms or less. (Note that jitter buffers are sometimes included in the design of a VoIP system to compensate for variability of packet transmission times – however such buffers will introduce further delay, which may not be acceptable if the delay budget is already under pressure.)

Similarly, the end-to-end packet loss ratio L_{tot} can be established from the contributions L_P, L_i, L_c, L_r, L_{buffer} as:

$$L_{tot} = 1 - (1 - \mathrm{L_P}) \times (1 - L_i) \times (1 - L_c) \times (1 - L_r) \times (1 - L_{buffer}).$$

In this case there is no burst buffer, thus L_{buffer} goes to zero and all losses, with the exception of L_r, will be negligible. It will be necessary to establish a worst-case link budget that ensures a BER corresponding to an acceptable packet loss over the air interface.

The final provisioned QoS parameter set is given in Table 7.9.

7.3.3.3 Required QoS parameter set for web browsing

Web browsing will use the extended real-time polling service (*ertPS*) flow designation and so it is necessary to define:

Table 7.9. Provisioned QoS parameter set for VoIP

Application: VoIP (non-VAD)	
Service class	Unsolicited grant service
Maximum latency tolerance	20 ms
Maximum sustained rate	27 kbps
Jitter tolerance	5 ms
HARQ	On

- Maximum latency tolerance,
- Jitter tolerance,
- Maximum sustained rate,
- Minimum reserved rate,
- Traffic priority.

The parameters for the provisioned QoS set will now be evaluated.

Maximum latency tolerance

For the web-browsing application it will also be necessary to consider the delays introduced by buffering of traffic in the network, as illustrated in Figure 7.12. Buffering is necessary to minimise packet loss, if bearers below the peak rate are to be used, since the traffic produced by web browsing is bursty in nature, unlike the constant bit-rate traffic for VoIP. The system designer must, therefore, specify the appropriate dimensions for the application buffer and evaluate the contribution to the overall delay budget which it will introduce.

The web-browsing application specified in Table 4.1 calls for a maximum delay of 6 seconds, which will be the sum of all components:

$$D_{\text{total}} = D_{\text{s}} + D_{\text{i}} + D_{\text{c}} + D_{\text{buffer}} + D_{\text{r}},$$

where:

- D_{s} is the server delay,
- D_{i} is the delay due to transmission over the internet,

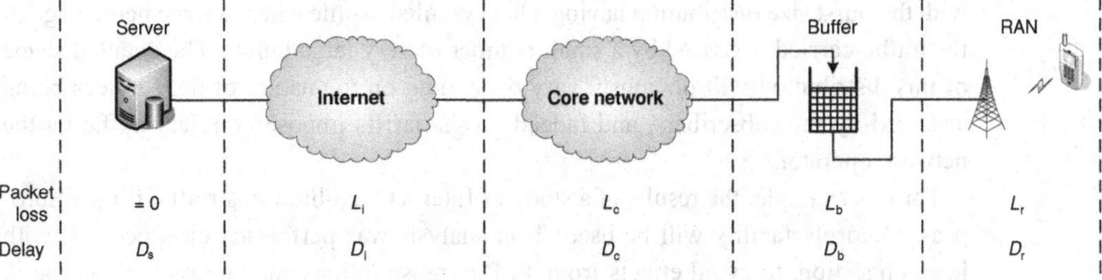

Figure 7.12 System budget components for web browsing (downlink)

- D_c is the delay due to transport over the core network,
- D_{buffer} is the delay introduced by the buffer in the buffer–pipe smoothing process,
- D_r is caused by scheduling and transmission over the air interface.

In practice, as with any statistical distribution, it is not possible to define an absolute maximum, and a delay value within which 95% of packets are delivered will be used.

D_{buffer} will be calculated when the minimum bearer rate requirement is estimated, as the two have a strong interaction. D_i has already been discussed and the server delay, D_s, will be set to 0.5 s, a representative delay for a significantly loaded shared server. The remaining component D_r can be specified as the maximum latency tolerance in the QoS parameter set, once D_{buffer} is known. D_{buffer} will be different in the uplink and downlink directions; a typical web session is characterised by occasional clicks from the user (sent on the uplink) followed by variable-sized bursts of data sent to the browser from the server in the downlink. Since the volume of data sent in the uplink is very small, it is usually possible to neglect the contribution from buffering in the uplink. In this example, only the delay due to buffering in the downlink direction will be considered.

Jitter tolerance

Web browsing using http is quite robust to jitter. For the WiMAX air interface, jitter is likely to result from HARQ and ARQ retransmissions when the scheduler has moved to an over-aggressive modulation and coding scheme in an attempt to maximise capacity. The jitter tolerance should be tuned after deployment based on observed field measurements, but an initial value of 100–200 ms is appropriate for this parameter, well within the recommended range from Table 4.1. It is desirable to keep the jitter delay reasonably small such that TCP does not begin retransmission or congestion control.

Maximum sustained rate and minimum reserved rate

Care is needed in specifying the minimum reserved rate; too low a value will lead to unacceptable queuing delays and buffer overflows, too high a value will result in an inefficient system unable to meet the overall capacity goals due to wasteful allocation of resources. The requested rate must match the application buffer size to ensure that the overall application QoS requirements are met.

Web browsing can lead to a high maximum data rate requirement. Various studies have shown [37] that the size of data bursts generated by a user click is highly variable, with the burst size distribution having a heavy-tailed profile where a large percentage of the traffic carried is caused by a small number of very large bursts. The detailed shape of this distribution will obviously vary depending on the nature of the services being accessed by the subscribers, and indeed on the tariffs imposed on data traffic by the network operator.

For this example, the results of a study of Internet web-browsing traffic from a European Motorola facility will be used. This analysis was performed on a network with low congestion, to avoid effects from TCP retransmissions and congestion control. A burst was defined to be the aggregated size of a number of consecutive http requests originating from a single user activity – effectively the traffic generated by the browser, as a result of a single mouse click. Although the majority of bursts were reasonably small, there was a long tail caused by a small number of very large bursts. This distribution was found to be approximated by a Pareto distribution (discussed in Chapter 3). The complementary cumulative distribution function for the three-parameter form of Pareto distribution, which describes the area under the 'tail', may be written as:

$$F(x) = (1 + b(x - c))^{-a}.$$

The best fit parameterisation was found to be

$$a = 1.329,$$

$$b = 1.923 \times 10^{-5},$$

$$c = 2864.$$

This parameterisation of the distribution allows for an infinite burst size; in practice there is invariably a maximum burst size associated with the application, which represents the largest burst that can be admitted onto the system without truncation. Usually, operators will limit the maximum burst size that can be sourced by the server; in this example, 1 Mbyte will be used as the upper limit. To estimate the load offered to the system it is also necessary to specify the mean burst arrival rate – this can be considered to be the average 'reading time' associated with a web page – the time between two 'clicks' during a web-browsing session. Weinreich *et al.* [38] monitored this and found a median time between clicks of 9.4 seconds, with a somewhat larger mean, owing to the heavy tail of the distribution; in this example a mean reading time of 30 seconds will be used, resulting in a burst arrival rate of $1/30 = 0.033$ bursts/second and a Poisson burst arrival distribution is assumed (although this may not be representative in all cases).

Using this description, it is now possible to evaluate the bandwidth and application buffer size requirements to support the traffic QoS requirements. The graph in Figure 7.13(a) generated by an analytical modelling tool (in this case the Motorola 'ARC' tool) shows how the packet loss decreases as bearer rate increases, for a fixed buffer size of 700 kbytes. After an initial fall, the packet-loss rate is seen to be insensitive to further increases of bearer rate beyond about 50 kbytes/s; the packet loss at higher rates is almost entirely a result of truncation of large bursts due to a finite buffer size.

Figure 7.13(b) shows how the packet loss varies as the buffer size is increased to accommodate larger bursts. An acceptable packet loss (within the design guidelines for the scenario) is reached with a buffer size of about 700 kbytes. Note that once the buffer size is increased to the maximum burst size of 1 Mbytes, the packet loss ratio becomes vanishingly small.

A similar exercise, using the ARC tool, was carried out to investigate how the 95th-percentile delay time altered as the buffer size was increased, for two different bearer rates (400 kbits/s and 800 kbits/s). It was found that a bearer rate of 800 kbits/s and a buffer size of 5.6 Mbits (0.7 Mbytes) resulted in a 95th-percentile delay of 4.2 s, which should enable the overall delay design criteria for this application to be met.

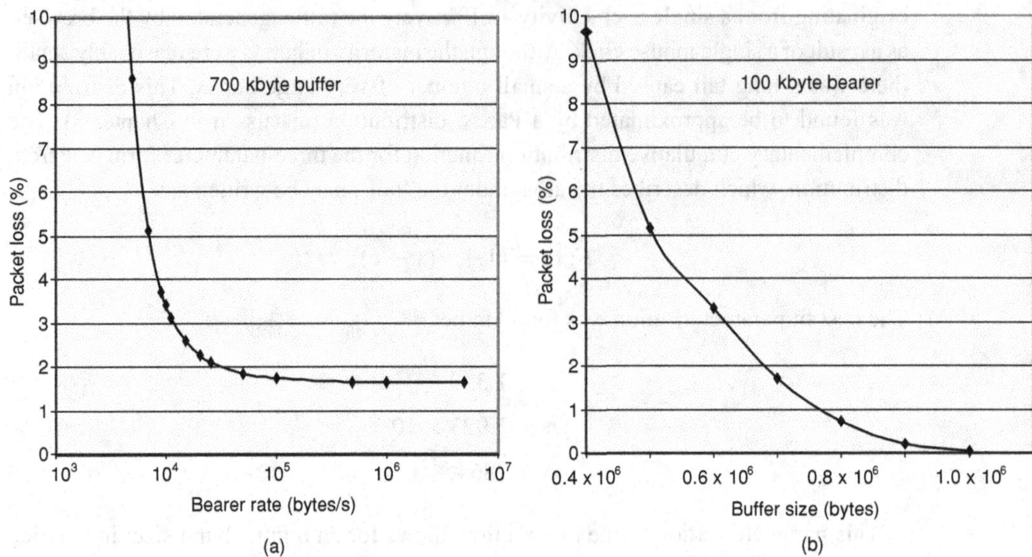

Figure 7.13 Packet loss vs. bearer rate and buffer size for heavy-tailed web-browsing traffic

The ertPS maximum tolerated latency is, therefore, found as follows:

$$D_{\text{tot}} = D_{\text{s}} + D_{\text{i}} + D_{\text{c}} + D_{\text{buffer}} + D_{\text{r}},$$
$$6\,\text{s} = 0.5 + 0.07 + 0.01 + 4.2 + D_r,$$
$$D_{\text{r}} = 1.22\,\text{s}.$$

From the above it can be seen that the ertPS maximum sustained rate should be set to 800 kbits/s and the delay D_{r} set to 1.22 s.

Traffic priority

The traffic priority (alternatively called the 'user priority') can be used to allow tagging of packets emerging from the RAN with different VLAN priorities. This field provides a 3-bit value (0–7), which can be used to assist prioritisation of traffic in the back-haul network if required. Typically, the highest priorities (5–7) are automatically allocated to unsolicited grant (UGS) flows; values 2–4 can be used for low-to-high priority data services, and 0–1 can be allocated to BE flows. In this example, therefore, a traffic priority of 4 is appropriate. The final provisioned QoS parameter set is shown in Table 7.10.

7.3.4 Estimation of maximum cell size

To estimate the maximum cell size it is appropriate to study the heaviest-coded modulation scheme, as this is the most resilient to interference and hence will ensure operation at the poorest C/I values. (Note that higher coding rates will, of course, be used away from the cell edge). From Table 7.6, it can be seen that a 10 MHz QPSK 1/2 rate channel offers a theoretical maximum rate of 4.71 Mbps on the uplink at a 40:60 UL/DL split, comfortably greater than the minimum throughput requirement for a single instance of

Table 7.10. Provisioned QoS parameter set for web-browsing application

Application: web browsing	
Service class	Extended real time polling service
Maximum latency tolerance	1.22 s
Jitter tolerance	200 ms
Maximum sustained rate	800 kbits/s
Minimum sustained rate	800 kbits/s
Traffic priority	4
HARQ	On

VoIP or web browsing at the cell edge. To provide ubiquitous coverage for the selected applications, it is, therefore, necessary to ensure that the received signal and CINR values at each location are adequate for QPSK to be used. As discussed previously, this is done using a radio planning tool with a suitable terrain–clutter model and the appropriate propagation model for the frequency domain, tuned for the particular deployment environment. QPSK 1/2 rate needs a C/I of around 6 dB, and this is the level for which initial coverage should be planned. To perform the planning task, it is necessary to carry out a detailed link budget analysis, which will not be addressed in this example (for an illustration of the use of a detailed radio link budget, see the worked example for UMTS in Chapter 6). Studies performed using Motorola's frequency planning tools indicate that a cell radius in the region of up to 2 km will give 95% coverage at the required CINR in dense urban environments, provided that suitable transmitter locations can be found.

7.3.5 Estimation of system capacity

System capacity will obviously vary, depending on such factors as cell radius, channel bandwidth, frequency reuse patterns and the radio environment. For this example, the system is modelled using the parameters in Table 7.11.

Planning tools of the type referenced in Chapter 4 indicate that this configuration provides a sector data capacity of up to 12 Mbits/s (UL and DL combined) or 36 Mbits for the tri-sectored site. What is the most efficient UL:DL split for the TDD frame, and how many users of WWW and VoIP services can be supported in each cell? The application attributes and targeted application mix are shown in Table 7.12. In Chapter 3, a two-stage planning approach was proposed for dimensioning mixed-service traffic, and this will be used in this example. The burstiness of the WWW traffic is accommodated by providing an application buffer of appropriate size to meet the delay and packet-loss requirements; subsequently the service-class maximum latency tolerance and minimum sustained bitrate have been specified in the ertPS service flow. It is assumed that the scheduler will always provide the WWW application with this QoS (although unused capacity may be reused for other applications) – consequently the flow can be treated as a virtual circuit, and subscriber capacity can be estimated using the Erlang-B queuing model. The web-session bandwidth of 800 kbits/s (100 kbytes/s) is the minimum sustained rate specified in the ertPS parameter set, and was calculated with the analytical model for

Table 7.11. Key system simulation parameters

<table>
<tr><th></th><th>Downlink</th><th>Uplink</th></tr>
<tr><td>Site-to-site distance</td><td colspan="2">1.74 km</td></tr>
<tr><td>Number of sectors</td><td colspan="2">3</td></tr>
<tr><td>Frequency reuse pattern</td><td colspan="2">1 × 3 × 3</td></tr>
<tr><td>Sub-carrier permutation</td><td colspan="2">PUSC</td></tr>
<tr><td>Carrier frequency</td><td colspan="2">3.5 GHz</td></tr>
<tr><td>Channel bandwidth</td><td colspan="2">10 MHz</td></tr>
<tr><td>Propagation law</td><td colspan="2">Erceg B</td></tr>
<tr><td>Building penetration loss</td><td colspan="2">6 dB</td></tr>
<tr><td>Antenna gain</td><td>14 dB</td><td>7 dB</td></tr>
<tr><td>Antenna height (m)</td><td>30 m</td><td>1.5 m</td></tr>
<tr><td>Transmit power (dbW)</td><td>0</td><td>−3</td></tr>
<tr><td>MIMO</td><td>No</td><td>No</td></tr>
</table>

Table 7.12. Application attributes

<table>
<tr><th>Application</th><th>WWW</th><th>VoIP</th></tr>
<tr><td>Bearer rate (Mbits/s)</td><td>0.8</td><td>0.027</td></tr>
<tr><td>Session duration (seconds)</td><td>180</td><td>90</td></tr>
<tr><td>Erlang demand</td><td>0.05</td><td>0.025</td></tr>
<tr><td>Busy hour call attempts</td><td colspan="2">1</td></tr>
<tr><td>Target application mix (%)</td><td>10</td><td>90</td></tr>
</table>

the bursty application. The web-session duration of 180 seconds, together with a BHCA of 1, is used to calculate the busy hour Erlang demand for web browsing of 0.05E (the session duration divided by 3600). VoIP Erlang demand per subscriber is evaluated with the mean call duration of 90 seconds. The detail of the capacity evaluation process is illustrated in Table 7.13.

The design scenario requires a mixture of 90% VoIP sessions and 10% WWW sessions, and a subscriber density of 5000 users/km^2. Although the 75%:25% downlink to uplink allocation almost meets the application mix requirement at 9.4% WWW (the paramount consideration), the achieved target subscriber density (1376/km^2) falls considerably short of the target. This is primarily a consequence of:

- The large allowed burst size for mobile web traffic – 1 Mbyte,
- The tight QoS delay target put on worst case response times (6 s).

These two considerations are the driver for the 800 kbits/s bearer required by the WWW service. Capacity demand for this service relative to VoIP is further aggravated by a session duration of 180 seconds versus 90 seconds for VoIP.

Despite these considerations, the subscriber capacity for the application mix might still, on first appearance, seem to imply lower capacity for WiMAX than expected.

Table 7.13. System capacity modelling for a mixed service WiMAX cell with VoIP and low delay www services

	Notes	See key below											
Sector area	[a]	0.79 km²											
Sector capacity	[b]	12 Mbits/s											
DL:UL split (%)	[c]	75: 25				60:40				50:50			
System overhead (%)	[d]	7				6.7				6.5			
Capacity		Downlink		Uplink		Downlink		Uplink		Downlink		Uplink	
	[e]	8.93		2.23[f]		7.61		3.58[f]		6.73		4.49[f]	
Application		WWW	VoIP	WWW	VoIP	WWW	VoIP	WWW	VoIP	WWW	VoIP	WWW	VoIP
Allocated capacity	[g]	8.04	0.89	0.02	0.89	6.85	0.76	0.02	0.76	6.06	0.67	0.02	0.67
Number of bearers	[h]	10.04	33.07			8.56	28.2			7.57	24.93		
Erlangs	[i]	5.1	24.6			3.63	20.15			2.94	16.63		
Erlang demand		0.05	0.025			0.05	0.025			0.05	0.025		
Number of instances		102	985			73	806			59	665		
WWW (% of total)		9.4				8.3				8.1			
Subscribers per sector		1087				879				724			
Subscribers per site		3261				2637				2172			
Subscriber density		1376				1113				916			

[a] One of three sectors giving a site area of 2.38 km²

[b] From radio planning tool for the particular propagation region, penetration loss, mobility profile, etc.

[c] The percentage split reflects the allocation of radio-frame symbols to downlink or uplink; it does not equate directly to capacity allocation

[d] Percentage of WiMAX frame used by MAP, CQICH, ACKs, ranging, etc. The MAP overhead will show some increase as the number of users increases. Note, in this table it also includes an allowance for scheduler inefficiency for low delay services

[e] Usable radio-frame capacity in each direction once overhead has been considered. See also note ([c])

[f] Note that not all capacity has been used in the uplink – this is more pronounced as the downlink-dominated aggregate application traffic is fitted to an increasingly symmetric downlink and uplink. This is a reflection of the currently standardised uplink/downlink ratios

[g] Usable frame capacity in the downlink and uplink allocated to each allocation. This is a reflection of the currently standardised uplink / downlink ratios. This is policed by admission control

[h] Effective number of bearers estimated by dividing allocated capacity or application by bearer rate. Note that an allowance for scheduler inefficiency on low delay bearers has been included in the system overhead

[i] Traffic capacity in Erlangs for 2% blocking

Another perspective on this can, however, be obtained by observing that about 25 VoIP users are supported simultaneously in 0.89 MHz for the 75:25 downlink/uplink configuration. This means that about five times as many users can be supported per sector if VoIP were the only service (even allowing for an increase in MAP overhead). This compares favourably with 3G systems, and this is without VAD active in the codec.

7.3.6 How can the overall system capacity requirements be met?

If the subscriber density goals in this example are to be met, what avenues are available to the system designer? At least three options are available.

- Reducing the cell radius from the current value of 870 m,
- Leaving the cell radius unchanged and introducing a MIMO antenna configuration,
- Limiting the maximum burst size to less than 1 Mbyte (or degrading guaranteed QoS).

Reducing the cell radius to about 450 m will allow the subscriber density goals to be met and these cells are still larger than the 200 m voice-only cells deployed in 2G and 3G. This is probably the preferred solution, especially recognising that this high quality WWW service would probably be deployed only in business areas for some time. Nevertheless, the other two options also represent viable ways of meeting the subscriber density goal; as ever, the particular solution chosen will reflect a given operator's goals and long-term ambitions.

7.3.7 Worked example – summary comments

It is doubtless apparent by now that planning for guaranteed QoS is complicated – even though it is not difficult. To ensure that the process can be followed straightforwardly using a simple two-dimensional table (Table 7.13) a number of simplifying assumptions have been made. Most of these have been addressed via the notes associated with the table but it is appropriate that the main issues should be repeated here.

- The main vehicle used for allocating resources to the downlink or uplink is the number of radio-frame symbols assigned to each link. In reality, the coding schemes in use on the uplink and downlink may be different and static overheads may form a larger proportion of one link than the other and thus, as is evident from the table, the *actual* downlink/uplink ratio and capacities will not reflect the simple symbol allocation ratio.
- To provide a simple and visible relationship between the available downlink or uplink capacity and the number of bearers that can be supported, the artificial process of dividing the link capacity by the bearer rate has been used. Whilst this might be reasonable if scheduling delay was unimportant, the reality is that for these two services quite stringent limits have been placed on the worst-case delay across the radio interface. This will require the MAC scheduler to leave unfilled payload regions to enable it to guarantee delivery within the time limit. In practice, the number of bearers of a given QoS that can be supported within a given capacity is dependent on the vendor

scheduler and will form part of the data provided by the manufacturer. For this example, a scheduler inefficiency factor has been added to the 'overhead' line.

- In reviewing the 'allocated capacity' row of Table 7.13, it will be apparent that very considerable capacity is being wasted in the uplink, because there is not significant associated uplink traffic from WWW; this is particularly true for the 50:50 symbol allocation, where almost 4 Mbits/s are wasted. In practice a number of measures might be taken to minimise this occurrence:
  - Support 'best-effort' traffic, if this is available,
  - See whether non-standard downlink-to-uplink ratios can be configured, if this is a key service,
  - Ensure that the cost of this idle uplink capacity is reflected in the WWW service cost.

Hopefully, it is now apparent that there is a viable system design approach that will enable user applications to be delivered with full assurance of correct operation through the establishment of guaranteed QoS bearers matched to the application. It is expected that this process will become the norm, as 3GPP LTE and WiMAX enter into volume deployment and enable operators to access new premium-revenue streams through delivery of applications rather than 'data pipes'. Although this example has only dealt with two applications, the technique employed above can be extended to multiple applications, provided the application burst-size distributions are known. Identifying the best partition of system resources to meet the design guidelines can become a rather complex and iterative task, best performed in a software tool embodying the calculations described above.

References

1 3GPP, *Evolution of 3GPP System (Release 6)*, 3GPP TR 21.902 version 6.0.0 (2003).

2 3GPP, *All-IP Network (AIPN) Feasibility study (Release 7)*, 3GPP TR 22.978 version 7.1.0 (2005).

3 3GPP, *Service Requirements for the All-IP Network (AIPN); Stage 1 (Release 8)*, 3GPP TS 22.258 version 8.0.0 (2006).

4 3GPP, *Proposed Study Item on Evolved UTRA and UTRAN*, RP-040461, TSG-RAN Meeting #26, Athens, Greece, 8–10, December, 2004.

5 3GPP, *Feasibility Study for Evolved Universal Terrestrial Radio Access (UTRA) and Universal Terrestrial Radio Access Network (UTRAN) (Release 7)*, 3GPP TR 25.912 version 7.1.0 (2006).

6 3GPP, *Requirements for Evolved UTRA (E-UTRA) and Evolved UTRAN (E-UTRAN) (Release 7)*, 3GPP TR 25.913 version 7.3.0 (2006).

7 3GPP, *Physical Layer Aspects for Evolved Universal Terrestrial Radio Access (UTRA) (Release 7)*, 3GPP TR 25.814 version 7.1.0 (2006).

8 3GPP, *3GPP System Architecture Evolution: Report on Technical Options and Conclusions (Release 7)*, 3GPP TR 23.882 version 1.9.0 (2007).

9 3GPP, *3GPP System Architecture Evolution: GPRS enhancements for LTE Access; Release 8*, 3GPP TS 23.401 version 0.2.0 (2007).

10 3GPP, *3GPP System Architecture Evolution: Architecture Enhancements for Non-3GPP Accesses; Release 8*, 3GPP TS 23.402 version 0.2.0 (2007).

11 S. Gundavelli *et al.*, Proxy Mobile IPv6 draft-sgundave-mip6-proxymip6–02.txt, IETF NETLMM WG (March 05, 2007).

12 IEEE Computer Society/Local and Metropolitan Area Networks, *802.21 – Standard for Media Independent Handover Services*, http://www.ieee802.org/21/.

13 P. Carlsson *et al.*, Delay performance in routers, *Proceedings of the 2nd International Working Conference (HET-NETs '04)*, P15 (2004) pp. 1–10.

14 3GPP, *Evolved Universal Terrestrial Radio Access (E-UTRA) and Evolved Universal Terrestrial Radio Access Network (E-UTRAN); Overall description; Stage 2 (Release 8)*, 3GPP TS 36.300 version 1.0.0 (2007).

15 R. Love, K. Stewart, R. Bachu and A. Ghosh, MMSE equalization for UMTS HSDPA, *IEEE Vehicular Technology Conference 2003-Fall*, **4** (2003) 2416–2420.

16 K. Baum, P. Sartori and V. Nangia, System-level study of OFDM and CDMA with adaptive modulation/coding for a 5 MHz channel, *IEEE Wireless Communications and Networking Conference, 2004*, **4** (2004) 2131–2136.

17 L. Hanzo, W. Webb and T. Keller, *Single and Multi-Carrier Quadrature Amplitude Modulation* (John Wiley and Sons Ltd, 2000) pp. 423–436.

18 J. Romberg, *Circular Convolution and the DFT*, http://cnx.rice.edu/content/m10786/latest/, version 2.8 (July 2006).

19 Y. Sun, R. Love, K. Stewart *et al.*, Cellular SFN broadcast network modelling and performance analysis, *IEEE Vehicular Technology Conference 2005-Fall*, **4** (2005) 2684–2690.

20 Y. W. Blankenship, P. J. Sartori, B. K. Classon, V. Desai and K. L. Baum, Link error prediction methods for multi-carrier systems, *IEEE Vehicular Technology Conference 2004-Fall*, **6** (2004) 4175–4179.

21 L. Hanzo, W. Webb and T. Keller, *Single and Multi-Carrier Quadrature Amplitude Modulation* (John Wiley and Sons Ltd, 2000) pp. 452–454.

22 Motorola, *Single Carrier Uplink Options for E-UTRA: IFDMA/DFT-SOFDM Discussion and Initial Performance Results*, 3GPP TSG RAN WG1 #42, London, United Kingdom (Aug 29–Sept 2, 2005) document R1–050971.

23 Motorola, *EUTRA Uplink Numerology and Design*, 3GPP RAN1#41bis, Sophia Antipolis, France (June 20–21, 2005) document R1–050584.

24 3GPP, *Physical Channels and Modulation (Release 8)*, 3GPP TS 36.211 version 1.0.0 (2007).

25 B. Classon, P. Sartori, V. Nangia, X. Zhuang and K. Baum, Multi-dimensional adaptation and multi-user scheduling techniques for wireless OFDM systems, *IEEE International Conference on Communications*, **3** (2003) 2251–2255.

26 Y. Sun, W. Xiao, R. Love *et al.*, Multi-user scheduling for OFDM downlink with limited feedback for evolved UTRA, *IEEE Vehicular Technology Conference 2006-Fall* (2006) 1–5.

27 B. Classon, K. Baum, V. Nangia *et al.*, Overview of UMTS air-interface evolution, *IEEE Vehicular Technology Conference 2006-Fall* (2006) 1–5.

28 P. Sartori *et al.*, Broadband mobile MIMO-OFDM cellular system capacity and link experiments, *Proc. 39th Annual Allerton Conference on Communications, Control, and Computing* (Sept. 2001) 553–562.

29 P. J. Sartori, K. L. Baum and F. W. Vook, Impact of spatial correlation on the spectral efficiency of wireless OFDM systems using multiple antenna techniques, *IEEE Vehicular Technology Conference 2002-Spring*, **3** (2002) 1150–1154.

30 The WiMAX Forum, *Mobile WiMAX – Part I: A Technical Overview and Performance Evaluation* (August 2006) www.wimaxforum.org/home/.

31 IEEE, *IEEE Standard for Local and Metropolitan Area Networks, Part 16: Air Interface for Fixed Broadband Wireless Access Systems*, IEEE Standard 802.16 – 2004, http://standards.ieee.org/getieee802/download/802.16-2004.pdf.

32 IEEE, *IEEE Standard for Local and Metropolitan Area Networks, Part 16: Air Interface for Fixed and Mobile Broadband Wireless Access Systems, Amendment 2: Physical and Medium Access Control Layers for Combined Fixed and Mobile Operation in Licensed Bands and Corrigendum 1*, http://standards.ieee.org/getieee802/download/802.16e-2005.pdf.

33 CableLabs, *Data-Over-Cable Service Interface Specifications DOCSIS 2.0, Radio Frequency Interface Specification*, CM-SP-RFIv2.0-I07–041210, pp. 219–223, www.cablelabs.com/specifications/archives/CM-SP-RFIv2.0-I07-041210.pdf.

34 IEEE, *Channel Models for Fixed Wireless Applications, IEEE 802.16 Broadband Wireless Access Working Group*, IEEE 802.16.3c-01/29r4, http://ieee802.org/16.

35 J. G. Andrews, A. Ghosh and R. Muhamed, *Fundamentals of WiMAX* (Prentice Hall, 2007) pp. 385–386.

36 ETSI, *Transmission and Multiplexing (TM); Speech Communication Quality From Mouth to Ear for 3,1 kHz Handset Telephony Across Networks*, ETR 250 (July 1996).

37 M. E. Crovella and A. Bestavros, Self-similarity in world wide web traffic: evidence and possible causes, *IEEE/ACM Transactions on Networking*, **5** 6 (1997) 835–846.

38 H. Weinreich, H. Obendorf, E. Herder and M. Mayer, Off the beaten tracks: exploring three aspects of web navigation, *Proceedings of the 15th International Conference on World Wide Web* (Edinburgh, Scotland, May 23 – 26, 2006) (ACM Press, 2006) 133–142. doi = http://doi.acm.org/10.1145/1135777.1135802.

8 Mesh network planning and design

8.1 Principles of mesh networking

As discussed earlier, one method for increasing capacity in cellular networks is to make the cell sizes smaller, allowing more subscribers to use the available radio spectrum without interfering with each other. A similar approach is used in 802.11 (Wi-Fi) – either to provide public 'hot-spot' broadband connections or to link to broadband connections in the home via a Wi-Fi router. The range of such systems is limited to a few tens of metres by restricting the power output of the Wi-Fi transmitter. The goal of a *wireless mesh* is to extend the 802.11 coverage outdoors over a wider area (typically tens of square kilometres) not simply by increasing the power, but by creating contiguous coverage with dozens of *access points* (APs) or *nodes*, separated by distances of 100–150 metres. For such a solution to be economically viable, the access points themselves need to be relatively cheap to manufacture and install, and the back-haul costs must be tightly managed. To address this latter requirement, only a small percentage of APs (typically 10–20%) have dedicated back-haul to the Internet; the other APs pass their traffic through neighbouring APs until an AP with back-haul is reached. At the time of writing, the IEEE 802.11s standard for mesh networking is still being drafted, and various proprietary flavours of mesh networks exist. The following discussion outlines the principal characteristics and properties of most commercially available mesh networks, which will be embodied in 802.11s when it is finalised.

Conceptually and physically, a mesh network can be considered to consist of two distinct layers – the *access layer* and the *gateway* layer. The access layer consists of a number of nodes, which provide wireless LAN coverage to subscribers, who access the WLAN via suitably equipped devices, such as wireless laptop computers, WLAN VoIP phones, PDAs, and so on. Typically the access layer provides 802.11 access at 2.4 GHz, although other access protocols and frequencies are possible. The gateway layer consists of nodes, which have back-haul connectivity to the service-provider network and the Internet beyond. Nodes in the gateway layer can themselves act as access nodes in most mesh configurations, but their most important function is to act as gateways to back-haul for the access layer. The nodes in the access layer communicate wirelessly with the gateways – either directly or through intermediate 'hops' to other access nodes.

The advantage of this approach is that not all nodes need to have potentially expensive back-haul provision, thus reducing the overall cost of deployment of the network. Node APs in the access layer only require a suitable mounting point, a power supply and a

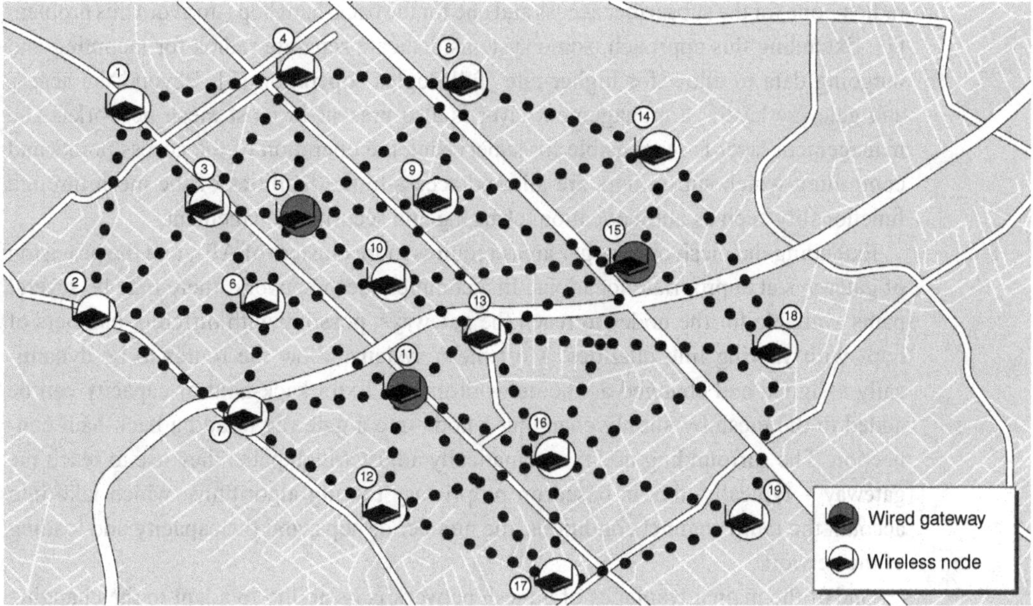

Figure 8.1 Example mesh network

radio link to another AP in the mesh. Additional coverage can be provided simply by adding more nodes to the access layer – the only requirement is that adjacent nodes in the access layer must be within range of each other so that a wireless link of suitable quality can be established between them. As the network expands and traffic increases, the gateway layer can be augmented simply by converting access nodes to gateways by providing them with back-haul links.

To illustrate this concept, an example mesh network is shown in Figure 8.1. Nineteen APs are to provide wireless LAN coverage to a small area of a city. Three of the nodes (AP5, AP11 and AP15) have been provisioned with DSL back-haul and are termed *gateway nodes*, or simply *gateways*. Nodes AP3, AP4, AP6, AP9 and AP10 are within range of AP5 and can pass their data wirelessly to AP5 for back-haul. AP1 and AP2 are beyond the range of AP5 but can see AP3, AP4 or AP6 and they can route their data to AP5 via the intermediate APs. A wireless link between neighbouring APs (gateways or nodes) is termed a *hop* – as can be seen in the diagram, several hops may be required to route traffic to the gateway, for example for nodes AP19, AP17, AP2, etc. A group of nodes sharing a common gateway is called a *cluster*.

In the simplest mesh networks, each node has only one radio transceiver. The radio channel used to provide access for the subscribers is the same one that is used for the wireless hop to the gateway or neighbouring node. Consequently, each hop required introduces a drop in capacity, with the loss increasing as the number of hops n increases; this must be taken into account when designing the mesh. Additionally, more hops can mean longer latency for delivery of packets, which can be important for applications such as VoIP. More recently, mesh nodes have been developed with two (or more) radios – one for the subscriber access and one for the back-haul hop – to avoid this problem [1]. Extending this approach, some systems dedicate separate radios for incoming and outgoing data to allow for higher-rate 'full duplex' operation [2]. Beyond the access and gateway layers, a management layer is also present in most mesh networks. The management layer is responsible for supervising the operation of the access points and controlling which subscribers are allowed access to the facilities of the mesh through functionality such as authentication, charging and security management.

Extending the mesh over larger areas requires more clusters of APs, and the provision of gateways at appropriate locations. In a complex mesh network, there may be several paths available for the nodes to reach the gateways, possibly with different numbers of hops with varying link qualities. Most mesh systems allow the routes to be dynamically assigned and changed as the network grows. Extra back-haul or capacity can be added to the mesh by simply changing a node into a gateway to add a back-haul connection. The surrounding nodes automatically adjust which paths they use to reach the gateway – typically, this is based on proprietary routing algorithms, which take into account the radio channel conditions, the number of hops, and the capacity and loading of the network.

One of the main advantages of a mesh network is its ability to adapt to the changing environment, such that users do not experience any interruption in service. In Figure 8.1, suppose the link between wireless node AP18 and gateway AP15 has been broken, for example by an obstacle such as a large truck being placed in the path between them. Then AP18 can reconfigure its path to the gateway via the intermediate neighbour node AP14. Although this introduces another hop into the path, it does ensure that network service can continue without interruption. When the obstacle is removed from the original path, the node will detect that the link quality to AP15 has improved and it will revert to the direct link. Hysteresis is built into this process to ensure that mesh linkages do not 'flip flop' due to short-timescale obstructions, such as passing traffic.

8.2 Mesh linking protocols

Because of the largely non-standardised status of mesh networking, a wide variety of protocols is used for forming mesh linkages between APs. The simplest use route-selection techniques similar to those used in standard IP routing, such as open shortest path first (OSPF) [3], whereby nodes exchange information between each other about available routes and their 'cost'. More sophisticated algorithms have been proposed and implemented, which take into account specific characteristics of radio links, such as variable packet-loss rate and congestion.

8.2.1 Handover, routing and roaming in MESH networks

It is perfectly possible for gateway nodes to use ADSL as the back-haul medium. However, it is also common for wireless point-to-point or point-to-multipoint links to be

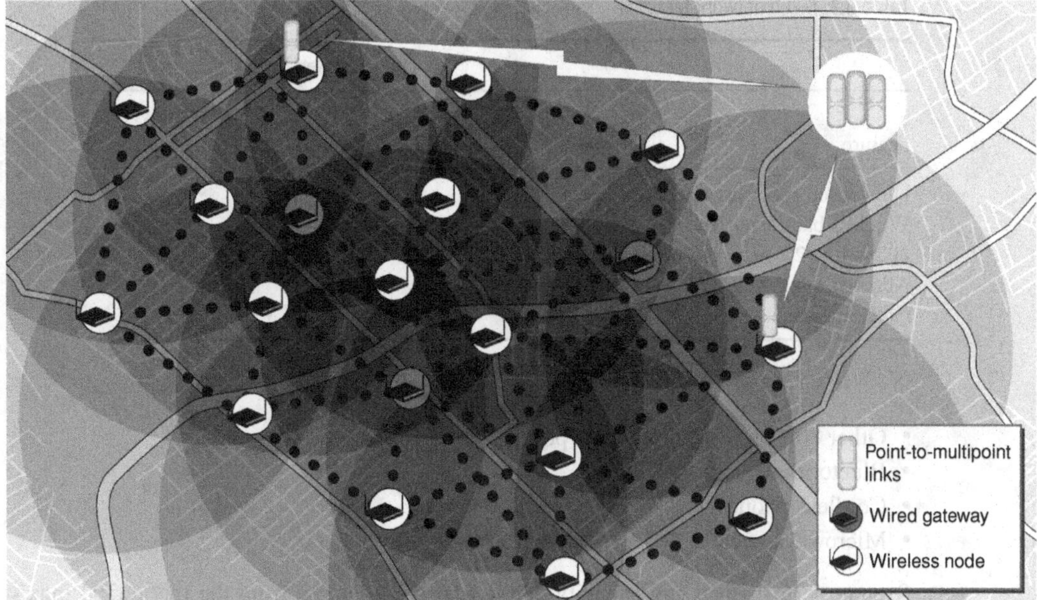

Figure 8.2 Wireless point-to-multipoint back-haul for capacity injection

used for this purpose. As well as being potentially cheaper to install and more flexible, they have greater uplink/downlink bandwidth symmetry. Depending on local regulatory requirements, these links may be operated in licensed or unlicensed spectrum, usually at 5.4 GHz or 5.7 GHz. The gateway uses a directional antenna to establish a back-haul link to some central location, usually a so-called 'high site', where the 5.4 or 5.7 GHz link is terminated. In point-to-multipoint configurations, several gateways access the same high site and the back-haul is aggregated at this central location. This is illustrated in Figure 8.2. The point-to-point nature of the links makes them resilient to interference even in unlicensed bands; however this does place stringent line-of-sight requirements for the gateways.

8.3 Mesh network planning example

8.3.1 Mesh-network air interface features

Mesh network features are summarised in Table 8.1.

8.3.1.1 Unlicensed spectrum operation

Most wireless mesh-network products are designed to operate in the 2.4 GHz ISM band. This avoids the costs associated with acquiring radio spectrum, but has a significant disadvantage – interference from other users is not under operator control. Interference in the ISM band comes from a number of possible sources:

Table 8.1. Mesh network features

Feature	Benefits: Capacity	Coverage	QoS	Deployment flexibility
Unlicensed spectrum operation				✓
Wireless back-haul	✓	✓		✓
802.11b/802.11g mixed-mode operation		✓		✓
Adaptive routing		✓	✓	✓
QoS support			✓	

(✓) Primary benefit (✓) Secondary benefit

- Other 802.11b/802.11g access points,
- Bluetooth devices,
- Cordless phones,
- Microwave ovens.

System performance and capacity can depart considerably from the theoretical maximum, and it is, therefore, not possible to plan for guaranteed throughput and latency performance for specific applications (contrast this with WiMAX or LTE operating in licensed spectrum). Instead, the goal for mesh networks is to provide a wideband data pipe, which can be shared by subscribers. The key parameter that influences system dimensioning is the contention ratio. This ratio is a direct measure of the number of subscribers who might be sharing the service at any one time. Thus, if a system has a bearer rate of 1 Mbit/s, 400 subscribers and a contention ratio of 20, the theoretical *average* service rate is 50 kbits/second (20 users will be sharing the bearer), but in practice only a few users are active at any instant, so the *peak* rate is closer to 1 Mbit/s. This type of system is well matched to bursty applications that can tolerate variable delay, such as email downloading.

8.3.1.2 Wireless back-haul

As discussed earlier in this chapter, one of the fundamental design concepts of mesh networks is that not all access points need to have dedicated back-haul provisioned, as they can route their traffic to other 'visible' nodes until egress from the network at a gateway node. In single-radio systems, the system capacity is reduced in proportion to the number of hops required to reach the gateway node. Overall system capacity is limited by the aggregate allocated total back-haul capacity; this can be increased by converting more access nodes into gateways. As a rule of thumb for design purposes, approximately 10% of the mesh nodes should be gateway nodes with back-haul.

8.3.1.3 802.11b/802.11g mixed-mode operation

Although 802.11g offers higher data rates (up to a theoretical maximum of 54 Mbps), there is still a significant number of legacy WLAN 802.11b-only clients, which need to be supported. Equally, while the higher rates available with 802.11g will increase overall

Table 8.2. Scenario for mesh-network worked example

Environment	Population density (/km^2)	Market penetration	User data rate (kbps)	Contention ratio
Urban or campus	1000	20%	512	20:1

system throughput, the more robust modulation of 802.11b is more useful in determining the effective cell radius.

8.3.1.4 Adaptive routing

During system design, an ideal mesh-routing topology might be identified, whereby the access nodes reach the gateways using the smallest number of hops. In practical systems, the mesh routes to the gateways will change over time with the radio environment. The system designer should endeavour to ensure that at least one and preferably two or more nodes are visible from each AP; isolated nodes or 'orphans' will be unable to mesh into the network, and will only be able to offer service if they are gateways with dedicated back-haul. If each AP has a choice of mesh routes, the network resilience to interference, obstructions and node failures will be increased.

8.3.1.5 QoS support

Since the mesh system operates in unlicensed spectrum, guaranteed QoS is not possible, as service may be degraded at any time by interferers. Most networks offer 802.11e QoS support, which provides various mechanisms for prioritising traffic associated with particular clients or applications. The standard provides a useful mechanism for prioritising VoIP traffic over best-effort Web-browsing traffic, for example. However, the unpredictable nature of interference means that detailed QoS design to deliver specific latency and throughput targets is usually futile.

8.3.2 Planning – preparation

8.3.2.1 Scenario sub-set selection

In this example, the scenario outlined in Table 8.2 will be examined. The scenario is a typical case, where an operator wishes to provide Wi-Fi service to a student campus or industrial estate. The *contention ratio* R_c is a measure of expected demand and is equal to the number of subscribers divided by the average number of subscribers expected to be active at any instant. A contention ratio of 20 would thus reflects an expectation that only 1 in 20 users would actively be transferring data at any one time. Many ISPs use a contention ratio of about 20:1 although higher ratios up to and beyond 50:1 are also found, with a corresponding degradation in end-user service quality. This number is sometimes also referred to as the *activity ratio*, or *over-subscription*.

8.3.2.2 Key questions to address

The system design activity for a mesh access network must answer the following questions:

- What propagation models are appropriate?
- How many nodes are required to deliver coverage?
- What radio channel plan should be used?
- How much back-haul is required to deliver the required capacity?
- Which nodes should be configured as gateways?
- What can be done to improve performance and capacity?

Each of these questions will now be addressed in turn.

8.3.3 Propagation model selection

Radio-frequency propagation modelling for mesh systems in the unlicensed ISM band is viewed in some quarters as a 'waste of time', since the interference from other users is, by definition, unknown. For rapid, large-scale and low-cost deployments it may well be possible to provide near-ubiquitous 802.11 coverage by simply placing an AP every 150 yards or so along the streets, on available lampposts. However, more careful modelling is appropriate if some or all of the following conditions apply:

- Wireless point-to-point back-haul is used – in this case LOS studies may be necessary,
- Terrain and clutter is highly non-uniform, and the street layout is not a regular grid,
- Coverage requirements are contractually committed to a network operator (e.g., guaranteed RSSI levels across 95% of the system).

Planning for mesh networks is somewhat different from that for traditional cellular systems, as there is an additional need to ensure that the mesh gateways can 'see' each other, as well as verifying coverage provided to subscribers. Most of the APs in a mesh network will be placed below roof height on suitable street furniture. If they are to mesh effectively, each node should be within the coverage area of at least one neighbouring node. In urban environments, the easiest way to ensure this is by making sure that there is line-of-sight coverage between the nodes. For small area deployments, the best way of verifying this is by performing a site visit; for large metropolitan systems, a detailed three-dimensional model of the buildings and terrain is required; most RF planning tools provide a facility to determine if LOS exists between two points, and the degree to which the first Fresnel zone has been impaired for point-to-point wireless back-haul. Ray-tracing models are more likely to model accurately the reflections and 'canyon' effects along streets that can occur when both the transmitter and receiver are below roof height. Wider-area coverage over less dense rural and suburban areas can be modelled with a COST231/Okomura–Hata model, tuned using measurements performed in the target environment. For more examples of propagation models at 2.4 GHz, consult [4].

Table 8.3. Sample link budget for cellular 802.11b [5]

	Indoor	Outdoor	Comment
Thermal noise (dBm/Hz)	−174		(Noise spectral density)
Channel bandwidth (dB)	73.4		(dB relative to 1 Hz)
Noise factor (dB)	5		
Noise power (dBm)	**−95.6**		
Interference margin (dB)	3		
Minimum SINR (dB)	0		
(*a*) Minimum Rx signal power (dBm)	**−92.6**	**−92.6**	
(*b*) Transmitter EIRP (dBm)	30	30	
(*c*) Sector gain (dBi)	6	9	
(*d*) Shadowing margin (dB)	−8	−8	
(*e*) Building penetration (dB)	−15	0	
(*f*) **Allowable path loss (dB)**	**105.6**	**123.6**	$[(b) + (c) + (d) + (e) - (a)]$
Median path loss @100 m (dB)	**80**	**80**	
Excess link budget @ 100 m (dB)	25.6	43.6	
Propagation exponent	4	4	
Cell radius (km)	**0.4**	**1.2**	

8.3.4 How many nodes are required to deliver coverage?

An estimate of the range of a cellular-style 802.11 network has been performed by Leung *et al.* in [5]. This considers in-building and outdoor coverage in a lightly wooded, flat environment with an access point antenna mounted at 30 m above ground – this is useful as a best-case scenario for the mesh network, whereas the propagation conditions are likely to be harsher. The link budget based on Leung *et al.* for 802.11b is shown in Table 8.3.

The maximum range is calculated for 1 Mb/s modulation on 802.11b, which requires a minimum SINR of 0 dB. The example shown assumes a transmitter EIRP of 30 dBm (1 W), which is the maximum in the USA; in Europe this figure is considerably reduced to 20 dBm (100 mW), which brings the typical AP maximum radius down to 150–200 metres, if good meshing is to be achieved.

An inter-node separation of 200 metres gives a node density of 20–25 nodes/km^2. Note that this number could vary considerably depending on radio propagation and interference conditions.

8.3.5 What radio channel plan should be used?

The 802.11b and 802.11g standards in the 2.4 GHz band divide the spectrum into 14 channels (11 in the USA) of 22 MHz bandwidth at 5 MHz separation – these channels overlap, and if adjacent channels are used on neighbouring access points there will be considerable interference. For in-building systems, channels 1, 7 and 13 are often used as these do not overlap each other. For outdoor systems this technique is also possible, but it does not allow for the fact that other users of the spectrum will be generating interference at unpredictable levels. For this reason, most commercial WLAN mesh systems use automated techniques to select the AP frequency. During quiet periods, or at a scheduled time in the night, the AP scans the spectrum to see which frequencies are being used most heavily, and selects the quietest channel on which to operate. If a single radio mesh product is being used, all nodes in the cluster need to be tuned to the same channel.

8.3.6 How much back-haul capacity is required?

Referring to the scenario in Table 8.2, the total number of subscribers to be offered service per square kilometre is simply:

$$\begin{aligned} N_{\text{subs}} &= \text{population/km}^2 \times \text{market penetration} \\ &= 1000 \times 0.2 \\ &= 200. \end{aligned}$$

The aggregated back-haul requirement (μ_{tot}) is, therefore, found by multiplying the number of users by the contention ratio (R_{c}) and the user data rate:

$$\begin{aligned} \mu_{\text{tot}} &= N_{\text{subs}} \times R_{\text{c}} \times 0.5\,\text{Mbps} \\ &= 200 \times 1/20 \times 0.5 \\ &= 5\,\text{Mbps}. \end{aligned}$$

This capacity can be met by provisioning four links of 1.5 Mbps for each km^2 of the campus area in question; at least four of the APs should be connected to these back-haul links and configured as gateways. Note that the aggregate capacity of the radio-access layer of the nodes is greater than the back-haul capacity. Each individual node is typically capable of delivering >1 Mbps (uplink and downlink) to individual users, who will then contend for back-haul through the gateway nodes.

8.3.7 Which nodes should be configured as gateways?

The choice of gateway nodes may be dictated by the ability to provision back-haul at particular locations. If wireless back-haul (for example using point-to-point links in the 5 GHz range) is being used, the selected nodes should have good line-of-sight coverage to the back-haul aggregation cluster. If fixed-line back-haul links are provisioned, then the gateway candidates should be mounted on or near a suitable building for the connections to be terminated.

In choosing the gateway nodes, the mesh topology is studied and a configuration selected that minimises the total number of hops in the system. This will provide the maximum aggregated throughput and lowest average latency. Clearly, for a system of >100 nodes the number of possible permutations is very large, and software tools can help in identifying optimal configurations.

8.3.8 What can be done to improve performance and capacity?

As data volumes grow, a number of actions can be taken to increase capacity, including:

- Adding more back-haul capacity at gateway nodes,
- Adding extra mesh nodes,
- Converting access nodes into gateways by providing back-haul.

If the higher rates of 802.11g are to be fully exploited it may also be necessary to consider making the network 802.11g only, and denying access to legacy 802.11b users – the feasibility of this will depend on the market in which the deployment is to take place.

References

1 Motorola Inc., *Motorola's MOTOMESH Series of Products Now Offer More Choices for Municipal Wireless Networks*, http://www.motorola.com/mesh/pages/newsroom/press_releases/ 2007_07_24.htm.

2 Strix Systems Inc., *Solving the Wireless Multihop Dilemma*, 2006.

3 IETF, Open Shortest Path First (OSPF), *RFC 2328*, April 1998.

4 R. A. Santos, O. Álvarez and A. Edwards, Experimental analysis of wireless propagation models with mobile computing applications, *2nd International Conference on Electrical and Electronic Engineering (ICEEE) and XI Conference on Electrical Engineering* (CIE 2005).

5 K. K. Leung, M. V. Clark, B. McNair *et al.*, Outdoor IEEE 802.11 cellular networks: radio and MAC design and their performance, *IEEE Transactions on Vehicular Technology*, **56** 5 (2007).

9 Core network and transmission

This chapter aims to provide an overview of the role of the *core network* and *transmission* in wireless solutions. Insight is given into the factors that have influenced network evolution from early cellular architectures, such as GSM Release 98, through to systems currently being standardised for the future, exemplified by Release 8. The chapter will conclude with a worked example illustrating the dimensioning of an IP multimedia system (IMS) transmission for a system supporting multiple applications.

It is useful to establish a common terminology before discussing networks in more detail. In the early 1990s, ETSI proposed the convention shown in Figure 9.1 [1], to distinguish between two distinct types of *circuit* service that a network might provide, namely *bearer services* and end-to-end applications, which it called *teleservices*. In the case of bearer services, a wireless network is 'providing the capability to transmit signals between two access points'. Support of teleservices, however, requires the provision of 'the complete capability, including terminal equipment functions, for communication between users according to protocols established by agreement between network operators'. Defining teleservices in this way has standardised the details of the complete set of services, applications and supplementary services that they provide. As a consequence, substantial effort is often required to introduce new services or simply to modify the existing one (customisation). This makes it more difficult for operators to differentiate their services.

In 1999, therefore, a second service framework was introduced, to run in parallel with the circuit service framework, whose aim was to 'standardise service capabilities and not the services themselves' [2]. Service capabilities consist of bearers defined by QoS parameters and the mechanisms needed to realise services. The mechanisms include the functionality provided by various network elements, the communication between them and the storage of associated data. The intention was that these standardised capabilities should provide a defined platform, which would enable the support of speech, video, multimedia, messaging, data, teleservices, user applications and supplementary services and enable the market for services to be determined by users as well as operators.

9.1 Core network evolution

9.1.1 Factors driving core network evolution

Table 9.1 shows the introduction of major feature functionality by 3GPP release. In GSM phase 1, the originally specified services covered full rate speech (FR) and

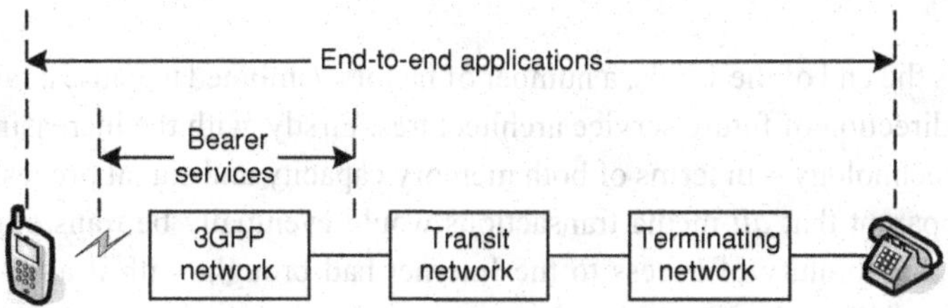

Figure 9.1 Description framework for telecommunications services

circuit teleservices in the form of SMS and fax. Releases 96 and 97 saw the introduction of two key features, CAMEL (customised applications for mobile enhanced logic) and GPRS-bearer services, along with other data and voice services. The significance of GPRS has already been discussed in Chapter 2 but IN (intelligent networking in the form of INAP (IN applications protocol) and CAMEL) is important because, for the first time, services that the subscriber is familiar with in the home network can operate in exactly the same way when they are roaming. CAMEL is used extensively for the implementation of prepaid services and relies on *event triggers*, occurring in the MSC of the network supporting roaming, to trigger services hosted by the CAMEL server in the home network. This functionality is discussed in detail in [3].

Release 98 represented the final release focused solely on the GSM air interface and introduced the *adaptive multi-rate* (AMR) codec [4] which provides a family of speech codecs of different rates that can be selected according to the prevailing C/I at the air interface. Additionally, *location services* [5] were introduced to provide a standardised enabling technology for the introduction of location-based applications.

Table 9.1. System features by major release

	3GPP release						Key functionality
	Ph1/2	96/97	98	99/4	5/6	7/8	
Basic services	✓						Speech, SMS, fax
HSCSD		✓					Circuit-switched data (4.8 to 57.6 kbits/s)
CAMEL		✓					Support of home network services – even when the user is roaming
Voice services		✓					Large number of largely voice-focused supplementary services introduced
GPRS		✓					Introduction of packet radio service
AMR			✓				Adaptive multi-rate codec and associated adaptation functionality introduced
Location			✓				Introduction of location service to support higher-layer applications
UMTS				✓			Introduction of 3G services
IMS					✓		Support of IP based multimedia services
MBMS					✓		Introduction of multimedia broadcast and multicast services
E-UTRAN/EPC						✓	Evolution of the 3GPP system

Towards the end of the 1990s, a number of factors combined to cause a radical rethink about the direction of future service architectures. Firstly, with the increasing capability of digital technology – in terms of both memory capacity and signal processing – it was already apparent that *all* media transactions would eventually be transmitted digitally. Secondly, the ubiquity of access to the Internet had brought with it a de-facto global acceptance of IP as the layer 3 routing protocol. Finally, the increasing penetration of broadband to the home as well as offices was moving service expectations from 'dial-up' access times of tens of seconds to 'always-on' latencies of 100 ms or so.

The first industry response to this trend is generally perceived to have been the introduction of UMTS [6] in Release 99. The reality, as discussed in Chapter 6, is that the performance of UMTS Release 99 reveals its origins in the mid 1990s when architectures to support low-access latency and effective QoS implementation were not foreseen as important by operators or vendors.

Arguably, the industry's first response to the new multimedia world was the introduction of IMS [7] in Release 5. For the first time, an all IP framework was established, which enables:

- Support of applications not developed within 3GPP,
- Common service capability independent of the access network,
- Support of multiple (and different) multimedia applications concurrently within one user session,
- Negotiation of QoS for each IP multimedia application at the time of application establishment.

The rollout of IMS has been fairly limited at the time of writing, largely because of the significant session set-up times caused by the time taken to transport the relatively large SIP (session initiation protocol) messages (used for session set-up between the mobile and IMS) over the narrow signalling bearers of UMTS. The final piece of the multimedia solution will, therefore, fall into place with the introduction of evolved UMTS in Release 8. The low latencies of the control and bearer plane in E-UTRAN and EPC combined with the higher control bearer rates [8] will for the first time deliver a mobile multimedia experience comparable to that for the fixed network.

9.1.2 The GSM Release 98 core network

In Chapter 2, the components of a Rel.98 GSM core network were discussed and the way in which they inter-work to support call establishment was described; it is not intended to expand further on this here. Rather, the intention is to discuss the functionality of the MSC, since it is the single most complex element of the core network, with a view to understanding how to dimension this key element. For convenience, the core network aspects of Figure 2.1 are reproduced here as Figure 9.2.

The Release 98 MSC is based on the ISDN exchange (then in common use). It has, for call control, the same interface as the fixed network exchanges towards the legacy PSTN and new functionality in the form of *mobility management* (MM), *connection management* (CM) and *call control* (CC) towards the radio-access network. The CC

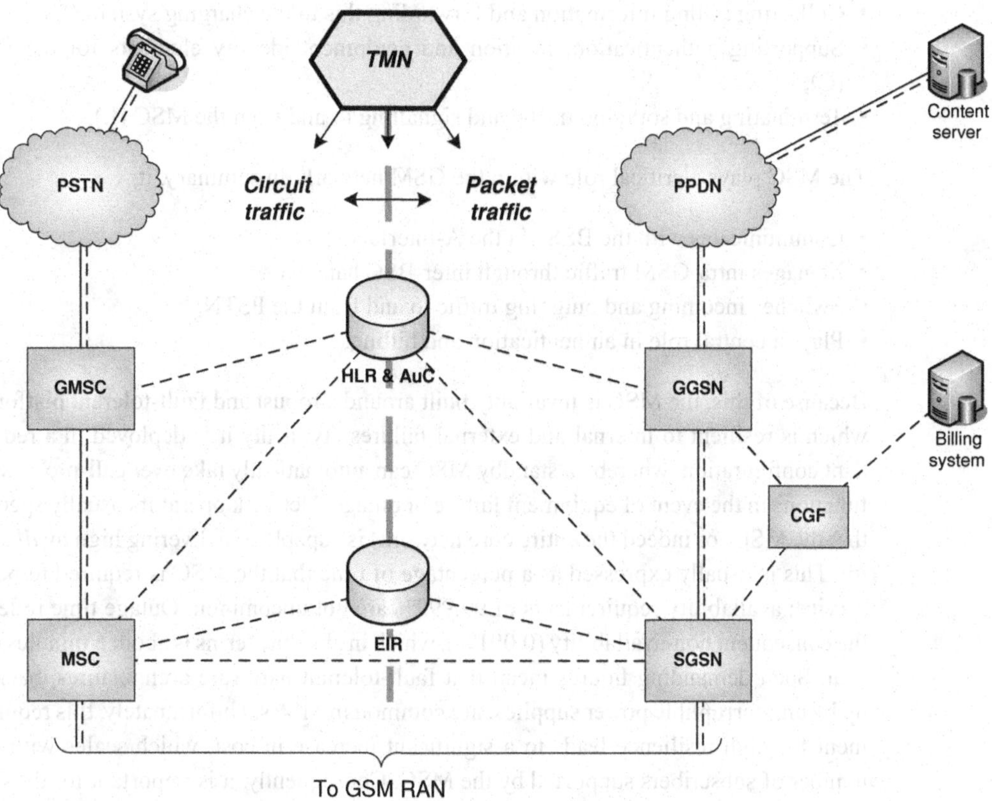

Figure 9.2 The Release 98 core network

signalling protocols (discussed further in the GSM technical specifications) are related to the signalling system No 7 user parts TUP (telephone user part) and ISUP (integrated services user part (of ITU signalling system No 7)) associated with the circuits used for incoming and outgoing calls. The major component parts can be summarised as:

- 64 kbits/s circuit switch fabric (S),
- Control processor unit (C),
- Line cards (L).

The key functions of an MSC [9] are set out below and annotated with either an 'S', a 'C' or an 'L' to indicate which element of the MSC architecture is primarily responsible for its realisation:

- Connecting incoming voice calls from the PSTN to mobile on the RAN (S), (C),
- Routing outgoing voice calls from the mobile to the fixed-line PSTN user (C), (S),
- Routing voice calls from mobile users to other mobile users (C), (S),
- Providing access to other services, e.g., voicemail for the cellular subscriber (C),
- Managing handovers from one BSC to another BSC (C), (S),

- Collecting billing information and forwarding this to the charging system (C),
- Supporting authentication, location and equipment identity elements for the BSS (C),
- Terminating and sourcing traffic and signalling to and from the MSC (L).

The MSC plays a critical role within the GSM network. In summary, it:

- Communicates with the BSS via the A-interface,
- Manages intra-GSM traffic through inter-BSC handover,
- Switches incoming and outgoing traffic to and from the PSTN,
- Plays a central role in authentication and billing.

Because of this, the MSC is invariably built around a robust and fault-tolerant platform, which is resilient to internal and external failures. Typically it is deployed in a redundant configuration, whereby a standby MSC can automatically take over call-processing functions in the event of equipment failure or outage. Network operators usually specify that the MSC (or indeed the entire core network) is capable of delivering high *availability*. This is usually expressed as a percentage of time that the MSC is required to be in service; availability requirements of 99.999% are not uncommon. Outage time reflects the consequent non-availability (0.001%), which in absolute terms is about 5 minutes per year. Such demanding figures mean that fault-tolerant hardware architectures, backed up by uninterruptible power supplies, are common in MSCs. Unfortunately, this requirement for high resilience leads to a significant increase in cost, which scales with the number of subscribers supported by the MSC. Consequently, it is important for the network designer to ensure that the MSC is adequately designed to cater for the expected traffic and subscriber volumes, but is not over-dimensioned, as this would lead to unused capacity and hence unnecessary capital expenditure.

In designing the RAN, key parameters have been the number of subscribers per km^2 and the total area to be covered. In considering the core network design, it is important to recognise that an increasing proportion of traffic is *between* mobile subscribers rather than from a mobile to a 'fixed' network subscriber. This directly impacts the bearer path dimensioning of the MSC and BSC as more traffic stays within the RAN and can potentially reduce the BSC/MSC transport needs. Other key parameters to establish include:

- Number of subscribers per MSC,
- Number of busy hour call attempts (BHCA),
- Traffic in mE/sub,
- Proportion of traffic mobile-to-land, land-to-mobile and mobile-to-mobile,
- Services to be offered: voice, voicemail, SMS, CS data, etc.,
- Number of pages per successful mobile termination,
- Number of handovers per call,
- Number of location updates,
- Number of authentication, EIR check requests,
- Number of SMS, voicemails, CS data per session, etc., per subscriber.

The amount of MSC equipment required to support the planned traffic will vary from manufacturer to manufacturer, depending on the relative capabilities of the hardware and software architectures. Most manufacturers will provide an equipment planning guide – either a document describing how to calculate MSC capacity, or more commonly a spreadsheet or automated tool. The network designer inputs the design parameters as listed above, and the tool will provide a complete parts list and pricing estimate.

An important observation to make in the context of core network evolution is that in the Release 4 architecture [10], ETSI specified a distributed MSC architecture comprising an MSC server and a circuit-switched media gateway (CS-MGW). In this new architecture, the MSC server is responsible for the management of connection, mobility and call control for circuit-switched calls. It terminates the user–network signalling and translates it into the relevant network–network signalling. The MSC server also contains a VLR to hold the mobile subscriber's service data and CAMEL-related data. The CS-MGW operates under the control of the MSC server and may terminate bearer channels from a switched circuit network and media streams from a packet network (e.g., from an IP network). The CS-MGW also provides media conversion, bearer control and payload processing (e.g., codec, echo canceller, conference bridge functionality) for support of different service and transport options for CS services (AAL2/ATM-based as well as RTP/UDP/IP-based).

The thinking behind the Release 4 distributed MSC option was that it would become increasingly cost effective as:

- The MSC bearer and control requirements ceased to scale together.
- More traffic stayed local to the mobile network and small CS-MGWs could be sited close to the BSS to provide inter-BSC switching, eliminating back-haul to and from the remote MSC.

Whilst the scalability arguments are realised (with networks architected accordingly) it is, in fact, unusual for networks to be configured with MGWs sited at remote BSC or RNC sites. Instead, it is more usual to find the RNC/MSC server and MGW co-sited in more centralised network offices. By locating these three entities in the same room, there are very large savings in transmission and reduced travel and labour costs for maintenance. The network MSC server function can be provided by a very small number of very high availability platforms. Typical MGW:MSC server ratios are 1:1 for very small networks and 3:1 for larger networks.

Although not addressed here, dimensioning of the SGSN and GGSN will follow a similar approach, again relying on the manufacturer's equipment planning guide. In this case, though, the key parameters will be different and include maximum values for:

- Number of attached subscribers,
- Number of active subscribers,
- Number of mobility events per second,
- Ciphering requirements,
- Expected data generation per user (uplink),
- Expected received data per user (downlink).

9.1.3 3GPP Release 8 core network

Until recently, communication systems were designed with a specific form of content in mind. The earliest systems were designed to support data (Morse code) and voice (in analogue format) and, as recently as the 1980s, new PSTN networks were being designed around specific codecs, such as G711. However, with the dramatic rise in the power of digital technology discussed earlier, it became possible to envisage a time when all content would be conveyed over digital transmission systems with the concomitant requirement for a more flexible architecture. Concurrent with this realisation, the success of the World Wide Web made a de-facto world standard of the Internet Protocol (IP) and it became possible to envisage a new communications platform based on broadband IP transport, where users would request the resources required to sustain the QoS necessary for their communication session. A communications platform built in this way would be able to support applications that are as yet unforeseen because the resource requirements are defined in a generic set of attributes. This concept is captured in two key IETF (Internet Engineering Task Force) RFCs (requests for comment): RFC 3261 [11] and RFC 4566 [12]. These describe the session initiation protocol (SIP) and session description protocol (SDP). These are at the heart of IMS, first defined in 3GPP Release 5, and central to the Release 8 end-to-end solution. These protocols will be briefly discussed next.

9.1.3.1 SIP and SDP

Anyone associated with the Internet development community quickly becomes aware that the IETF does much more than specify transport protocols, such as IP. The IETF's charter is, 'Identifying, and proposing solutions to, operational and technical problems in the Internet.' It largely achieves this through the development and selection of standards to form the Internet protocol suite. In this context, SIP was originally proposed as a means of establishing audio and video conferencing sessions between multiple locations. However, the open and flexible nature of SIP means that it is equally suited to the support of general IP multimedia flows. The SIP signalling messages can contain an SDP payload, which gives details of the type of service being requested.

SIP is used between two *user agents* (UA) wishing to communicate over some medium. Figure 9.3(a) shows a simplified set of SIP transactions that might occur in establishing a call [13]. User agent 1 sends an invite message to user agent 2, which immediately responds with a 'trying' message. The request is queued whilst the target user deals with other calls, and eventually user agent 2 responds with an OK message to user agent 1, which immediately sends back an acknowledgement (ACK). At this point the session has been established and communication continues until user agent 1 terminates the call with a BYE message, acknowledged by user agent 2 with an OK.

User agents, by definition, reflect the capabilities of the device they represent – perhaps user agent 1 is representing a PC whilst user agent 2 is reflecting the capabilities of a multimedia mobile device. Extensions to the SIP protocol [14] allow the SIP 'contact' header field to contain parameters reflecting the agent (and hence device) capability. SIP invitations contain an SDP message, which is used to describe the multimedia session

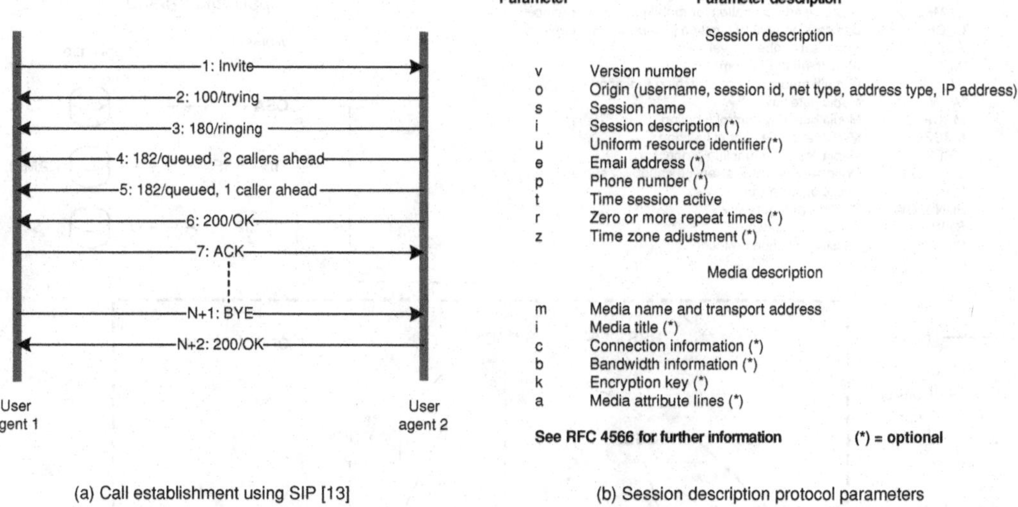

(a) Call establishment using SIP [13]

(b) Session description protocol parameters

Figure 9.3 SIP and SDP protocols

requested, for the purposes of session announcement, session invitation and other forms of multimedia session initiation. This information, together with the UA capabilities descriptions, allows participants to agree on a set of compatible media types. SIP makes use of elements called proxy servers to help route requests to the user's current location, authenticate and authorise users for services, implement provider call-routing policies and provide features to users. SIP also provides a registration function that allows users to upload their current locations for use by proxy servers.

Figure 9.3 (b) shows the format of the SDP protocol. It comprises a sequence of strings consisting of a 'type' field of 1 character, an = sign, and a value field, whose format depends on the type. The first few lines of the descriptor comprise the *session-level* section, giving details of the session such as name, originator, time-zone, etc. These are followed by one or more *media-level* sections which describe the media flows that are to be established during the session. The IETF defines a small number of media types: *audio*, *video*, *text*, *application* and *message* (other types can be proposed to IANA should they become necessary). The media descriptions (beginning with 'm =') contain details of media type, port numbers, protocols to be used and other formatting information. Thus, an example format could define the first media stream to be of type audio, on port 49170, and using the RTP protocol, with default encoding parameters. A second stream of type video might also be specified on port 51372, again using RTP, but this time additional attributes might be included via the 'a =' *attributes* descriptor; the encoding type could be explicitly specified – e.g., h263–1998. By listing multiple attributes for the media stream, the participant in the SIP session can list a variety of encoding types that can be supported for the stream, and the other participant can decide which encoding method to use based upon its capabilities. Further SDP lines describe characteristics such as bandwidth, encryption, quality and many other media-dependent attributes to allow a rich description of the multimedia sessions [12].

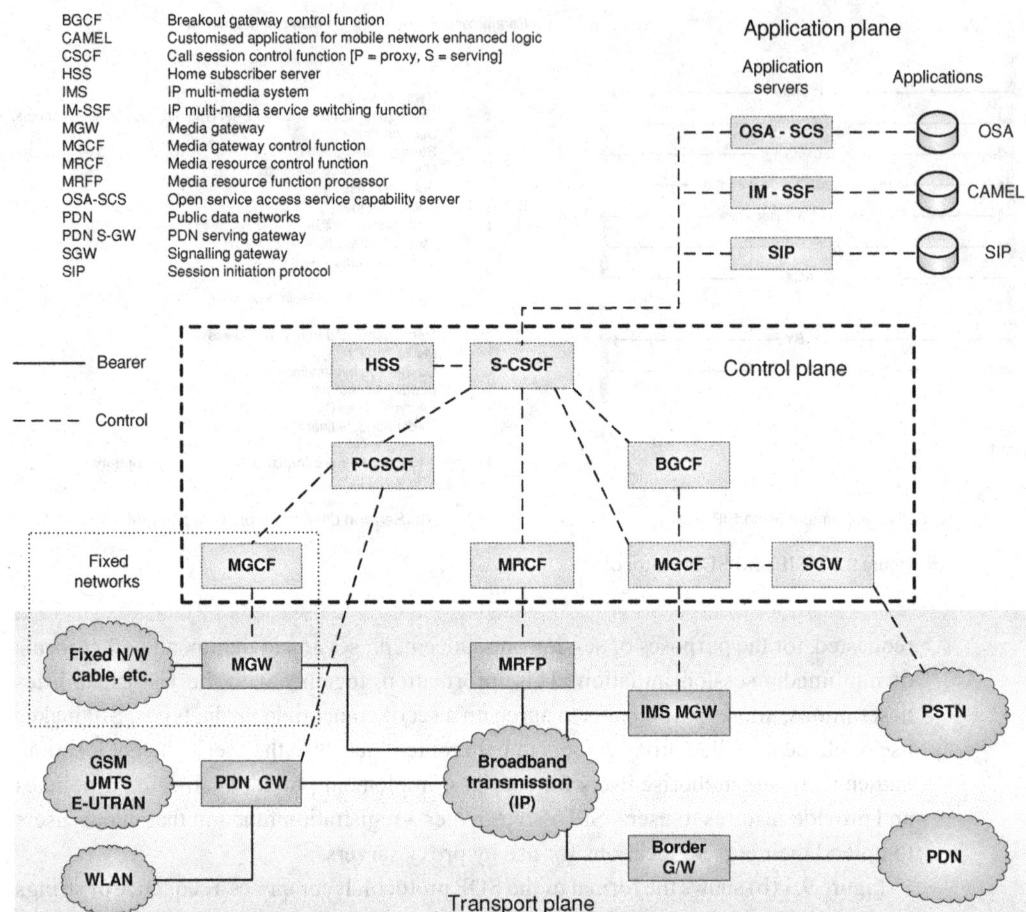

Figure 9.4 Expected Release 8 core network architecture

9.1.3.2 Release 8 core network

The network architecture for 3GPP Release 8 [15] defines the core network as logically divided into a CS domain, a PS domain and an IM sub-system. The CS domain refers to the set of all the CN entities offering 'CS type of connection'. The focus of this section will be on the delivery of multimedia services via the packet domain, as this will reflect the majority of new network deployments. Note, though, that there will be increasing inter-working of CS and PS domains via a shared IMS platform until the time comes when 2G devices are no longer supported, although this point is some way in the future.

Figure 9.4 shows the expected architecture of the Release 8 core network supporting packet-switched (PS) services via an E-UTRAN and evolved packet core. The SAE gateway (not shown) and the PDN gateway form the new EPC, as discussed in Chapter 7, and, in a similar way to the SGSN and GGSN, hide terminal mobility from the rest of the network using a dedicated mobility-management entity (MME). The IMS sub-system is seen to comprise three distinct layers or 'planes': the *application plane* (sometimes called the service layer), the *control plane* and the connectivity or *transport plane*.

The application plane is responsible for delivering the applications to the end user – voice services, messaging, video clips, music downloads or combinations of all these and more. *Application servers* (AS) are used to deliver this content, which may be drawn from one or more AS and combined as necessary.

Access to these services is managed by the control plane, which is responsible for authenticating, routing and coordinating requests for services with the appropriate AS. Within the control plane is the HSS, which is the master database for all users. It contains enhanced HLR authentication information (familiar from the Release 98 architecture) and new subscription-related information to support the network entities actually handling calls or sessions [15]. The primary function of the control plane is session establishment, which is accomplished by network elements called *call session control functions* (CSCF); the name is somewhat misleading as they control far more than 'calls' in the traditional telephony sense. Essentially, these are nothing more than SIP servers and proxies, which set up the appropriate media streams with the AS and negotiate QoS between end points. As SIP and SDP are being used to set up the various media streams, the CSCF has knowledge of both the bearer characteristics and the application content being delivered to the end user. It can then forward this information to the billing and charging functions so that users are billed for the resources they use in a flexible manner. In addition to the CSCF, another set of SIP/SDP instances is hosted in the proxy CSCF (P-CSCF). The proxy CSCF is the first contact point within the IM CN sub-system for users seeking to set up a session. Its address is discovered by UEs during 'call' set-up and it behaves like a proxy (as defined in RFC 3261 [11]), i.e., it accepts requests and services them internally or forwards them for further processing. The P-CSCF supports a number of key functions including:

- Maintaining a security association between itself and each UE,
- Generation of call detail records,
- Forwarding SIP messages received from the UE to the SIP server (e.g., S-CSCF),
- Forwarding SIP requests or responses to the UE,
- Authorisation of bearer resources and QoS management.

The breakout gateway control function (BGCF) processes requests for routing from an S-CSCF for the cases where the S-CSCF has determined that the session cannot be routed using DNS or ENUM/DNS. When traffic is destined for the PSTN, it manages the IMS media gateway (IMS-MGW) and signalling gateway (SGW) via the media gateway control function (MGCF). The IMS-MGW transforms IP traffic to an appropriate format for the PSTN, and the SGW transforms the IP control protocols used within IMS to the SS7 protocols on the CS network. Further control of transport-plane functionality is afforded by additional instances of the MGCF and the media resource control function (MRCF).

Below the control plane is the *transport plane*, responsible for interfacing the control and data streams to the communication infrastructure. Key resources in the transport plane are the IMS-MGW, the media resource function processor (MRFP) and IP transport. The MRFP is a key enabler for several services including:

- Audio mail,
- Audio conferencing,
- Video conferencing,
- Media stream processing (between different formats or codecs),
- Media stream source (announcements, etc.).

The final and most important elements of the transport plane are the transmission resources themselves; these comprise the IP connectivity access network (IP-CAN) – in this case, the E-UTRAN and EPC – and IP bearer services (IP-BS) in the IMS IP transport network. These resources are policed by the policy charging enforcement function (PCEF) found in the PDN S-GW or GGSN. The PCEF has the capability of policing packet flow into the IP network, and restricting the set of IP destinations that may be reached from or through an IP-CAN bearer according to a packet classifier. This policy 'gate' function has an external control interface that allows it to be selectively 'opened' or 'closed' on the basis of IP destination address and port. The 'gate' control is performed by a policy charging and resource function (PCRF), which is a logical entity of the P-CSCF.

During establishment or modification of a SIP session, the P-CSCF will use the SDP contained in the SIP signalling to derive the session information that is relevant for policy and charging control and forward it to the PCRF. The PCRF will use the received information to calculate and authorise the required QoS resources for IP-CAN and IP-BS. The authorisation will be expressed in terms of the IP resources to be authorised and include limits on IP packet flows, and may include restrictions on IP destination address and port [16, 17, 18].

In concluding this overview of IMS, it is appropriate to note that this open services framework has proved very attractive to other standards bodies, as it provides a vehicle for sharing a common application set across many access media. ETSI TISPAN (telecoms and Internet converged services and protocols for advanced networks) has included IMS as one of the key components of the next-generation networking (NGN) architecture [19]. CableLabs, responsible for defining standards for the cable TV industry, have included IMS in their PacketCable standard, allowing IP-based multimedia content to be delivered over bidirectional domestic cable connections [20]. The convergence of fixed-line telephony, Internet access and broadcast video is sometimes described by marketing departments as *'triple-play'*; the addition of wireless services leads to the concept of *'quad-play'*. From a commercial perspective, the convergence of the underlying wired and wireless worlds has brought about a blurring of the distinctions between cellular network operators, Internet service providers, fixed-line telcos and broadcast and cable TV operators. Increasingly, users of these services are able to buy integrated services offerings from one provider with a single bill for all their communications requirements.

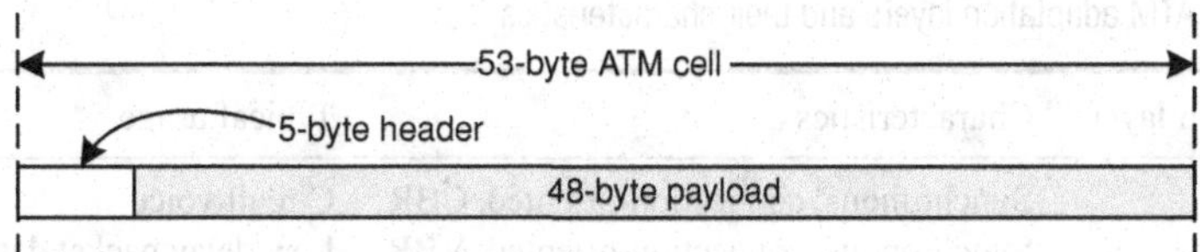

Figure 9.5 Structure of an ATM cell

9.2 Transmission systems

The R99 CN made two recommendations for the transport protocol architecture: asynchronous transfer mode (ATM) for CS transport and IP transport for the control plane and optionally for the user plane. By Release 5, the increasing dominance of IP as the networking protocol of choice led 3GPP to propose that IP or ATM were equally valid choices for both the user plane and the control plane. This change in transport architecture allows an all-IP core, which can radically reduce the number of protocols to be supported within the CN, with corresponding opportunities for major operational cost savings. It is important to note, however, that the IP protocol has its roots in the computer-networking community, where communications between nodes were largely based on a best-effort basis, and over-provisioning of links was the main mechanism for satisfying QoS requirements. In the telecommunications world, where back-haul costs are a major component of operational expenditure, such practices are not ideal – especially when they do not even result in a guarantee that QoS requirements will be met.

In this section, an overview of the key advantages and disadvantages of two current transport solutions – ATM and IP-based 'packet' networks – will be provided, along with a brief commentary on some of the mechanisms used to manage QoS in such networks.

9.2.1 Asynchronous transfer mode

Asynchronous transfer mode (ATM), proposed by the ITU, was standardised in the early 1990s, and was designed to provide guaranteed performance over high-speed links to support packetised data flows for real time and non-real time applications. To ensure the low latency and jitter required by demanding interactive applications, such as voice, traffic is divided into small fixed-size chunks called cells (equivalent to small packets) consisting of a 48-byte payload and 5-byte header (Figure 9.5). These cells are transmitted over a virtual circuit negotiated between two end points of the ATM network. The small size of the cells ensures that voice traffic is not subject to unacceptable delays if larger bursts of traffic (for example from video streams) are presented to the network, as the voice traffic can still be efficiently interleaved.

ATM defined a set of service classes, which can be used to differentiate the characteristics of the traffic being carried. From an application perspective, the key characteristics of the service classes are:

Table 9.2. ATM adaptation layers and their characteristics

Adaptation layer	Characteristics	Typical usage
AAL1	Synchronous, connection-oriented, CBR	Circuit voice
AAL2	Synchronous, connection-oriented, VBR	Low delay packet data
AAL3/4	Synchronous, connectionless, VBR	Connectionless packet data
AAL5	Asynchronous, connectionless, VBR	IP over ATM

Burstiness Constant bit rate (CBR), variable bit rate (VBR) or unspecified (UBR),
Connection Connection-oriented or connectionless,
Synchronisation End points synchronised or unsynchronised.

Using these definitions, a series of service classes was proposed to meet different application requirements, such that network designers would be able to map their transport requirements onto the ATM bearers efficiently. The way in which the incoming packets are disassembled into cells is defined by one of a set of ATM adaptation layers (AALs). The original intention of the ITU was that each service class would be mapped onto its own AAL; however, over time this distinction became blurred and it is more instructive to simply refer to the AALx characteristics rather than the service classes themselves. These are summarised in Table 9.2, along with typical applications, where relevant.

The advantage of fixed, small ATM cells is that they can be very rapidly switched in hardware by the core network routers, allowing the bandwidth of high-speed links to be efficiently utilised. However, one key disadvantage of ATM is its complexity; the segmentation and reassembly (SAR) of the incoming and outgoing packet streams via the AAL poses a significant processing overhead, and as data links have become ever faster, the SAR has become a bottleneck. In addition, the emergence of very-high-speed transmission links (capable of speeds up to almost 10 Gbits/s for OC-192) has meant that the need for ATM to remove delay and jitter is fast disappearing in these systems. ATM is, however, ideally suited for more limited bandwidth DSL links, with rates of around 2–8 Mbit/s, and hence its use is likely to continue for some time.

9.2.2 IP packet networks

IP-based networks, like ATM, employ a packet-based transport approach. Messages sent over a packet-switched network are first divided into packets containing the destination address. Then each packet is sent over the network, with each intermediate router in the network determining where the packet goes next. A packet does not need to be routed over the same links as previous related packets. Thus, packets sent between two network devices can be transmitted over different routes in the event of congestion at one node or a link failure. Unlike ATM, the number of bytes per packet is not fixed and there are no predetermined QoS configurations. This has a number of advantages in that the payload size of the IP packet can be optimised to minimise the header overhead for a given service but it therefore means that other mechanisms must be used to manage QoS.

To address the problem of IP QoS, the IETF has proposed two candidates solutions; *integrated services* (IntServ) [20] and *differentiated services* (DiffServ) [21]. IntServ uses a resource reservation protocol (RSVP) along a path between two end points. Intermediate nodes, such as routers, must ensure that the path QoS requirements can be met, or they must reject the reservation request. Whilst IntServ does indeed provide a workable mechanism for QoS reservation, it requires all nodes supporting it to maintain and refresh the state of the various paths, which makes it unwieldy for large-scale IP networks and, consequently, its adoption has been limited. In the alternative DiffServ model, packets are classified and marked with a *per-hop behaviour* (PHB) depending on the stringency of the QoS requirements. For example, packets associated with an application with a low-latency requirement (e.g., VoIP) can be flagged as such and are forwarded quickly, whilst packets associated with background traffic classes can be delayed or dropped altogether in the worst case. The DiffServ PHBs can be summarised as follows:

- *Expedited forwarding (EF)* Traffic in this category requires low latency or delay, low packet loss and low jitter. These are commonly the requirements for highly interactive services, such as voice [22].
- *Assured forwarding (AF)* This category attempts to guarantee delivery of packets unless a particular traffic rate is exceeded, otherwise packets will become candidates for being dropped. Assured forwarding supports four classes, with differing delay and bandwidth characteristics [23].
- *Default* Traffic not allocated to other classes is dealt with on a best-effort basis.

The concept of DiffServ precedence classes maps reasonably well to UMTS CN QoS classes, and, hence, the adoption of DiffServ within the CN is common (see for example [24] and [25]). The expedited forwarding PHB maps to the conversational class, and the remaining UMTS QoS classes map to the different available AF classes. Nonetheless, DiffServ does not help much in situations where the network is experiencing significant congestion or where the incoming traffic is highly bursty in nature, as the QoS requirements may still not be met.

9.2.3 Multi-protocol label switching

Many of the concepts from ATM and DiffServ have been adopted in *multi-protocol label switching (MPLS)* [26], defined by the IETF, which integrates the IP and ATM data-link protocols into one networking domain, as illustrated in Figure 9.6.

All multilayer switching solutions, including MPLS, are composed of two distinct functional components; a control component and a forwarding component. The control component uses standard routing protocols to exchange information with other routers to build and maintain a forwarding table. When packets arrive, the forwarding component searches the forwarding table maintained by the control component to make a routing decision for each packet. Specifically, the forwarding component examines information contained in the packet's header, searches the forwarding table for a match and routes the packet from the input interface to the output interface across the system's switching fabric.

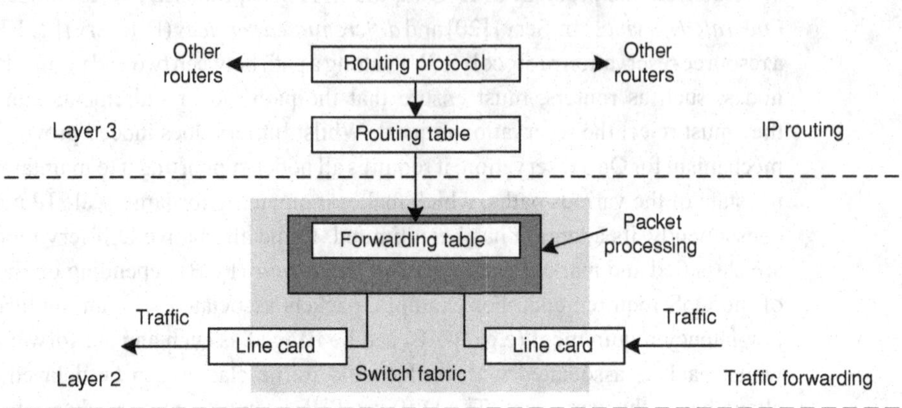

Figure 9.6 MPLS integrates layer 2 and layer 3 protocols

At the system level, MPLS is a packet-forwarding protocol, which only applies between routers and not between the end-user applications. The basic topology of MPLS is shown in Figure 9.7.

Incoming traffic arrives at a *label-edge router (LER)*, and packets are forwarded over *label-switching routers (LSR)* via a pre-determined *label-switching path (LSP)* to the outgoing LER. Packets arriving at the incoming LER have a label attached to them prior to being forwarded to the next LSR in the path. The destination LSR uses the label to look up the next LSR, updates the label field and forwards the packet to the next LSR. This process continues until the outgoing LER is reached, at which point labels are removed and the packet is sent onwards from the MPLS domain. Clearly, MPLS can only work if the routing tables and paths are set up appropriately and maintained to provide the required QoS; MPLS uses RSVP or *label-distribution protocol (LDP)* for this task. The label path is analogous to the virtual circuit created in ATM networks, providing traffic isolation as a mechanism to enable traffic engineering. MPLS is essentially

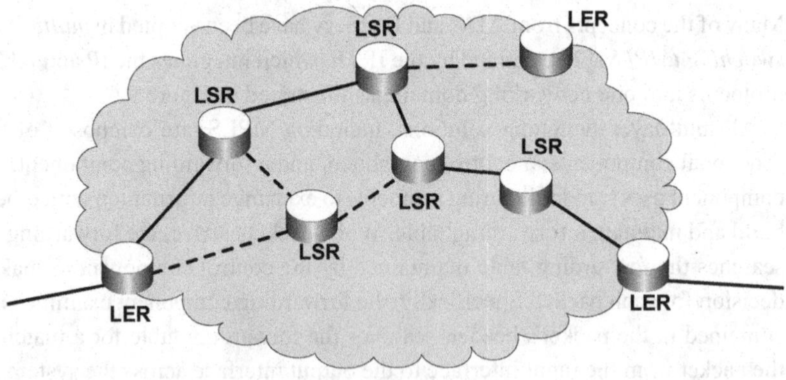

Figure 9.7 MPLS network architecture

protocol-agnostic, in that it only cares about delivering incoming data packets to the destination, and not about the internal structure of the packets beyond the label, or the protocol associated with the traffic. It can be used to carry IP or ATM traffic equally well, and is therefore increasingly widely used for provisioning of large-scale transport networks. Although MPLS and ATM networks can coexist, the reduced cost and complexity of MPLS is beginning to encourage network operators to migrate their ATM networks towards MPLS.

It is appropriate to note that all of the techniques discussed in Section 9.2 concerned with the management of transmission focus on the *prioritisation* of traffic. This, in itself, is not sufficient to provide any QoS guarantees. To ensure QoS, traffic prioritisation *and* network dimensioning are required. Dimensioning of packet transport networks to support many independent media streams, each with their own (different) QoS needs, is discussed in the next section.

9.3 Worked example: EPC transmission planning

The intention of this worked example is to illustrate a dimensioning approach that may be used to determine the transmission bandwidth needed between the SAE and PDN gateways of the EPC and, more generally, the transport of any arbitrary set of mixed applications over common packet transport. The SAE gateway concerned is sourcing and aggregating traffic to and from a number of e-node Bs, which are deployed to cover an area of some 40 km^2. The traffic to be transported in the downlink is summarised in Table 9.3.

It is first appropriate to highlight a significant difference between the worked example of Chapter 7, and the question being addressed here. In the Chapter 7 example, there was a finite amount of capacity and the question was to determine how many users could share this resource for a given QoS performance and level of blocking. In this case, the question is to *determine* the transmission capacity required to guarantee to transport, concurrently, a maximum (and different) number of instances for each of several applications, whilst preserving the QoS specified for each application. Figure 9.8 shows the configuration that is being modelled to address this question.

For the purposes of this model, the PDN gateway can be considered as a number of signal sources. Each signal source will originate traffic that is to be transported within a narrow window of QoS performance, in terms of bearer rate, maximum delay, packet loss, etc. The reality is that in most cases the traffic will be in response to specific mobiles previously interrogating or interacting with specific servers or sources. The QoS information relevant to this flow and accompanying the resulting packet stream will have directed the traffic to a port associated with the required transport QoS. A total of six QoS classes is implied in the diagram; six is not a magic number! If there are only a small number of QoS classes, the transport attributes must match the requirements of the most demanding service directed to this link, with all the capacity wastage that implies. At the other extreme, complexity will rise if there is too large a number of classes.

Table 9.3. Back-haul traffic to be transported

Chapter	System demand	Busy hour call attempts/ subscriber	Application Usage (% 'calls' during the busy hour)							System characteristics			
			1	2	3	4	5	6	7	8	9	10	11
			Voice	Mobile web	Interactive gaming	Video	VoIP	Web browsing	Messaging	Population density (km^2)	Penetration (%)	Propagation environment	Mobility (km/hour)
9	–	0.5	35	0	5	10	30	15	5	5000	100	Urban	–

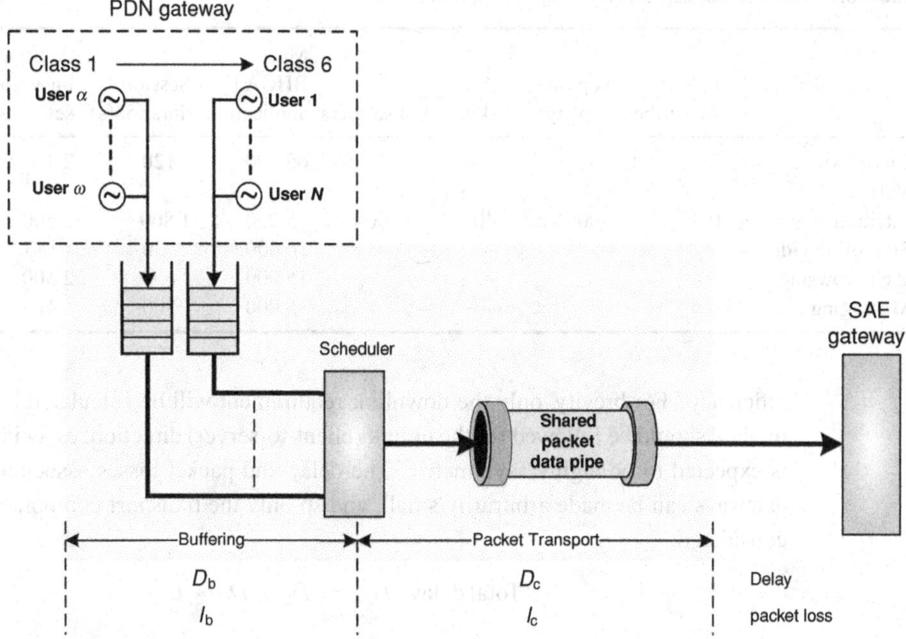

Figure 9.8 Back-haul transport model

The second component of the solution comprises buffers, associated with each service class, a scheduler and multiplexer, and, finally, the broadband packet transport network. Although only one transport link exists, the question to be addressed is, 'What are the bearer rates and buffer sizes of the (in this case six) buffer–pipe combinations that will enable the QoS needs of the traffic in each class to be met?'

The notionally separate transport pipes are, of course, realised by scheduling and de-multiplexing traffic onto the one link of an appropriate bandwidth. However, recall the discussion of Chapter 3, where discussion of 'self-similar' traffic, such as web browsing, revealed that, unlike many statistical distributions, there is not convergence towards the mean as the number of instances increases. The bandwidth of the shared pipe must, therefore, equal the sum of the bandwidths of the individual pipes, unless it can be guaranteed that no 'self-similar' traffic will be carried.

The final part of the transport solution is the SAE gateway, which has to de-multiplex packets relating to one mobile (but different applications and thus QoS classes) and direct them towards the appropriate cell where the targeted mobile is located.

9.3.1 Back-haul transmission principles

Reference to the buffer–pipe element shown in Figure 9.8 reveals the questions to be answered. Specifically, for each of the different service classes: what are the bearer rates, and what are the buffer sizes that allow both the required QoS delay and packet loss rate, for the flow of data in each class, to be met, whilst delivering reasonable transport

Table 9.4. Back-haul session dimensioning by application

	BHCA/ subscriber	Population density	Area (km^2)	Subscribers	Mean BHCA by application	Session duration(s)	Average busy hour sessions	95% sessions
Circuit voice	–	–	–	–	65 000	120	2 167	2 244
VoIP	–	–	–	–				
Interactive gaming	0.5	5000/km^2	40	200 000	5 000	1 800	2 500	2 583
Streaming video	–	–	–	–	10 000	30	83	98
Web browsing	–	–	–	–	15 000	600	2 500	2 583
Messaging	–	–	–	–	5 000	300	417	451

efficiency? For brevity, only the downlink requirement will be calculated but a similar method should be followed for the uplink (client-to-server) direction, especially if traffic is expected to be highly asymmetric. The delay and packet losses associated with the gateways can be made arbitrarily small, and so only the transport components need be considered:

$$\text{Total delay, } D_{\text{tot}} = D_b + D_c \sim D_b$$

and

$$\text{packet loss } L_{\text{tot}} = 1 - (1 - l_b) \times (1 - l_c) \sim l_b.$$

Since l_c is negligible for modern broadband packet transport and with well managed intra-company links, packet loss in ordinary operation can be ignored and the core transport delay can be managed to be a few milliseconds. The whole of the QoS budget can thus be reserved for the buffering operation designed to reduce the transport bandwidth.

It is also necessary to consider the application specifications for jitter set out in Table 4.1; VoIP traffic is particularly sensitive to this effect and significant attention was paid to HARQ in the WiMAX example to meet this requirement. Fortunately, broadband packet transport clock speeds will mean that even the 10 millisecond requirement is trivial to meet in this case.

9.3.2 Back-haul application scenario

Table 9.4 summarises the process of session dimensioning that is reflected in the examples. Table 9.3 repeats here, for convenience, the traffic scenario proposed in Table 4.2. It advises that 0.5 sessions can be expected in the busy hour for each subscriber and indicates the probability that any one of these sessions is voice, gaming, etc. From the introduction, the area to be covered is known to be 40 km^2 and this may be combined with the population density and subscriber penetration to give a subscriber total and number of instances by application. Since the application session durations are known, it is possible to estimate the average number of sessions active at a given time; this is reflected in the 'Average busy hour sessions' column in Table 9.4. Further, since the mean session arrival rate is known to follow a Poisson distribution, it is also possible to determine an upper bound for the number of active sessions, which will only be exceeded 5% of the time, and this, too, is evaluated. It should be noted that the choice of '5%' should be consistent with the operators' overall QoS outage objectives. Stringent targets tend to be inexpensive in the context of back-haul dimensioning here, because the number of sessions tends to be high, and the penalty for setting more stringent goals (e.g., 2% or 1%) can be quite small because of the narrow skirts of the Poisson distribution.

9.3.3 Back-haul capacity evaluation

For each application, it is necessary to calculate the bearer rate and logical buffer size required to support the aggregated traffic from all users. For applications with a variable bit rate, this is done by characterising the burst size distribution (which will then apply for each instance of the application) and calculating the aggregated burst arrival rate due to all instances of the distribution. The resulting aggregated traffic can be considered to be one super application with a much enhanced burst-arrival rate and a burst size distribution which is unchanged from that for a single application instance. Crucially, this super application is always present. This should be contrasted with the technique used in Chapter 7, where the question to be addressed was whether a *transient* application of finite duration could be served on demand and in competition with a population of similar transient services. In that case, it was appropriate to evaluate the capacity of the WiMAX cell by calculating the number of bearers using an Erlangian model and evaluating the number of users based on Erlang demand.

It is also useful to note that the buffer–pipe estimation process falls into two broad cases.

- The first type is characterised by applications that are distinguished by having a *substantially constant burst size* and typically have a Poisson burst-arrival rate.
- The second type is characterised instead by having a *burst size distribution* that may exhibit wide variation and may often, but not always, have a Poisson burst-arrival rate.

For simplicity, and because it is often representative of application statistics involving human interactions, a Poisson burst-arrival rate will be assumed in each of the example applications.

9.3.3.1 Application classes with near-constant burst sizes

VoIP

Examination of Table 9.3 shows that there are in fact *two* distinct voice applications; one originating from a legacy PSTN circuit system and a second originating as VoIP voice traffic. Reference to Table 4.1 indicates that the circuit codec may have a rate anywhere between 4 kbits/s and 64 kbits/s. Inter-working with legacy systems now takes place at the media gateway (MGW) where resources are available to translate the legacy speech to a modern G729 codec; with VAD off, this is now identical to the characteristics for VoIP shown in the same table. Thus, the aggregate number of busy-hour call attempts for circuit and VoIP is 65 000.

In this example, it is assumed that the G729 codec is used without VAD active. In this configuration, the codec gives rise to constant frame size bursts. (Note that if VAD was active, it would be necessary to obtain a statistical description of the resulting burst size distribution for a representative profile of talkers. These data would then be used as the input to the second class of back-haul estimation, discussed in Section 9.3.3.2.)

To calculate the bandwidth allocation required, the mean number of simultaneous active sessions during the busy hour must first be found:

$$\begin{aligned} N_{\text{mean}} &= (\text{BCHA}_{\text{VoIP}} + \text{BCHA}_{\text{circuit}}) \times (\text{session duration}) / 3600 \\ &= 65000 \times (1/30) \\ &= 2167 \text{ simultaneous sessions.} \end{aligned}$$

The mean data rate of this traffic is:

$$\begin{aligned} \mu_{\text{mean}} &= 2167 \times 27 \text{ kbps} \\ &= 58.5 \text{ Mbps.} \end{aligned}$$

To ensure that the probability of exceeding the design back-haul capacity being available is small, say 5%:

$$P(N > N_{\text{max}}) < 0.05.$$

Assuming Poisson arrival with a mean of 2167 sessions, this is achieved with:

$$N_{\text{max}} = 2244 \text{ (when the cumulative Poisson is 0.95).}$$

Hence the bearer allocation μ_{max} should be

$$\begin{aligned} \mu_{\text{max}} &= 2244 \times 27 \\ &= 60588 \text{ kbps} \\ &= 60.6 \text{ Mbps.} \end{aligned}$$

A buffer may be allocated to deal with aggregate traffic arrival, but VoIP traffic is sensitive to delays greater than 100–150 ms. To ensure that the maximum queuing delay T_{queue} is less than 100 ms, the maximum buffer size should be:

$$\begin{aligned} \text{Buffer} &= 60.6 \times (T_{\text{queue}}) \\ &= 60.6 \times 0.1 \\ &= 6.06 \text{ Mbits} \\ &= 0.75 \text{ Mbytes.} \end{aligned}$$

Note that this buffer is for the aggregated VoIP traffic from all users. This process is summarised in the intermediate results of Table 9.5.

Interactive gaming

The data rates for this category will vary depending on the nature of the games being supported on the network and the expected number of simultaneous participants in each games session; for this example, a traffic model for the multiplayer combat game 'Quake3' is used [27]. Here, one packet is sent from the server to the client every 50 ms, with a packet length of about 120 bytes (for four users). This gives approx. 20 × 120 bytes = 2400 bytes/s = 19.2 kbps per user. The burst size can be considered constant in this example and from Table 4.1 a maximum queuing delay of 50 ms can be tolerated. A similar process to the VoIP case is followed, resulting in the following bearer rate and buffer requirements:

- $N_{mean} = 2500$,
- $\mu_{mean} = 48.0$ Mbps,
- $\mu_{max} = 49.6$ Mbps,
- Buffer = 310 kbytes.

It should be stressed that, depending on the particular game and its associated burst size distribution, either the constant or variable planning process may need to be used.

9.3.3.2 Application classes with a burst size distribution

Streaming video

Traffic generated by this application will vary considerably, depending on such factors as the video codec being used, the type of image being transmitted, scene duration and the video quality required. For this example it is assumed that the variable bit-rate MPEG-2 codec is being used to transmit reasonable resolution images of a low-motion scene to a remote user (for example, remote security surveillance). The codec generates a sequence of frames, termed a GOP (group of pictures) which consists of an initial image followed by a sequence of frames that describe changes to the original image. A typical GOP duration is 12–15 frames (at a rate of 25 frames/second). Hassan *et al.* [28] found that the GOP size distribution followed the following distribution:

GOP duration: 12/25 s,
GOP size distribution: log–normal,
GOP size mean: 141 kbytes,
GOP standard deviation: 46 kbytes.

From Table 4.1, the packet loss requirement is <1% and the delay should be 1–5 s. To model the aggregate traffic from all users, the aggregate burst arrival rate must be found. The mean number of simultaneous video sessions is:

$$N_{mean} = \text{BHCA} \times (\text{session duration})/3600 = 10\,000 \times 30/3600 = 83.3 \text{ sessions.}$$

Taking N_{mean} as the mean of the Poisson-distributed session activation rate, the 95th-percentile of the Poisson distribution is

$$N_{max} = 98.$$

Hence, the aggregate burst arrival rate λ is:

$$\begin{aligned}\lambda &= 98 \times 1/(\text{GOP duration})\\ &= 98 \times 1/(12/25)\\ &= 204\ \text{bursts/s.}\end{aligned}$$

The burst size in this case is not constant and, therefore, must be modelled using ARC or similar tools. The input parameters to the model comprise: the aggregate burst arrival rate just calculated, the allowable latency and BER and, finally, the log-normal distribution, which characterises this streaming video codec. The results of this modelling activity give the buffer size and aggregate bearer rate required:

- $\mu_{\text{mean}} = 203$ Mbps,
- $\mu_{\text{max}} = 221$ Mbps,
- Buffer $= 23$ Mbytes.

Web browsing

An analysis of this application is included in Chapter 7, which used the following application characterisation:

- Burst size distribution: Pareto (1.329, 1.923 $\times 10^{-5}$, 2864),
- Burst arrival rate/user: 0.033 bursts/s,
- Maximum burst size: 1 MB.

Using data from Table 9.4, the mean number of simultaneous web-browsing sessions for the scenario is:

$$N_{\text{mean}} = \text{BHCA} \times (\text{session duration})/3600 = 15\,000 \times 600/3600 = 2500.$$

Again, assuming Poisson-distributed session activation, the 95th percentile number of session activations is 2583. The burst arrival rate λ_{max} is, therefore:

$$\lambda_{\text{max}} = 2583 \times 0.033 = 85.2\ \text{bursts/s.}$$

Using data from Table 4.1 and the bursts/second result obtained above as inputs to the ARC tool gives the following results:

- $\mu_{\text{mean}} = 58.5$ Mbps,
- $\mu_{\text{max}} = 64$ Mbps,
- Buffer $= 5.2$ Mbytes.

Messaging

Traffic associated with text messaging is usually dealt with by allocating large buffers (sometimes disk-based) to cope with surges in demand, with no guaranteed delivery time offered. In this scenario, the designer has to ensure that messages are delivered within 30 seconds and so a study of the queuing is required. The application is characterised from Table 4.1 as follows

Table 9.5. Back-haul aggregate bearer and buffer dimensioning

	Notes	95% sessions	Single instance (bursts/s)	Service flow (bursts/s)	Burst size distribution	Mean burst size (bytes)	Aggregate bearer rate (Mbits/s)	Latency (ms)	Buffer size (MB)
Voice	a,b,c,d	2 244	50	112 200	Constant	67.5	60.6	100	0.75
Interactive Gaming	b,c,d	2 583	20	51 660	'Constant'	120	49.6	50	0.31
Streaming Video	b,e	98	2.1	204	Log-normal	0.14M	221	1 000–5 000	23
Web Browsing	b,e	2 583	0.033	85.2	Pareto	<1M	64	1 000–6 000	5.2
Messaging	b,e	451	0.016	7.2	Exponential	<10k	.058	30 seconds	0.2

[a] Burst arrival rate for all applications is assumed to conform to a Poisson distribution
[b] Service flow in bursts/s calculated by multiplying single instance rate by the number of sessions within the 95th percentile
[c] Aggregate bearer rate determined by service-flow burst rate multiplied by mean burst size
[d] Buffer size set by maximum latency multiplied by aggregate bearer rate
[e] Buffer size and aggregate bearer rate to support required delay and packet loss limits evaluated using ARC or similar tool

- Maximum burst size: 10 kbytes,
- Maximum latency: 30 s,
- Burst arrival rate: 1 (bursts/min)/user = 0.016 (bursts/s)/users,
- Burst size distribution: exponential with mean of 1 kbyte.

The mean number of active sessions is:

N_{mean} = BHCA × (session duration) / 3600 = 5000 × 300/3600 = 417 sessions,
95th percentile of Poisson distribution = 451.

Hence, the aggregate burst arrival rate to be provisioned for is:

$$\lambda = 451 \times 0.016$$
$$= 7.22/\text{s}.$$

Since the maximum latency is quite high, the bearer rate can be specified as slightly greater than the product of the mean burst arrival rate and the mean burst size:

$$\mu_{max} \sim \lambda \times 1000 = 7.2\,\text{kbytes/s} = 58\,\text{kbps}.$$

Modelling of this application showed that a bearer rate of 58 kbps and an application buffer of 200 kbytes gives a packet loss ratio of 0.005 and a 95th percentile response time of 20 s – well within design limits.

- μ_{mean} = 53.3 kbps,
- μ_{max} = 58.0 kbps,
- Buffer = 200 kbytes.

Table 9.5 summarises the key results from these worked examples and, together with Table 9.4, illustrates the process and shows intermediate results.

References

1 ETSI TC-SMG, *European Digital Cellular Telecommunications System (Phase 2); Principles of Telecommunication Services Supported by a GSM Public Land Mobile Network (PLMN)*, GSM 02.01 version 4.6.0.
2 3GPP, *Service Aspects; Service Principles (Release 8)*, 3GPP TS 22.101 8.5.0 (2007).
3 3GPP, *Customised Applications for Mobile network Enhanced Logic (CAMEL); Service Description; Stage 1 (Release 7)*, 3GPP TS 22.078 version 7.6.0 (2005).
4 ETSI, *Digital Cellular Telecommunications System (Phase 2+); Adaptive Multi-Rate (AMR) Speech Trans-coding* (GSM 06.90 version 7.2.1 Release 1998), ETSI EN 301 704 version 7.2.1 (2000).
5 3GPP, *Location Services (LCS); Service Description, Stage 1 (Release 1998)*, 3GPP TS 02.71 version 7.3.0 (2001).
6 3GPP, *UMTS Phase 1 (Release 1999)*, 3G TS 22.100 version 3.7.0 (2001).
7 3GPP, *Service Requirements for the IP Multimedia Core Network Subsystem (IMS); Stage 1 (Release 5)*, 3GPP TS 22.228 version 5.7.0 (2006).

8 3GPP, *Service Requirements for Evolution of the 3GPP System (Release 8)*, 3GPP TS 22.278 version 8.2.0 (2007).

9 ETSI, *Digital Cellular Telecommunications System (Phase 2+); Network Architecture*, GSM 03.02 version 7.1.0 Release 1998.

10 3GPP, *Network Architecture (Release 4)*, 3GPP TS 23.002 version 4.8.0 (2003).

11 IETF, *SIP: Session Initiation Protocol*, RFC 3261.

12 IETF, *SDP: Session Description Protocol*, RFC 4566.

13 CableLabs Inc., *PacketCable 2.0*, PKT-SP-23.218-I02–061013, www.packetcable.com/specifications/specifications20.html (2005).

14 IETF, *Indicating User Agent Capabilities in the Session Initiation Protocol (SIP)*, RFC 3840.

15 3GPP, *Network Architecture (Release 8)*, 3GPP TS 23.002 version 8.0.0 (2007).

16 3GPP, *IP Multimedia Subsystem (IMS); Stage 2 (Release 8)*, 3GPP TS 23.228 version 8.1.0 (2007).

17 3GPP, *Policy and Charging Control Over Gx Reference Point (Release 7)*, 3GPP TS 29.212 version 7.1.0 (2007).

18 3GPP, *Policy and Charging Control Over Rx Reference Point (Release 7)*, 3GPP TS 29.214 version 7.1.0 (2007).

19 ETSI, *Telecommunications and Internet Converged Services and Protocols for Advanced Networking (TISPAN) NGN Release 1*, www.etsi.org/tispan/ (2005).

20 IETF, *The Use of RSVP with IETF Integrated Services (Proposed Standard)*, RFC2210.

21 IETF, *An Architecture for Differentiated Services*, RFC 2475.

22 IETF, *An Expedited Forwarding PHB (Per-Hop Behaviour)*, RFC 3246.

23 IETF, *Assured Forwarding PHB Group*, RFC 2597.

24 R. Ben Ali, R. S. Pierre and Y. Lemieux, UMTS-to-IP QoS mapping for voice and video telephony services, *IEEE Network*, **19** 2 (2005) 26–32.

25 S. Maniatis, E. Nikolouzou and I. Venieris, QoS issues in the converged 3G wireless and wired networks, *IEEE Communications Magazine*, **40** 8 (2002) 44–53.

26 IETF, *Multi-Protocol Label Switching Architecture*, RFC 3031.

27 G. Armitage, P. Branch and T. Lang, A synthetic traffic model for Quake3, *ACE'04*, June 3–5 (2004) Singapore.

28 H. Hassan, J-M Garcia and O. Brun, Generic modeling of multimedia traffic sources, *HET-NETs '05 Third International Working Conference, Performance Modelling and Evaluation of Heterogenous Networks* (2005).

10 Network operation and optimisation

The life-cycle of any wireless telecommunications network broadly follows the process illustrated in Figure 10.1. As discussed in Chapter 3, the initial planning makes technology choices to meet the overall business goals. This leads to the design phase, where detailed studies are made of system capacity and coverage to ensure that the network performance criteria are likely to be met. Once the design is complete, the network infrastructure is ordered, installed and commissioned – depending on the scale of the network, this phase can take many months or even years. Once the network has been built and commissioned, subscribers are given access to the network, which is then said to be operational. The performance of the network is subsequently routinely monitored to ensure that any equipment failures, software problems or other issues are quickly identified. Any problems that do occur are fixed by operational support engineers or automatically dealt with by equipment redundancy and fault correction procedures. Finally, the performance of the stable network is examined, and its configuration is fine-tuned to maximise capacity or optimise the quality of service delivered to the subscribers. If necessary, the cycle is repeated, as new network expansion is planned and the network grows to cope with growth in the subscriber base.

In practice, the phases of the network life-cycle are rarely as distinct as this, and there is much overlap and iteration around the cycle. With a green-field network or when a new access technology is being deployed, it is normally useful to begin with a small trial network, where design choices can be evaluated and results compared in a controlled manner. Errors in the design phase can be very costly to remedy once a full network has been built, so the trial network is intended to reduce risk and increase the probability of successful network rollout. Additionally, trial networks can be used to evaluate and compare the performance of equipment from different manufacturers, and to ensure interoperability with existing systems, such as core network or transmission infrastructure. A further benefit of trial networks is that they can continue to be used once the main network is operational, to validate new software releases or new network configurations; this avoids exposing the operational network to the risk of disruption or outages.

The activities of the wireless network life-cycle need to be integrated and synchronised with the operator's core business processes, leading to additional dependencies and complexity. A useful framework for these and other activities has been developed by the *TeleManagement Forum*, called the enhanced telecom operations model (e-TOM) [1], and illustrated in Figure 10.2. An overview of the major elements of this model will

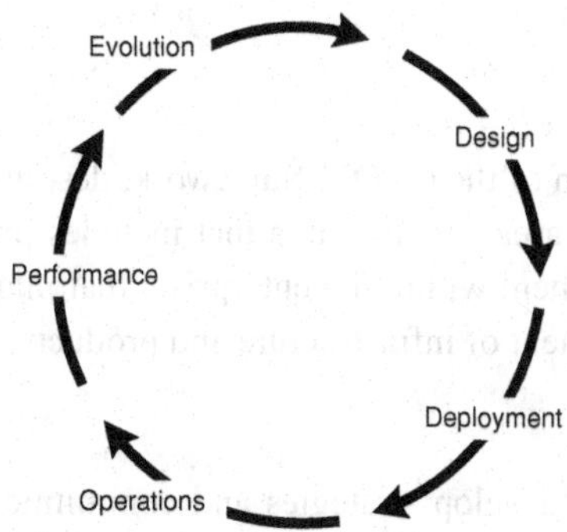

Figure 10.1 Wireless network life-cycle

follow in Section 10.1, identifying the way in which the key activities for the wireless network operator map into this model. Subsequent sections will discuss the methodology and techniques for one specific part of this model – radio network performance optimisation. This will be addressed in turn for GSM, GPRS and UMTS radio-access technologies.

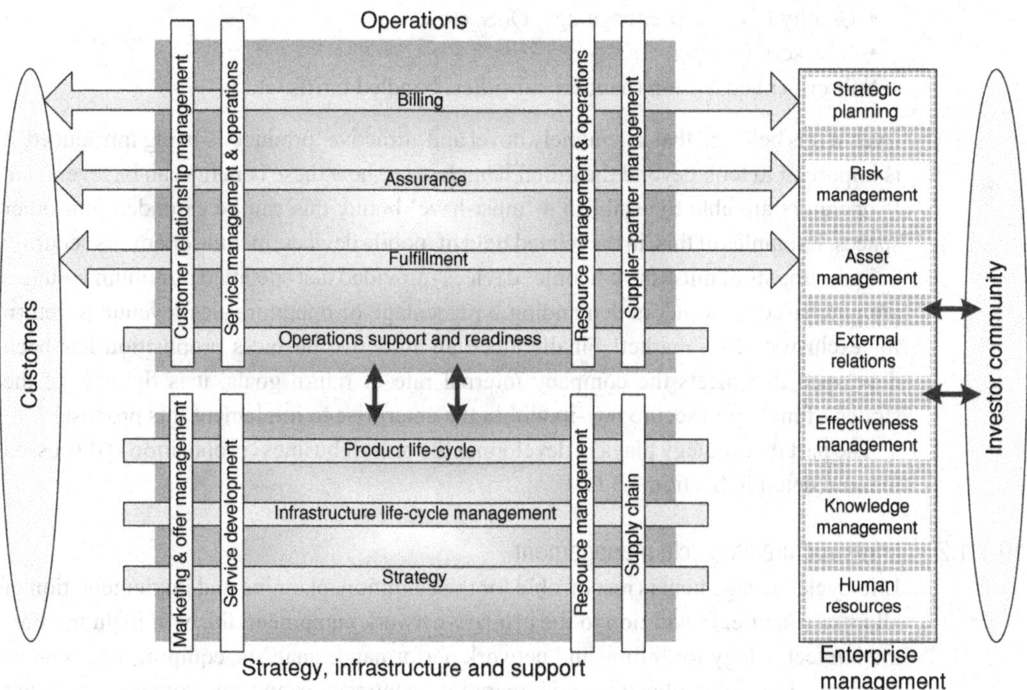

Figure 10.2 Enhanced telecom operations model (eTOM)

10.1 The network life-cycle

10.1.1 Strategy, infrastructure and product

The ITU, in recommending the adoption of the e-TOM framework, describes the strategy, infrastructure and product process area as, 'The area that includes processes that develop strategies and commitment to them within the enterprise; that plans, develops and manages the delivery and enhancement of infrastructure and products.' [2]

10.1.1.1 Strategy

The strategy activity is undertaken to 'develop strategies and commitment to them'. The business development unit will have identified candidate regions or countries, that the operator currently does not serve, or perhaps areas that are already served but seem likely targets for the deployment of new applications. Market research will have provided an estimate of the ARPU expected for the services and this can be compared with an estimate of the OPEX cost per user over the project period, based on historic data for the particular air interface to be used.

In putting together such business propositions, strategy teams will additionally consider a number of other factors including:

- Number of service providers already in the market,
- Entry strategy,
- Positioning of the new product against current (competing) services,
  - Value based (novel tariffs, etc.),
  - Quality based (better coverage, QoS, etc.),
  - New service type,
- Subscriber acquisition (handset subsidies, bundled tariffs, etc.).

Even if it is believed that a genuinely novel and attractive 'product' is being introduced, it is important to look beyond the initial launch to see how these benefits can be leveraged. If operators are able to establish a 'must-have' brand, this can be extended into other spaces. Examples of this, in the related field of mobile devices, include operators securing unique variants of must-have 'iconic' devices (provided that specified minimum volumes are purchased) or vendors demanding a percentage of operator sales revenue in return for exclusivity in a market! Finally, once an attractive business proposition has been developed that meets the company internal rate of return goals, it is the role of the strategy function to secure buy-in within the enterprise to implement this proposal.

The part that strategy plays in developing the overall business proposition is discussed in more detail in Section 3.3.6.

10.1.1.2 Infrastructure life-cycle management

Life-cycle management is responsible for the definition, planning and implementation of all infrastructure. In addition to the primary network equipment, this will include information technology for billing and network performance analysis, equipment to support the deployment of applications and operations centres to support customer relationship management. Provisioning for a large-scale wireless network is a major programme, which requires extensive planning and comprehensive project management if it is to be successful. Some network operators or service providers choose to outsource this activity completely, simply specifying key programme milestones and a service-level agreement (SLA) with an infrastructure provider to guarantee network performance targets. Other operators coordinate this activity within their own operations, particularly if they are purchasing equipment from many vendors, or if the wireless network is intended to be integrated seamlessly with an existing network. In either case, the provisioning activity will require the following non-exhaustive list of tasks:

- Generation of equipment lists from designs,
- Collecting pricing information from vendors,
- Placing orders,
- Managing inventory, spares, materials and warehousing,
- Logistics, shipping and import–export management,
- Site acquisition, preparation and readiness,
- Installation planning,
- Subcontractor management,
- Quality management,
- Regulatory compliance and licensing,
- Installation,
- Commissioning and integration.

A key consideration is the delivery of the desired level of service availability. In the RAN, a major operating expense is transmission, which tends to be more expensive in rural areas where microwave links are often required to back-haul traffic from base stations to the BTS controller. A 'daisy-chain' configuration, which progressively consolidates traffic from the most remote BTS with traffic from sites closer to the controller can reduce transmission costs, compared with a 'hub-and-spoke' arrangement, which would require multi-hop microwave links to each base site. However, this cost reduction is delivered at the expense of lower system availability (because all sites towards the end of the daisy chain are lost if one of the links near the controller fails).

Similarly, in the core network, ever-increasing concentration of traffic occurs as it is transported towards the BTS controller and MSC (or its packet equivalent – the serving or public data network gateway). End-to-end system analysis will be required to ensure that appropriate availability specifications are placed on transmission and platforms. This will sometimes require that redundancy and 'hot standby' are provisioned for key network elements.

It has long been standard practice to invite vendors to provide a network solution to support an operator network scenario, addressing subscriber growth over a specified deployment region with time, and reflecting an evolving service deployment. However, because the component of OPEX associated with infrastructure equipment has progressively reduced with time, it is increasingly common to ask vendors to consider the minimisation of total life-cycle costs in their submission. Operators also often ask vendors to guarantee certain network performance levels as a function of time, which result in penalties if not met.

10.1.1.3 Product life-cycle management

The product life-cycle management process helps the operator's *products* (primarily end-user services) to achieve required profit and loss margins, customer satisfaction and quality commitments, as well as assisting in the delivery of new products to the market. The activity comprises two key sub-processes, *product management* and *product development*. The product management process is required to understand the business environment, customer requirements and competitive offerings, in order to design and manage products that succeed in their specific markets. The product development activity is predominantly a project-oriented process that develops and delivers new products to customers, as well as new features and enhancements for existing products and services.

10.1.2 Operations

Typically, the majority of staff are employed in this area. Operations is responsible for processes that support the customer, network operations and management. As such, it is usually regarded as the heart of the enterprise. Failure to deliver against performance goals in this area can quickly result in loss of revenue and market share. Note, also, that it is only in the *operations* area that three of the four sub-processes directly touch the customer (fulfilment, assurance and billing, or FAB). For this reason, the FAB processes are typically extended directly into enterprise management, where they provide visibility of cycle times and life from a customer perspective.

The whole area of *operations* is rapidly becoming more complex. The conventional activities of service management, fault isolation and performance management associated with the current – essentially single (voice) service – network require a small number of performance metrics, such as service (area) coverage, blocked call, dropped call and handover success. These metrics will increasingly become transformed into multidimensional parameters where 'success', currently measured in a binary manner for each of the above metrics, will expand into: did it meet its 'call set-up time', guaranteed minimum data rate, packet delay and loss requirements, etc. The need to guarantee these more complex metrics arises because, increasingly, high margin revenue will be generated by delivering applications that require such parameters to assure correct end-user operation.

The key activities of fault, configuration, performance and security (FCAPS) [3] management are largely responsible for delivering the customer-facing FAB roles. FCAPS management, in turn, is mainly delivered either directly or indirectly from the network management system (NMS). At its highest level, an NMS is a set of software and hardware tools to enable the network to be monitored and managed from a central point. It embraces many disciplines, such as element installation and integration, as well as the coordination of hardware, software and human resources for the monitoring, test, configuration, analysis and evaluation of network performance. The final goal is to control the network in real time, knowing its resource utilisation and performance relative to established goals. At its heart is the NMS core. This is responsible for keeping records of network element, interface and data-link performance, along with a database recording subscription and other information for users, user groups, administrators and, finally, private management information base (MIB) elements reflecting current and fallback network configuration information. The core also processes and analyses all failure and intermediate events collected from any of the elements in the entire network.

10.1.2.1 Fulfilment

Fulfilment is the process responsible for providing customers with the products they have requested in a timely manner. It translates the customer's request into a solution, which can be delivered using the specific products in the enterprise's current portfolio. The process must also inform customers of the status of their purchase order, forecast completion and quickly update them in the event of some unforeseen delay. It will, in practice, require online real-time links between the HSS and the customer-facing 'front office', so that the database can be kept up to date with all subscribed services for each customer. The NMS will also need visibility of the total number of subscriptions by service so that it can provide timely advice of the need to upgrade particular network elements to ensure sufficient provisioned capacity.

10.1.2.2 Assurance

The assurance process covers the execution of proactive and reactive maintenance activities to ensure that services provided to customers are continuously available and performing to SLA or QoS performance levels. It performs continuous resource status and performance monitoring to detect possible failures. It collects and analyses performance data to identify potential problems and resolve them without impact to the customer. This process manages the SLAs and, in the future, will need to provide reports of application QoS performance for individual customers if required. It will have to associate specific problem reports from the customer with action taken, and collect data to demonstrate that the reported problem has been corrected.

The assurance process will also initiate RAN *network optimisation* when required. Typically, networks are optimised immediately after commissioning and at intervals after this point – perhaps after major expansions or the deployment of a new frequency plan. The traditional approach requires several teams of field personnel to drive along predefined routes making calls. Each call is investigated and any potential problem is resolved. Most network operators rely on such 'drive-testing', even though the approach is slow and expensive. Worse, the network is really only optimised for subscribers who use the same routes as the drive-test teams.

Modern network optimisation collects data from the entire network, rather than data from drive-test routes alone. By using actual commercial subscriber data, the service helps optimise the network for the majority of users, not just those who make calls from the tested areas. Higher network quality after optimisation is typically delivered through the following improvements:

- Reduced number of dropped calls,
- More efficient spectrum utilisation to meet capacity demands,
- Optimal frequency allocation to ensure good call quality,
- Accurate neighbour topologies to ensure smooth handovers and call distribution,

A discussion of some of the specific techniques used to optimise GSM and UMTS networks is addressed in later sections of this chapter.

10.1.2.3 Billing

The billing process is responsible for the collection of appropriate usage records, and the generation of timely and accurate billing information for customers. As discussed elsewhere, either the MSC or the IMS sub-system automatically generates 'event triggers' when there is a significant change in the application state. Such information can be used to generate *call detail records* (CDRs), which, in aggregate, can be used to generate billing information in line with the operator's published tariffs. This process also handles customer inquiries about bills and supports prepayment for services.

Billing data can also be collected – either in aggregate or for specific user groups – to provide insight into user behaviour. Such information can be analysed by time of day, by application and, increasingly, by location. This can form the basis for establishing future network expansion requirements and to create more targeted tariffs.

10.1.2.4 Operations support and readiness

This process grouping provides management, logistics and administrative support to the FAB processes and ensures operational readiness in the fulfilment, assurance and billing areas. In general, processes in this area are concerned with activities that are less 'real-time' than those in FAB, and are typically concerned less with individual customers and services and more with ensuring that the FAB processes run effectively. A clear example of this type of process is staffing capacity management, which is used to ensure efficient operation of call centres. The existence of this category reflects a need in some enterprises to divide their processes between the immediate customer-facing and real-time operations of FAB and other operations processes, which act as a 'second-line' or 'operations management back-room'.

10.1.3 Enterprise management

This area includes those basic business processes required to run and manage any large business. They focus on both the establishment and delivery of strategic corporate goals and objectives, as well as providing the support services required throughout the enterprise. These processes are sometimes labelled 'corporate' functions – e.g., financial management and human resources management. Since enterprise management is aimed at general support, it may interface as needed with almost every other activity in the company, whether operational, strategic, infrastructure or product-focused.

During the discussion of operations, it was identified that the FAB processes were critical to the wellbeing of the company, but that, in addition, they provided key visibility of customer satisfaction and an opportunity to engage the customer. The enterprise processes have a similar dual role. At one level, the supervision provided by these is central to the establishment of strategic direction and the timely delivery of these goals. At a second level, these same processes can generate information with which to engage the company's other key customers . . . its investors. An active dialogue in this area is

Table 10.1. User metrics

	Attribute	Potential causes
Circuit telephony	Call set-up	Coverage area (%)
		Blocking (%)
	Call drop rate	Dropped on handover (same network) (%)
		Dropped on handover (multi-band) (%)
		Dropped on handover (multi-RAT) (%)
	Call quality	Low S/N or low C/I – speech (MOS)
Applications	Application set-up	Coverage area (%)
		Blocking (%)
	Call drop rate	Dropped on handover (same network) (%)
		Dropped on handover (multi-band) (%)
		Dropped on handover (multi-RAT) (%)
	Bearer rate	Meets or exceeds specified rate (kbits/s)
	Latency	Less than or equal to specified delay (s)
	Packet loss	Less than or equal to specified packet loss (%)
	Jitter	Less than or equal to specified delay variation (s)

essential to ensure that the enterprise's true value is reflected in its share price and thus provides appropriate returns to its shareholders.

10.2 Network optimisation

In the context of the e-TOM model of the previous section, *network optimisation* is a major component of *operations*, and *service assurance* in particular. Optimisation is key to ensuring:

- The best user application performance,
- Maximised network capacity,
- Maximised network availability.

Good performance in these areas is central to gaining and retaining subscribers and hence network profitability.

The first step in the optimisation process is not, in fact, optimisation at all but relates to equipment configuration checks. It is sensible to perform an initial verification of all the database parameters to ensure that no transcription errors have occurred in configuring the equipment. This will reference the original system design information. A performance baseline should then be established to understand the current state of key performance indicators (KPIs) and other critical performance measures. Some of the metrics that might be recorded in this process are included in Table 10.1. Consideration of KPIs at the level of the BTS or even the cell level will indicate where the worst performance is occurring and thus where problem-solving effort should be focused. Some localised problems often reflect wiring errors or out-of-box equipment failures; in other cases, antennas may have been incorrectly connected to the site or their configuration may not match what was intended. Excessively high numbers of dropped calls or unexpectedly low traffic levels are often symptomatic of hardware problems. These may be indicated by KPIs out of nominal ranges or by simply looking for the worst-performing areas. All such problems should be identified and corrected before system-level optimisation is attempted.

Optimisation will invariably be carried out following initial network deployment checks. It is important for the network operator to recognise that the deployed network is the result of plans based on predictions, not all of which reflect reality. At the planning stage, subscriber volumes and their geographic distribution are based on estimates; further, actual models of the clutter environments and resulting propagation loss may deviate significantly from results in the field. These variations will mean that the carefully crafted network developed during the planning stage will not in fact be ideal and these deviations from the target performance will have to be addressed through optimisation.

Once optimisation has been completed following initial deployment, it is important to recognise that no cellular network remains static. The subscriber base will grow throughout the life of the network and extra capacity will need to be provided through new carriers and new sites. Similarly, changes will be required to support additional services. Optimisation is part of managing this natural commercial growth and will occur at intervals throughout its life. This ongoing process can ensure that subscribers continue to experience a good service quality and will be triggered whenever statistics from the network fall below target values.

10.2.1 Optimisation – an end-to-end process

From an end-user perspective, the performance metrics of most interest are typically the quality of the 'call' and its price; if, for some reason, the call did not go through or was dropped the user is not interested in why – just the fact that it happened. If the blocked or dropped call experience happens sufficiently often, it may ultimately result in a decision to move to another operator. From an operator viewpoint, in addition to the 'call' quality there is a great interest in those (hopefully) few occasions when calls do not originate as expected or drop abnormally; this may be symptomatic of a more general problem, which occurs at some times or in some regions, and which may be experienced by many users and needs to be addressed. So, where might these problems originate? Table 10.1 provides a non-exhaustive set of metrics, which reflect, under the attribute column, the primary functionality that might be of interest to the subscriber. By way of example, the 'potential causes' column includes some of the root causes that might give rise to the corresponding primary attribute impairment.

In the case of circuit telephony, the dominant source of revenue for most operators at present, the number of user metrics is fairly small and for mature networks these metrics are managed within tight limits, which are largely acceptable to the user. For those few locations where performance consistently lies outside these limits, this is because operators have decided that further investment to improve performance cannot be justified by the potential additional revenue.

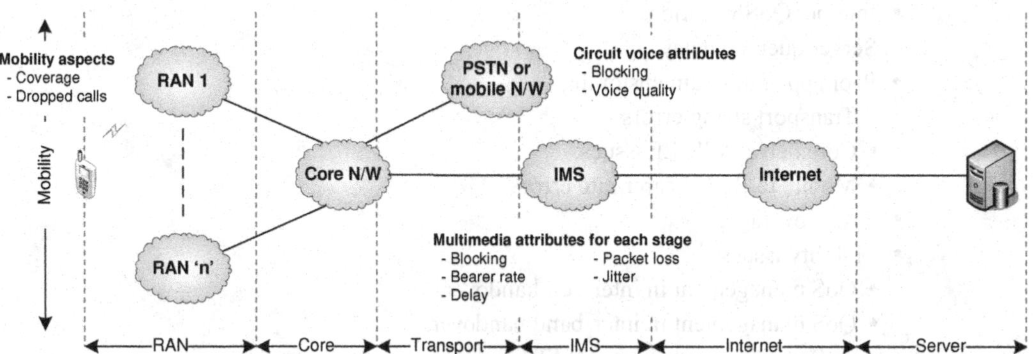

Figure 10.3 Components of end-to-end performance

Reference to Figure 10.3 makes it clear that, for circuit telephony, one or more RANs, the core network and a subscriber on the PSTN or, more usually, a subscriber on the same or another mobile network, are the elements involved in the complete call. Although problems can occasionally arise in the core or inter-working with other networks, the majority of issues in practice originate in the radio-access networks. As network demand grows and additional capacity has to be increased by means of new frequencies and perhaps new reuse strategies; cell neighbour lists need to be updated. Additionally, capacity may be increased by introducing networks operating at new frequency bands or utilising new air interfaces. These and other issues require a well-thought-out traffic management strategy if performance is to be maintained. Key considerations when planning, managing and optimising mobile telephony networks will be addressed in Section 10.3. Whilst this section is focused on GSM, similar considerations apply in the design of UMTS.

Moving from mobile telephony to networks supporting mobile applications, it is clear that network optimisation becomes more complex. Table 10.1 shows that support of data applications not only requires call set-up and dropped call performance at least as good as basic telephony also but that four metrics rather than the simple metrics for good voice quality are required if correct operation of applications is to be guaranteed.

As discussed in Chapter 7, an end-to-end system budget is established during design to ensure the delivery of this QoS performance between device and server. Each of the six segments from device to server will be allocated bearer rate, delay, packet loss and jitter budgets to assure correct operation of applications. Consequently, 'fault' diagnosis and resolution along with optimisation now becomes a far more complex process.

Under normal conditions, radio propagation continues to impact instantaneous performance for the user but when configured in line with the system design, the required RAN QoS will be delivered by the scheduler. Similarly, packet transport (within the RAN and between the RAN and core and between the core and IMS) and QoS metrics for the 'Internet' and 'server' segments will meet their requirements under normal conditions. However, when significant numbers of users complain about application performance, how is the operator to resolve the issue? It can no longer be assumed that the RAN is always the issue. Potential problems might include:

- Internet QoS variation,
- Server queuing delay,
- Poor application dimensioning;
  - Transport sizing errors,
  - Core network PCEF issues,
  - Mobile to RAN bearer rate errors.
- RAN coverage,
- Mobility issues;
  - QoS management in inter-cell handover,
  - QoS management in inter-band handover,
  - QoS management in inter-RAT handover.

This is not remotely an exhaustive list!

Central to the resolution of such complex issues is the analysis of user-to-server performance statistics from the mobile under the conditions in which performance impairment occurs. Once the impairment has been seen on an end-to-end basis, progressive analysis of examples of the particular call flow (or the same application made from an equivalent device) will enable the localisation of the QoS impairment to one or more of the six segments of Figure 10.3. It is to be expected that, from time to time, quite obscure and elusive impairments will be seen – for instance only provoked when sufficient users are active *and* using unforeseen application mixes that limit the available bandwidth on shared packet transmission links!

The worked examples of Chapters 7 and 9 should have made evident to the reader the methodology involved in designing packet transport links to support one or more bearers, each with its own arbitrary QoS. These same techniques can of course be used to analyse the behaviour of existing links or other equipment and identify elements that are not operating according to their specification. The remainder of this chapter will, therefore, discuss techniques that may be used to support the optimisation of GSM, GPRS and UMTS networks.

10.2.2 Optimisation – sources of data

The raw material of optimisation is information about current network performance. Where is this information to come from? In practice there are always at least three sources:

- RAN statistics,
- Third-party analysis tools,
- Drive testing.

These are presented in increasing order of the likely cost of information gathering.

The lowest cost method of learning about network performance is to make use of facilities built into equipment elements by the vendor. Statistics generated by trial or commercial users of the network are collected at the *operations and maintenance centre* for the RAN (the OMC-R). The data will normally include high-level KPIs, such as drop-call rate, call set-up success rate, handover success rate, usually broken down to BTS or cell level. Although this will vary from vendor to vendor, lower-level statistics may also be recorded to give visibility of system resource utilisation (frequencies, time slots, transmitter power, codes, etc.) and further background to the high-level KPIs – e.g., the reasons (by case) for dropped calls (low signal strength, low C/I, congested neighbour cells, etc.).

The vendor will also usually provide a feature known as 'call trace'. This enables the operator to collect information on calls exhibiting specific characteristics. The trigger for call trace may include the start and end times of the call, specific types of termination and a wealth of other events. When the trigger condition is detected, a comprehensive set of RAN measurements and states specific to the call is recorded – perhaps sufficiently detailed to include the evolution of all signalling messages that established the call – allowing detailed call-failure analysis to be performed. The processing required to capture the data invariably means that it may only be employed to sample a sub-set of the total calls made in the network but it is sometimes the only means of resolving obscure or intermittent problems.

Another approach that can be quite cost effective is the deployment of third-party tools. These provide a vendor-independent view of network performance by capturing and analysing data from the standardised 3GPP interfaces. It should be remembered that for 2G and many 3G standards, all of the mobile handset measurements used by the RAN to make decisions on handover and more general traffic management are transported over the A-bis or Iub links. Companies have developed tools to aggregate statistics and provide reports that can identify the worst-performing sites and cells and often have the capability to suggest corrective actions for specific problems. Again, facilities are usually provided to start collecting either a more comprehensive or specific data set following an 'event', which can be quite flexibly defined.

The most expensive source of data for the operator is drive testing. As its name suggests, the data are literally collected by deploying an engineering team to drive the network area under investigation. The drive test vehicle contains GPS-based equipment, which records appropriate KPIs and stamps them with the time and physical location where the data were gathered. A variation on drive testing is walk testing, which is common for tuning an in-building system. This typically uses the same equipment as the drive test, apart from GPS, which does not work indoors.

A variety of equipment configurations can be used by the engineering team performing the test. A BCCH or pilot scanner measures the network passively. It will test for adequate signal strength over the planned cell areas and measure the received strength of all such channels at given locations. In this way it will identify areas of BCCH or pilot pollution, where several channels are received at a similar level. For detailed problem investigation, one or more test mobiles will be deployed in addition to passive scanning receivers. These operate in dedicated mode and measure subscriber KPIs, such as set-up failures and dropped calls, together with any desired subset of associated information processed by the mobile. Because the 'test' mobile appears to the network exactly like any other subscriber, it can provide the operator with direct insight into the *actual* user experience. It can also measure call set-up times, the peak and average user throughput, voice quality, etc.

In summary then, the operator has a range of techniques for gathering data. Analysis of KPI data provided by facilities introduced by the equipment vendor is certainly the cheapest method. When supplemented by data gathering using specialist third-party equipment, this approach has the broadest scope, covering large areas of the network and making it possible to compare the performance of many cells simultaneously, and even providing recommendations for change. However some subscriber-specific quantities, such as throughput, may not be available in such systems. It can also be hard to locate geographically the network statistics at a resolution better than cell level.

Drive testing, on the other hand, is expensive and has long cycle times. It demands a team of experienced engineers, who must first drive to diagnose a problem, define a set of changes to deal with the problem and then drive again to verify the solution. Another drawback is that it can only focus on a relatively small cluster of cells at any one time. However, an over-riding advantage of drive testing is that it, alone, can identify areas where there is no coverage – drive testing can thus reveal problems of which the operator would otherwise be unaware. In practice therefore, a combination of the above techniques is adopted, with drive testing almost invariably employed during initial network deployment and for obscure problem resolution, and network statistics used to monitor day-to-day network performance.

10.2.3 Optimisation – where to start

To conclude this general introduction to optimisation, it is perhaps useful to add a few words on 'where to start'. The earlier discussion of Figure 10.3 made it clear that, as operators increasingly depend on applications as a major source of revenue and an even larger proportion of profits, delivering an optimised network is going to get more complicated. Limitations in the end-user performance can originate in at least six separate segments of the solution. How is the operator to home in on the problem areas?

In a modern network, it is often convenient to distinguish between four, orthogonal, dimensions, which can be used to assess network performance. These are shown conceptually in Figure 10.4 with each dimension represented by a circle. If the performance of the network is characterised by region A, it represents perfection; i.e., *all* of the services work at *any* time and *any* place for *any* user. Outside of regions A, B, C, D and E lies disaster . . . *some* of the time at least *some* of the services are not available to *some* of the people in *some* areas.

So far so obvious! For the operator to 'plot' where his network is on this framework there are at least four sources of information, although it is to be hoped that the last mechanism need not be exercised too often:

- OMC event reports (equipment failure, etc.),
- 'Call' detail records,
- 'Call' traffic levels by cell,
- Customer complaints.

If the data from these sources are structured according to Figure 10.4, patterns which give clues to the underlying issue may be discerned. For instance, supposing there are statistics

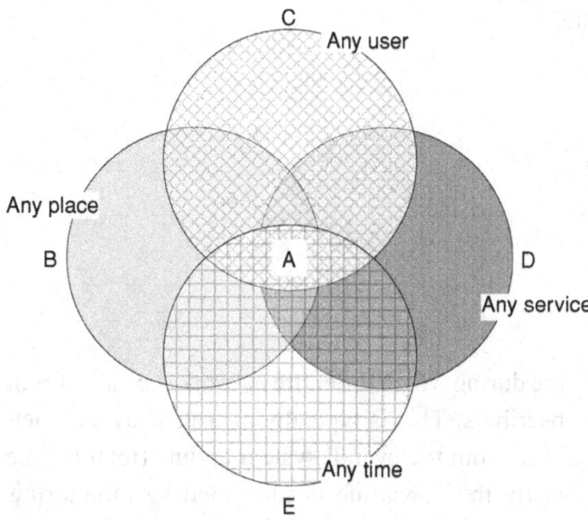

Figure 10.4 Orthogonal dimensions of optimisation

accumulating in the area common to B, D and E, indicating satisfactory performance, but *no similar statistics for area C*. The data in this case highlight that all of the services are available all of the time at any location – but not to all users. This could point to a problem with a particular mobile type, a common defect on a recent batch of mobile SIMs, etc. but it is almost certainly not a coverage, capacity or application-hosting problem.

Similarly, an accumulation of satisfactory statistics in a region common to zones B, C and E but *excluding* D might point towards a problem with one or more elements of the application server (depending upon the detail of the defects). If the problem, in fact, lay in the dimensioning of packet transport or the radio bearer assigned to the application, it is likely that correct operation of the application(s) would be seen from time to time under conditions of light load (late at night) or when the user was close to the BTS when surplus capacity could cause the application to work correctly.

The purpose of this short discussion is not to recommend this specific framework as *the* solution for network optimisation. Rather, it is to draw attention to the importance of the practitioner establishing an ordered framework that suits his or her method of operation in the resolution of what will become an increasingly complex problem. Optimisation thus moves further away from the simple task of providing a good quality voice service – any where, any time.

10.3 GSM network performance optimisation

10.3.1 Principles and key performance indicators

In this section, some of the most important KPIs for GSM networks are described, along with some achievable target values for these, based upon observed performance of operational networks. The following are some of the most important indicators:

- Network availability,
- SDCCH access success rate,

- Call set-up success rate,
- TCH blocking rate,
- Call volume,
- Dropped call rate,
- Mean time between drops,
- Handover success rate,
- Voice quality.

10.3.1.1 Network availability

This is the proportion of time during which the entire network (i.e., all cells in the system) is available for use by subscribers. This is typically measured by subtracting the time during which cells are 'down' from the overall system uptime (total number of cells × measurement period). Usually the downtime is calculated by monitoring events and alarms at the operations and maintenance centre. Cell outages are then further classified according to the root cause of the problem: human error, software failure, hardware failure, planned outage, etc. Network availability is improved by ensuring the following are in place:

- Routine maintenance procedures for all cell sites,
- Service level agreements with network equipment providers,
- Escalation and field call-out procedures,
- Fault resolution procedures,
- Spares management.

10.3.1.2 SDCCH access success rate

The SDCCH is used to initiate calls, and for SMS and location updates. When a mobile requests a channel on the RACH, the network attempts to assign an SDCCH channel to the mobile; if it does not receive communication from the mobile on the SDCCH after a specified time, it will be logged as a failure. Possible causes for such failures include high levels of interference, faulty radio equipment and incorrectly calibrated timers.

10.3.1.3 Call set-up success rate

This is the proportion of attempts to allocate a traffic channel that were successful; these are associated with requests to allocate a circuit – it does not include network accesses that do not require a circuit, such as location updates, SMS and supplementary services.

10.3.1.4 TCH blocking rate

High blocking on the traffic channels indicates that inadequate radio resources are allocated to the cell. This could be due to unexpected surges in demand (for example due to an emergency or traffic incident), or symptomatic of a longer-term increase in traffic, which may require capacity to be added in the area under investigation.

10.3.1.5 Call volume

This should be routinely monitored on all cells and compared against recent historical values for the monitoring period. Sudden drops in the number of calls have many possible causes, which could include equipment and antenna faults, or problems in the core network.

10.3.1.6 Dropped-call rate

This important statistic indicates the percentage of calls that were successfully established on a traffic channel, but terminated abnormally. Typical causes of dropped calls include: equipment or software failure, poor frequency planning, handover failures or coverage problems.

10.3.1.7 Mean time between drops

This is similar to the dropped call rate, but can be a more reliable indicator of performance on cells that handover a large number of calls very quickly – the large number of successful (but short) assignments can mask a high number of dropped calls in these cells.

10.3.1.8 Handover success rate

This measures the percentage of handover attempts that were successfully completed; usually this excludes handovers that failed owing to congestion on the target cell, which will be tracked using the KPIs for blocking. Poor handover success could indicate problems with coverage, or with neighbour lists.

10.3.1.9 Voice quality

Voice quality assessment is a difficult KPI to measure, and a wide variety of testing techniques are used. It is usually defined in terms of the mean opinion score (MOS), which is a number between 1 and 5 generated by combining the subjective assessment of voice quality by a panel of listeners, with 1 = bad and 5 = excellent. Clearly the use of a listening panel is not appropriate for routine network monitoring, and the ITU have defined a standard set of voice quality assessment tests to measure MOS in recommendations ITU-P.800 [4] and ITU-P.862 [5].

Table 10.2 presents target values for each of these KPIs.

10.3.2 Coverage optimisation

If the KPIs indicate that there could be problems with coverage, it will be necessary to identify where the coverage holes are, and what are the likely causes. The engineer leading the optimisation should firstly consult the RF propagation maps from the design team to identify possible black spots. Coverage holes can occur for many reasons, including:

- Shadowing from tall buildings or hills,
- Poor in-building penetration (for example in dense urban environments),

Table 10.2. Key performance indicator target values

Key performance indicator	Target value
Network availability	>99.9%
SDCCH access success rate	>95%
Call set-up success rate	>96%
TCH blocking rate	2% (busy hour)
Call volume	(network dependent)
Dropped-call rate	<1%
Mean time between drops	>7000 seconds
Handover success rate	>95%
Voice quality	>3.5 (MoS)

- RF attenuation by foliage and vegetation,
- Antenna misalignment or failure.

Coverage holes can be difficult to pinpoint due to the absence of radio information, and it may be necessary to perform drive tests in the suspect cell to determine where the problems may be. Measurement reports can be analysed to estimate subscriber location, which can sometimes highlight gaps in coverage. In some cases, a coverage hole may be surrounded by areas with high call volumes, caused by subscribers moving to locations where they can successfully make and receive calls.

In addition to detecting coverage holes, it is also necessary to ensure that cell coverage areas do not overlap by more than is necessary for successful handover, as this could result in increased interference. It may be possible to remedy coverage problems with some of the following techniques:

- Increasing or decreasing power (the former may adversely affect nearby cells),
- Changing antenna azimuth or tilt,
- Adding repeaters,
- Adding microcells to fill the coverage gaps,
- Adjusting power and alignment of neighbouring cells.

10.3.3 Neighbour-list optimisation

The initial neighbour lists for a GSM network are usually constructed from the 'best-server' plots produced by a radio planning tool. These will indicate which cells share a boundary with the serving cell, and should be included in the configured neighbour list. Over time, as cells are added to the network and the surrounding site configuration changes, it may become necessary to 'clean up' the neighbour list – this involves removing redundant legacy neighbours (which may result from cell splitting) and inserting new neighbours that were not in the original plan. Neighbour-list optimisation is usually performed by drive testing at locations where the handover success is low; test mobiles can report candidate neighbours seen at good signal strengths, which should be added to the list. Drive testing can also assist with optimisation of parameters for the handover algorithms – in some cases it may be necessary to force handovers to occur more rapidly, for example in situations where the serving-cell signal strength drops suddenly, owing to a moving mobile going behind a large building. A more cost-effective and comprehensive approach is to use automatic optimisation. In this technique, 'dummy' neighbours are periodically inserted into the mobile's neighbour list. The mobile will report the BSIC, BCCH and RXLEV of the dummy neighbours even though they are not genuine candidates for handover. Analysis of large volumes of measurement reports sampled from all mobiles in the system will identify neighbours that should be added to each cell's neighbour list for the entire system. Additionally, unused neighbours can easily be identified. Sophisticated automatic optimisation systems can autonomously make modifications to the system's neighbour lists without the need for manual operator intervention. Further, recommendations for changes to power control settings and operation can also be made, which will improve performance KPIs.

10.3.4 Frequency plan optimisation

In GSM, the optimisation of the frequency plan focuses on the BCCH, as these frequencies are permanently transmitting, unlike the other traffic carriers. The initially deployed BCCH frequency plan is based on propagation predictions from the radio planning tool. As discussed earlier, propagation models are at best a worthy attempt to model the radio environment based upon available terrain and demographic data. If the KPI analysis indicates that interference is an issue – for example, because dropped call rates are unacceptably high – then it will be necessary to optimise the frequency plan. In GSM networks, this is best performed using an automated optimisation tool, such as Motorola's *Intelligent Optimisation System (IOS)* [6]. Through the introduction of *dummy neighbours* described above, it is possible to gather detailed information on which surrounding base sites can be detected by mobiles on a given serving cell, irrespective of whether or not they are in the neighbour list. Analysis of large numbers of measurement reports allows creation of a system-wide carrier-to-interference matrix – this indicates the amount of interference which one site would cause to another, if they were to be assigned the same BCCH frequency. The C/I matrix can then be analysed with an automatic frequency planning tool to come up with an optimised BCCH plan, ready for deployment in the network. Extensive use of IOS in the field has shown that it consistently produces much better frequency plans than manual or prediction-based methods, and that the dropped call rate can be driven to well below 1% in a well-maintained network with good coverage.

10.3.5 Voice quality optimisation

Problems with voice quality can occur at any point between the mouth of the speaker and the ear of the recipient. Localising problems to specific parts of the transmission path is not easy, and automated voice quality assessment tools are often required (see, for example, [7]). A carefully selected sample of speech is repeatedly played over a circuit voice call, and the resulting speech sample collected at the other end is stored and digitally compared with the known original sample. A test station deployed with a

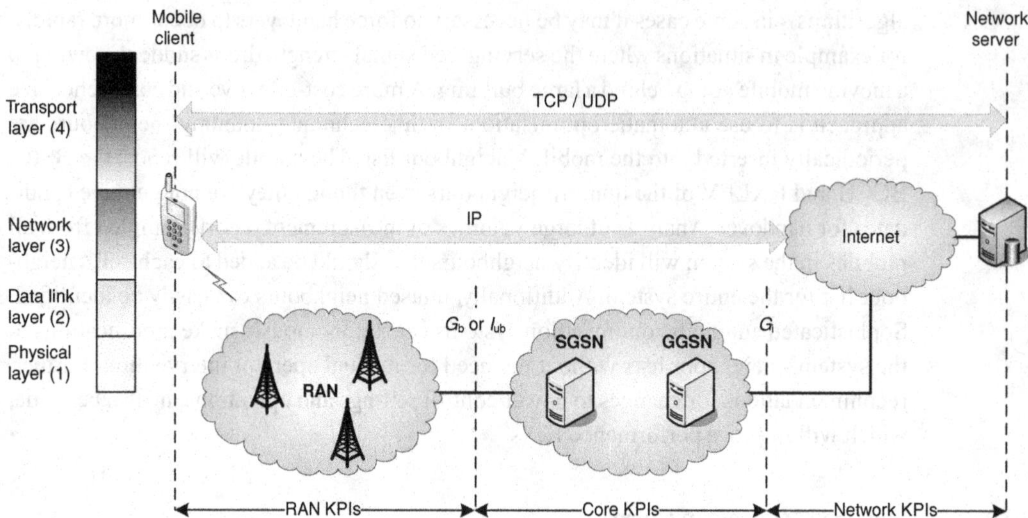

Figure 10.5 Mobile wireless system architecture

GPS unit attached can be driven around the network coverage area to identify cells with poor voice quality. Similar tests should be performed across all intermediate links in the transmission chain where possible – this will help identify problems in the transmission, switching and core network elements.

Voice quality in the RAN will be improved if the system-wide C/I is optimised, allowing the better-quality codecs within AMR to be employed; consequently, optimising the frequency plan and neighbour lists will have benefits for voice quality. Manufacturers of GSM infrastructure have also devised various methods for improving voice quality, for example by ensuring handovers are performed quietly and with minimal interruptions in speech.

10.4 GPRS RAN optimisation

In the next two sections, an overview of optimisation techniques used in GPRS and UMTS networks is provided. Most of the discussion will be centred on the diagnosis and resolution of problems in the radio access network, where most of the more obscure issues seem to be encountered, but always remember that the RAN is only one segment of the total solution delivering the end-to-end system performance that is experienced by the user.

10.4.1 An end-to-end view

Figure 10.5 shows a system view of application delivery to the user. Layer 1 and Layer 2 connectivity between the mobile device and applications server are provided by the GPRS RAN and core, as discussed at some length in Chapter 2, together with the underlying

Internet transmission system. At Layer 3, IP is used to route traffic between the mobile device and application server, with traffic flow managed at Layer 4 using transmission control protocol (TCP) or user datagram protocol (UDP).

During the original system design for a given application, an end-to-end QoS budget that *guarantees* correct operation of the application would have been formulated. Elements of the required QoS performance (in terms of bearer rate, delay, packet loss and jitter) would then have been allotted to each segment of the design so that, in aggregate, the required performance is delivered. In practice, for narrowband systems such as GPRS and EDGE with data rates heavily dependent on user location and with significant uncertainty as to when requested TBFs will be allocated, it is rarely possible to guarantee performance of real-time applications using budgets of the sort developed in the WiMAX example of Chapter 7. Nevertheless, a budget for the RAN, core and Internet components of the end-to-end solution *will* have been established, and it will be deviations from this planned performance that need to be corrected during any optimisation.

The operations staff charged with assuring the quality of the network on a day-to-day basis will normally be prompted, by some 'event', to consider network optimisation. This might be the completion of a network upgrade, the introduction of a new service or perhaps a downward trend in the end-to-end KPIs used to manage the network. Once a decision has been made to consider optimisation, operations staff should first consult the design specification for the network and establish the expected KPI levels; whenever possible, these design goals should be supplemented by the *actual* results obtained when the network was commissioned. These data provide a baseline against which current performance can be compared. Performance statistics from a representative cross-section of cell sites should then be accumulated and compared with the baseline established above. It should be possible to gather most of the data for this initial assessment using the in-built vendor facilities.

10.4.2 Assessment of end-to-end KPIs

The importance of end-to-end (E2E) metrics, such as bearer rate (or user throughput), delay, packet loss and jitter has already been established. How are these parameters to be determined in practice?

Throughput is an average (rather than instantaneous) parameter; applications depend on in the sustained throughput that a mobile link can deliver between the application server and the mobile client. Throughput can be measured straightforwardly by averaging the time it takes to transfer files of known size over the server–mobile link. The file transfer protocol (FTP) is commonly used for exchanging files over any network that supports TCP/IP; it is available on almost all operating systems and can be used for this purpose. Although FTP provides an easy means of determining link throughput, care nevertheless needs to be taken when making these measurements. Inappropriate TCP settings can cause TCP flow-control mechanisms to become active, reducing the apparent throughput below its true level.

In contrast to throughput, latency is a short-term metric and refers to the time taken for individual data packets to propagate from client to server or vice versa. It is an

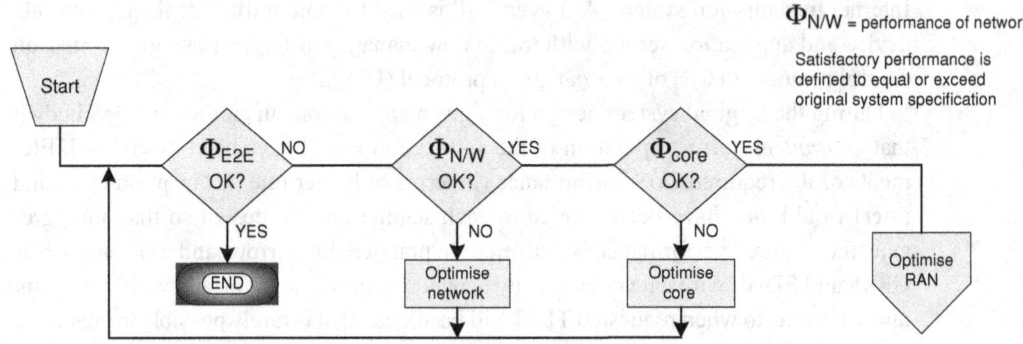

Figure 10.6 Diagnosis of the sub-system requiring optimisation

important metric for interactive and real-time applications. It is normally specified to be the one-way delay of a data packet. In practice, the measurement of one-way delays is complicated by the need to synchronise sender and receiver. For this reason, it is more common to measure the round-trip time of a data packet, which can be established with the PING command present in most operating systems. This same command can also be used to establish a packet-loss rate for the link.

The final metric of interest is jitter, which is a measure of packet-delay variability. Jitter is an important metric in ensuring the correct operation of real-time and streaming applications. Establishing system jitter is a more complicated task than assessing latency, as it is necessary to measure and form distributions from an ensemble of individual packet delay measurements. However, commercial tools are now available to perform these tasks automatically.

Figure 10.6 shows a process flow that represents an efficient optimisation methodology for GPRS and UMTS systems. The first step is to carry out end-to-end measurements of the four key metrics discussed above. If these metrics align with or exceed the original system design goals, optimisation is not necessary. If one or more of the end-to-end metrics are significantly outside of the system specification, optimisation should enable some improvement in performance.

Having established that optimisation is necessary, it is usually worthwhile to make more end-to-end measurements. During FTP measurements, the most obvious problem might be that the throughput is well below the design goal. In fact, this problem of low throughput could be the result of TCP responding to large round-trip times and high packet-loss ratios. It is important to carry out a thorough assessment of all four key parameters and, if necessary, eliminate the effects of TCP before representative link measurements can be recorded.

If any of the four key metrics is significantly outside its design goals, the next step is to determine where this limitation originates. In principle, it would be possible to start work on the GPRS RAN, core or network sub-systems; there is no particular diagnostic merit in choosing any particular one of the segments on which to start. In practice, gaining a comprehensive understanding of RAN performance is much more difficult. For one thing, performance varies depending upon the user's location! It is, therefore, much more sensible to check the performance of 'wired' segments progressively, only spending time on the RAN if the other sub-systems have been eliminated.

A protocol analyser, connected at the Gi interface and then the Gb interface, can be used to determine if the limitation is occurring within the network or core sub-systems. If a limitation is found within the network sub-system, dialogue will be required with the operator's interconnect supplier. If the problem is found to be for the GPRS network, then equipment statistics for the GGSN, SGSN and associated intra-company transmission need to be examined. Confirmation of within-specification performance for the network and core sub-systems almost certainly means that the problem is within the GPRS RAN.

10.4.3 GPRS radio-access network performance

In this concluding section on GPRS optimisation, it is assumed that network data have already been collected, a definite fall-off in performance against the original design goals has been identified and that, after further diagnostic work, suspicion has fallen on the GPRS RAN. Over the next few pages, KPIs associated with four areas critical to GPRS RAN performance are discussed:

- Radio-link quality,
- TBF performance,
- GPRS packet loss,
- GPRS blocking.

To assist in problem diagnosis, Table 10.2 provides target performance values, which should be achievable in most networks, for many of these key metrics.

10.4.3.1 Radio-link quality

The quality of the radio link directly influences the throughput of individual user connections and the capacity of the radio network. Two KPIs provide insight into the behaviour of the radio link; the block erasure rate (BLER) and the coding scheme usage.

The *BLER* is defined as the number of retransmissions out of the total number of blocks transmitted (including retransmissions). An overall BLER for each data transfer should be measured, although this may not determine whether there are radio-link problems. This is because of the presence of the link adaptation algorithm which, over a period of time, changes the coding scheme to that which is considered optimum for the C/I prevailing at the mobile location. If, after a number of BLER measurements, there are still concerns with the performance of the radio link it may be necessary to measure the BLER with the mobile at a number of different locations, chosen such that the link adaptation algorithm exercises each of the coding schemes. In this way the BLER performance of each scheme can be assessed. Figure 10.7 shows the downlink block error rate for the CS-2 coding scheme obtained from a live GPRS network.

Coding scheme usage, not unreasonably, is a measure of how often each coding scheme is used! One of the goals of optimisation is to maximise the usage of the highest-rate

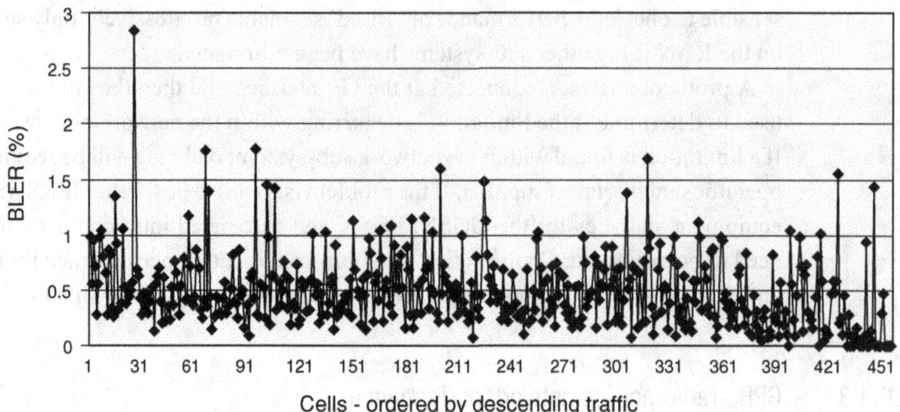

Figure 10.7 BLER performance measurements from a commercial GPRS network

(CS-4) scheme, consistent with an acceptable BLER, to provide the best possible end-to-end throughput. It is important to note that care must be exercised when interpreting the coding scheme usage KPI. On its own, it does not directly provide an indication of the quality of the radio link. This is because there is always a considerable use of CS-1 and CS-2 coding schemes in the network to ensure the delivery of control-plane information and because short data transfers might not exist for long enough to allow the radio-link adaptation algorithm to switch to CS-4. For example, if the majority of the GPRS traffic is originated by WAP transactions, there is unlikely to be a large percentage of CS-4 usage because the short duration of WAP transactions will not often allow time for link adaptation. The coding scheme usage KPI needs to be considered in conjunction with the BLER KPI.

10.4.3.2 TBF performance metrics

The *TBF establishment success rate* is a KPI analogous to the call set-up success rate in GSM. The uplink GPRS TBF set-up starts with the channel request or packet resource request sent by the mobile and ends with the transmission of the first data block or acknowledgement for the packet uplink assignment. The downlink GPRS TBF set-up might start with an immediate assignment if the network knows in which cell the mobile is, or with a packet downlink assignment if the mobile is in the ready state. For both the uplink and downlink it is necessary to evaluate the success rate of establishing the TBF, since it has a considerable impact on the end-to-end performance. Two KPIs are normally recorded:

- Downlink TBF establishment success rate,
- Uplink TBF establishment success rate.

When considering these KPIs it is important to remember that the number of TBFs established during a data transfer will have an impact on the end-to-end performance because of the time it takes to set them up. It is, therefore, often helpful to establish a counter for the number of TBFs established so that these may be minimised consistent with the number of independent 'calls'.

The *TBF drop rate* KPI is also analogous to its GSM counterpart – the drop-call rate KPI. An excessive drop rate, with its associated requirement for additional TBF establishments, can have an adverse impact on end-to-end performance. However, it should be noted that there is one circumstance when a dropped TBF has no impact. This occurs when the mobile reselects a new cell. In this case, the PCU loses contact with the mobile in the old cell and the TBF is terminated. Clearly, this is one occasion when the TBF may legitimately be dropped! To account for this specific case a network counter for the number of cell reselections should be implemented and correlated with the TBF drop rate KPI.

10.4.3.3 GPRS packet loss

The term GPRS packet loss has been used to distinguish it from the same phenomenon in the wide area network linking the G_i interface to the application server. Packet loss is one of the most important metrics of mobile-to-application server link quality. This is because TCP can respond to excessive packet losses with requests for data retransmission and may also restrict the flow rate of new data into the system. Within the GPRS radio network the following KPIs should be recorded:

- Loss of frame-relay frames,
- Loss of LLC frames.

The LLC frame-loss rate is a direct measure of the loss rate between the mobile and the SGSN. The frame-relay-loss rate is a measure of the losses occurring only between the SGSN and PCU and thus can provide further insight for problem resolution.

10.4.3.4 GPRS blocking

It will hopefully be apparent by now that GSM has been designed as a very flexible communications solution. The same physical resources can be used to convey GSM voice traffic or GPRS data traffic. Time slots in a cell can be configured to be used only by GSM traffic, only by GPRS traffic, or by either traffic type. In those time slots that are enabled to carry GSM or GPRS traffic it is normal that when both GSM and GPRS traffic is present and contending for the same channel, GSM traffic takes priority. When GSM traffic takes priority, the performance of GPRS data transfers will be impacted. For this reason it is important to assess the expected GSM and GPRS traffic to be supported by each carrier and, if possible, per time slot to ensure sufficient resources are available to ensure that the QoS requirements for both GSM and GPRS may be met.

It is also important to establish the congestion level of a GPRS network *before* any attempt is made at optimisation; if a network is fundamentally congested, no amount of optimisation effort is going to produce a significant improvement in performance. GPRS standards define congestion as the amount of time that all GPRS time slots are occupied. GPRS networks will normally operate at a congestion level less than 75%, depending upon blocking and other QoS goals.

Table 10.3. GPRS performance targets

	Parameter	Limit	Comments
End-to-end	Throughput	Q kbits/s	See note 1 below
	Delay	0.5–2.5 s	0.5 s for ~500 bytes and up to 2.5 s for ~1500 bytes
	Packet loss	0.1%	
GPRS RAN	TBF est.	>99%	(Percentage success of TBF establishment attempts)
	TBF drop	<1%	(Dropped TBFs as a percentage of established TBFs)
	BLER	<10%	
	Packet loss	<0.1%	Over BSS/SGSN and SGSN/GGSN transport (PDU loss)

Notes:
1) Mobile and signal strength dependent. For a link with a 4-time-slot mobile and assuming more than 4 time slots dedicated to GPRS in the network, the peak data rate seen by the user, Q kbits/s, is given by:

$$Q = 4 \times 20\,\text{kbits/s} \times 0.75 = 60\,\text{kbits/s}.$$

This assumes: a) Signal strength ensures selection of CS4 coding; b) 25% of payload for overhead.

Assuming that the GPRS network is not congested, it is also usually desirable to manage the GPRS blocking rate as a KPI. Even though GPRS is a packet system, the likelihood that a user will be granted access to the system is driven by the probability that the requested number of channels is available on demand, in a rather similar way to conventional Erlang calculations. It will, therefore, be down to the operator to provision sufficient GPRS time slots to ensure that the targeted blocking rate can be achieved for the average throughput goal. Target blocking rates will depend upon the GPRS service positioning of the particular operator but will normally be less than 5%.

Table 10.3 provides a summary of representative performance targets for some of the GPRS KPIs. It is important, of course, that these parameters are recorded and correlated at different mobile locations to give a view of how user performance varies in the cell. These data can then be compared with the expected performance, for instance checking the *actual* variation of coding scheme selected by the network with that which should be expected as the user moves from the centre to the cell edge.

10.5 UMTS network performance optimisation

As for GSM and GPRS, the quality of the user experience is very important for the operator. Users who experience poor quality of service are more likely to migrate to a competitor's network.

Table 10.4 summarises the key end-to-end performance attributes for a UMTS user. These are largely similar to GPRS with the addition of user throughput in the circuit bearer section, which may now be varied over a wide range to support streaming video and other applications. However, in a number of cases, the KPIs associated with the metrics have changed to reflect the impact of soft handoff and single frequency reuse.

Table 10.4. UMTS user experience and related KPIs

	Attribute	Relevant KPIs	Comments
Circuit bearers	Call set-up Success rate	Coverage area (%) Blocking (%)	Also affected by cell load
	Call drop rate	Intra-frequency handover (%)	Handover between cells
		Inter-frequency handover (%)	Handover within or between cells
		Inter-RAT handover (%)	Handover to a different air interface
	User throughput	Bearer rate (kbits/s)	Percentage of cell covered is key
	Cell capacity	(Mbits/s)	Affected by maximum cell radius
Applications	Application set-up	Coverage area (%) Blocking (%)	Also affected by cell load
	Call drop rate	Intra-frequency handover (%)	Handover between cells
		Inter-frequency handover (%)	Handover within or between cells
		Inter-RAT handover (%)	Handover to a different air interface
	Bearer rate	Meets or exceeds specified rate (kbits/s)	Because of the need for omnipresent power control with the attendant inefficiency and long set-up times, UMTS Release 99 packet bearers are not extensively used
	Latency	Less than or equal to specified delay (s)	
	Packet loss	Less than or equal to specified packet loss (%)	
	Jitter	Less than or equal to specified delay variation (s)	

As for GPRS, there are various components of system-level optimisation. Optimising the management of radio resources can preserve the capacity and health of the network. This includes tuning the parameters of the vendor-specific algorithms for power control, to ensure that excessive power in the system, and thus interference, is kept to a minimum. There will also be features in the UTRAN to control admission and react to congestion when this does occur. The parameters of these algorithms need to be optimally configured. As ever, the various components that make up mobility should be optimised to ensure that the user experiences no problems with call quality and stability. Finally, the coverage and capacity of the system should be balanced so that a good trade-off between supporting large numbers of subscribers in as many locations as possible can be reached.

10.5.1 Optimisation of mobility management

Correct configuration of neighbour lists is critical to the success of handovers and the health of the UTRAN in general. The neighbour list for each cell should include all the significant neighbours with which a mobile may benefit from going into soft handoff. The instances of insignificant cells (cells that cover small areas or carry little load) should be minimised. These extend the neighbour list so that handover measurements take longer, and the call may drop before the process is completed. The phenomenon of

Table 10.5. Events that can trigger intra-frequency measurement reports

Event name	Description
1A	A primary CPICH enters the reporting range
1B	A primary CPICH leaves the reporting range
1C	A non-active primary CPICH becomes better than an active primary CPICH
1D	Change of best cell
1E	A primary CPICH becomes better than an absolute threshold
1F	A primary CPICH becomes worse than an absolute threshold

pilot surprise is a case in point. Pilot surprise can happen when a mobile rounds a corner in a dense urban area and suddenly finds itself in the direct path of a cell that previously had not been received. Unless the mobile can decode that pilot, perform a measurement report and complete the channel set-up in a short time, the call may be lost to interference. It is, therefore, critical that the potential surprise pilot is in the neighbour list and that the neighbour list is short enough so that the UE has more chance of measuring and reporting the pilot to the UTRAN in time.

Changes to the active set of a UE are controlled by the RNC using reports of measurements made by the UE. These reports are triggered by certain events in the received signal strength of CPICH power. In the case of intra-frequency measurements for FDD, these are known as events 1A–1F. The definition of these events is given in Table 10.5. Different UTRAN vendors have proprietary mobility management approaches which operate on different subsets of these event-triggered messages. The exact sub-set of messages along with their thresholds, hysteresis and other parameters are all configurable at the UE by the RNC using appropriate messages. Of course, the operator must first ensure that these parameters are configured sensibly in the management information base (MIB).

A pilot that is not contributing to a call as a member of the *active set* is a source of interference. If that interference is too great it will degrade call quality and ultimately jeopardise call stability. Conversely, too many pilots in the active set cause loss of capacity in the back-haul as well as the air interface, for little system benefit. This situation can result if pilots that are no longer significant are not dropped from the active set fast enough. Hysteresis parameters that are too wide will result in this sort of delayed response. Alternatively, if these windows are too narrow, then the signalling load will rise and reduce spectral efficiency. Excessive dropped calls will be one of the key indicators of poor mobility management parameter configuration.

The UTRAN is free to construct the neighbour list as it pleases and it transmits this to each UE when required. Typically, the RNC will build the neighbour list for a UE based either on the membership of the active set, or just the dominant member. It can take into account factors such as the ranking or even the relative strengths of the received pilot powers at the UE to weight the contents and the ordering of the list. The precise mechanism used in a particular vendor's RNC is a consideration when constructing the list. The UTRAN will also control how the UE will report cell-strength measurements.

One approach to determining optimum neighbour lists is to define a cell-overlap matrix. This will be a triangular matrix with order equal to the number of cells. The matrix defines the strength of overlap between each pair of cells in the system. This will probably be a sparse matrix unless the other aspects of the radio are poorly configured, with excessive overlap or pilot pollution. Each element of the matrix will represent the strength of the relationship between the pair of sites defined by that row and column. Constructing the neighbour list for a cell will then be a case of looking at the row and column corresponding to that cell and taking the top neighbours as defined by the strength of that pairing. The neighbour list for that cell will comprise the top weighted candidates.

Constructing the matrix will require consideration of all geographical areas of the network and comparison of the predicted received signal strength of all cells received in each area. Pairs of cells that are received strongly in the same location will result in a stronger pressure to have a reciprocal neighbour relationship. The matrix may alternatively be based on real measurements collected from the network.

In addition to intra-frequency soft handovers, a network with multiple carriers or an underlying 2G network must be able to perform inter-frequency and inter-RAT hard handovers reliably. Again, this is controlled by neighbour lists, and handover decisions are made by the UTRAN, based on reports of events of received signal strengths for cells measured by the UE.

Mobility in idle mode is controlled by cell selection and reselection. As the UE moves between location areas it must perform location-area updates. Location areas (comprising local groups of cells) are designed to minimise the amount of signalling in idle mode. Therefore, it should be verified that there is not an excessive number of location-area updates being performed. As the size of location areas is increased, the number of location-area updates will reduce but the number of paging messages will increase. When the quantities of these messages are not well balanced, a location-area redesign should be performed to maintain RF efficiency. Tests for location area ping-pong should also be performed. If the situation arises that UEs often flip back and forth repeatedly between two location areas, that would suggest that the border between the two location areas has been poorly chosen and a re-bordering should be performed.

10.5.2 Coverage optimisation

Subscribers are normally assumed to be distributed non-uniformly. However, if this concentration of users is in the area between two sectors on a tri-sectored site it will increase the number of users in soft handover (SHO) and reduce the capacity of both the air interface and back-haul. Subscribers can also be non-uniformly distributed between the node Bs; a clump of users in the SHO areas will again reduce capacity.

One way to deal with this lack of homogeneity is to re-orientate antennas. However, knowing where the subscribers are located does not directly translate to an optimum antenna-orientation plan. Any proposals for changes to antenna direction must be backed up by a strategy for balancing the various conflicting objectives of the optimisation. In general, minimising the amount of SHO will minimise back-haul requirements. However, where the intention is to change the best server for some subscribers, care must be taken not to overload the back-haul of the recipient site. Minimising the SHO areas will, in general, minimise inter-cell interference and thus maximise capacity. However, cell shrinkage effects must be considered, which can adversely impact coverage. Further, what is optimal for downlink is not necessarily optimal for the uplink. Transmit power is more of an issue for the UE. If the propagation loss between the mobiles is increased, then reverse link transmit power will go up and battery life will go down, along with customer satisfaction.

Smart antennas are able to steer the transmit and receive beams dynamically to where the subscribers are actually located. This minimises interference and maximises received signal strength. However, such antennas require a physically large tower area (which may not be acceptable from an environmental perspective) and are expensive; thus, they are rarely viable from a return-on-investment perspective. A more attractive alternative is the use of antennas whose vertical or azimuth beams may be tilted or shaped electrically. Such configurations are particularly attractive at sites where the subscriber base may move during the day, perhaps from residential areas to motorways during commuting periods.

Where sites and antennas are shared with 2G systems, complete flexibility of the antenna bearing, beam width and down-tilt may not be possible because of different cell coverage areas. A compromise in these cases may be to use a combination of some antenna changes together with pilot-power optimisation. Tuning the pilot powers can control cell overlap and work to minimise the inter-cell interference and thus maximise capacity. Special care must be taken with the reverse link as there is a risk of subjecting the UE to high transmit-power demands if the pilot power non-uniformity is extreme. Indeed, for this very reason, many operators have a policy of uniform pilot power on macrocells.

Finally, it is easy to ignore the problems that can arise with scrambling codes. As there are 512 unique scrambling codes available, it is not hard at the planning stage to make sure that there is no ambiguity in cell identity. However, errors can creep in, especially at RNC borders or if the network has been expanded with extra cell sites added. Sites within measurement reporting distance sharing the same scrambling code can lead to confusion over which cell is being referenced by a UE measurement report. Excessive active-set-update failures on a given site may indicate a problem with scrambling code conflicts. These can be resolved manually using a map of the cells annotated with these codes.

10.5.3 Optimisation goals

When tasked to optimise a UMTS network it is important keep an eye on the big picture. In CDMA systems it is easy to improve the various KPIs, but the traffic carried may reduce! Similarly, the number of dropped calls could potentially be reduced significantly if the CPICH powers were increased to improve coverage. However, the associated reduction in system capacity, and thus traffic carried, would have an unacceptable impact on customer satisfaction.

Nevertheless, once the work has been completed, a well-optimised system will be characterised by good KPIs; what represents a good set of KPIs will depend to some extent on operator policy for the particular network. For instance, if the network is heavily loaded by design then admission control will often be triggered, reducing call set-up success rate. In this situation, unless new carriers or sites are added, it may not be possible to improve performance. However, unless there are limiting policy choices, it should be possible in a well-optimised network to achieve a call drop rate of <1%, a handover success rate of greater than 99% and blocking of around 1%–2%.

Finally, it is important to know your subscribers thoroughly as well as your network. There will probably be seasonal variations in the KPIs as people move about during holiday times, or there will be changes in foliation on trees, and thus changes in signal attenuation for example. Optimisation engineers should understand these characteristic variations and the associated variation in baseline performance, against which improvements should be measured.

References

1 The TeleManagement Forum, *Enhanced Telecom Operations Model*, www.tmforum.org.

2 ITU-T, *Enhanced Telecom Operations Map (eTOM) – The Business Process framework*, ITU-T Recommendation M.3050.1, www.itu.int/rec/T-REC-M.3050.1-200406-S/en.

3 A. Pathan, *The Functional Areas of Network Management*, http://trade.hamk.fi/~lseppane/courses/projects/FCAPS.htm.

4 ITU-T, *Methods for Subjective Determination of Transmission Quality*, Recommendation P.800, www.itu.int/rec/T-REC-p.

5 ITU-T, *Perceptual Evaluation of Speech Quality (PESQ): An Objective Method for End-to-End Speech Quality Assessment of Narrow-Band Telephone Networks and Speech Codecs*, Recommendation ITU P.862, www.itu.int/rec/T-REC-p.

6 Motorola, *Motorola Strategic Optimisation Services*, www.motorola.com/networkoperators/pdfs/strategic_op_svcs_brochure.pdf.

7 Ascom, *QVoice – The Testing Solution for Speech and Data*, www.qvoice.com/1-products/cpro_1000.html.

Acronyms

2G	Second generation
3G	Third generation
3GPP	3rd generation partnership project
8PSK	8 state phase shift keying
16QAM	16 state quadrature amplitude modulation
64QAM	64 state quadrature amplitude modulation
AAA	Authentication authorisation and accounting
AAL2	ATM adaptation layer 2
ADSL	Asymetric digital subscriber line
AF	Assured forwarding
AFP	Automatic frequency planner
AGCH	Access grant channel
AICH	Acquisition indicator channel
AIPN	All IP network
AMC	Adaptive modulation and coding
AMR	Adaptive multi-rate (codec)
AP	Access point
ARC	Application resource consumption (Motorola proprietary tool)
ARFCN	Absolute radio frequency channel number
ARPU	Average revenue per user
ARQ	Automatic repeat request
AS	Application servers
ATM	Asynchronous transfer mode
AuC	Authentication centre
BBH	Baseband hopping
BCCH	Broadcast control channel
BCH	Broadcast channel
BCIE	Bearer capability information element
BE	Best-effort (service)
BER	Bit-error rate
BGCF	Breakout gateway control function
BHCA	Busy-hour call attempts

BLER	Block error rate
BPSK	Binary phase shift keying
BSC	Base station controller
BSIC	Base station identity code
BSS	Base station sub-system
BSSGP	Base station system GPRS protocol
BTS	Base transceiver station
C/I	Carrier-to-interference ratio
CAD	Computer-aided design
CAMEL	Customised applications for mobile enhanced logic
CAPEX	Capital expense
CBCH	Cell broadcast channel
CBR	Constant bit rate
CC	Call control
CCCH	Common control channel
CCPCH	Common control physical channel
CCTrCH	Composite coded transport channel
CD	Compact disk
CD/CA-ICH	Collision detection/channel assignment indicator channel
CDF	Cumulative distribution function
CDMA	Code division multiple access
CDR	Call detail record
CGF	Charging gateway function
CINR	Carrier-to-interference + noise ratio
CIR	Channel impulse response
CM	Connection management
CN	Core network
Codec	Coder/decoder
COST	Co-operation in the field of scientific and technical research
CPCH	Common packet channel
CPE	Customer premises equipment
CPICH	Common pilot channel
CQI	Channel quality indicator
CQICH	Channel quality indicator channel
CRC	Cyclic redundancy check
CS	Circuit switched or Coding scheme
CSCF	Call session control function
CSICH	CPCH status indication channel
CSMA	Collision sense multiple access
CS-MGW	Circuit-switched media gateway
CTC	Convolutional turbo-code
CTS	Clear to send

D/A	Digital to analogue (converter)
dBm	Decibel relative to one milliwatt
dBW	Decibel relative to one watt
DCH	Dedicated channel
DECT	Digital enhanced cordless telecommunications
DEM	Digital elevation model
DFT-SOFDM	Discrete Fourier transform spread OFDM
DHCP	Dynamic host configuration protocol
DNS	Domain name server
DOCSIS	Data over cable service interface specifications
DPCCH	Dedicated physical control channels
DPCH	Dedicated physical channel
DPDCH	Dedicated physical data channel
DSCH	Downlink shared channel
DSL	Digital subscriber line
DTM	Digital terrain model
DTX	Discontinuous transmission
DVD	Digital versatile disk
E1	2 Mbits/s transmission circuit
E-AGCH	Enhanced access grant channel
EAP	Extensible authentication protocol
E_b/N_0	Energy/bit over noise power spectral density
E_c/I_0	the ratio of energy/chip (of the wanted code) to the total power spectral density, measured at the receiver input
EC	European commission
E-DCH	Enhanced dedicated channel
EDGE	Enhanced data rates for GSM evolution
E-DPCCH	Enhanced dedicated physical control channel
E-DPDCH	Enhanced dedicated physical data channel
EF	Expedited forwarding
EFR	Enhanced full rate
EGPRS	Enhanced GPRS
EGSM	Extended GSM (frequency band)
E-HICH	E-DCH hybrid ARQ indicator channel
EIR	Equipment identity register
EIRP	Effective isotropic radiated power
ENUM	Telephone number mapping protocol, defined in RFC 3761
EPC	Evolved packet core
ePDG	Evolved packet data gateway
E-RGCH	Enhanced relative grant channel
ERP	Effective radiated power
ertPS	Extended real-time polling service
e-TOM	Enhanced telecommunications operations model

ETSI	European Telecommunications Standards Institute
EU	European Union
E-UTRAN	Evolved universal terrestrial radio-access network
FAB	Fulfilment, assurance and billing
FAC	Final assembly code (of a mobile phone)
FACCH	Fast associated control channel
FACH	Forward access channel
FBI	Feedback information (bit)
FCAPS	Fault configuration performance and security
FCC	Federal Communications Commission
FDD	Frequency division duplex
F-DPCH	Fractional dedicated physical channel
FER	Frame-erasure rate
FFT	Fast Fourier transform
FH	Frequency hopping
FIR	Finite impulse response
FSK	Frequency shift keying
FTP	File-transfer protocol
FUSC	Full usage of sub-channels
G/W	Gateway
GERAN	GSM/EDGE radio access network
GGSN	Gateway GPRS support node
GIS	Geographic information system
GMSC	Gateway MSC
GMSK	Gaussian minimal shift keying
GOP	Group of pictures
GPRS	General packet radio service
GPS	Global positioning system
GSM	Global system for mobile communications
GSMA	GSM association
GSMFR	GSM full rate
GSN	GPRS support node
GTP	GSM tunnelling protocol
GUI	Graphical user interface
HA	Home agent
HARQ	Hybrid ARQ
HILI	High level interface
HLR	Home location register
HR	Half rate
HSCSD	High-speed circuit-switched data
HSDPA	High-speed downlink packet access

HS-DPCCH	High-speed dedicated physical control channel
HS-DSCH	High-speed downlink shared channel
HSN	Hopping sequence number
HSPA	High-speed packet access
HS-PDSCH	High-speed physical downlink shared channel
HSS	Home subscriber server
HS-SCCH	High-speed shared control channel
HSUPA	High-speed uplink packet access
HT100	Hilly terrain @ 100 km/hour
HTTP	Hypertext transfer protocol
IANA	Internet assigned numbers authority
IEEE	Institute of Electrical and Electronic Engineers
IETF	Internet Engineering Task Force
IFFT	Inverse fast Fourier transform
IMEI	International mobile equipment identity
IMS	IP multimedia system
IMSI	International mobile subscriber identity
IMSSF	IP multimedia service switching function
IN	Intelligent networking
INAP	IN applications protocol
IOI	Interference on idle
IP	Internet protocol
IP-BS	IP bearer services
IP-CAN	IP connectivity access network
IPR	Intellectual property rights
IR	Incremental redundancy
ISDN	Integrated service digital network
ISM	Industrial scientific and medical (frequency band)
ISO	International Standards Organization
ISP	Internet service provider
ISUP	Integrated services user part
ITU	International Telecommunications Union
KPI	Key performance indicator
LA	Link adaptation
LAN	Local area network
LDP	Label distribution protocol
LER	Label edge router
LLC	Logical link control
LMA	Local mobility anchor
LOS	Line of sight
LRM	Low-revenue market

LSR	Label switching path
LTE	(UMTS) long-term evolution
LTE	Linear transversal equaliser
MA	Mobile allocation
MAC	Media-access control
MAG	Mobile-access gateway
MAI	Multiple-access interference
MAIO	Mobile allocation index offset
MAN	Metropolitan area network
MAP	Media-access protocol
MBMS	Multimedia broadcast multicast service
MCHO	Mobile controlled handover
MCS	Modulation and coding schemes
ME	Mobile equipment
MGCF	Media gateway control function
MGW	Media gateway
MHA	Mast head amplifiers
MIB	Management information base
MIMO	Multiple-input–multiple-output (antenna system)
MLSE	Maximum likelihood sequence estimation
MM	Mobility management
MME	Mobility management entity
MO	Mobile originated
MOS	Mean opinion score
MPEG-2	Motion pictures expert group 2
MPLS	Multi-protocol label switching
MR	Measurement report
MRCF	Media resource control function
MRFP	Media resource function processor
MS	Mobile station
MSC	Mobile switching centre
MT	Mobile terminated
MTBF	Mean time between failures
MTTR	Mean time to repair
MVNO	Mobile virtual network operator
NAS	Non-access stratum
NGMN	Next generation mobile network group
NGN	Next generation networking
NLOS	Non-line of sight
NMF	Network management function
NMS	Network management system
NPV	Net present value

nrtPS	Non-real-time polling service
Ofcom	Office of Communication (UK Government Agency)
OFDM	Orthogonal frequency division multiplexing
OFDMA	Orthogonal frequency division multiple access
OMC	Operations and maintenance centre
OMC-R	OMC-radio
OPEX	Operating expense
OSA-SCS	Open-service access-service capability server
OSPF	Open shortest path first
OVSF	Orthogonal variable spreading factor
P&L	Profit and loss
PA	Proxy agent
PAGCH	Packet access grant channel
PAPR	Peak-to-average power ratio
PAR	(IEEE) project authorisation requests
PBCCH	Packet broadcast control channel
PC	Personal computer
PCCCH	Packet common-control channel
PCCPCH	Primary common-control physical channel
PCEF	Policy charging and enforcement function
PCH	Paging channel
PCM	Pulse code modulation
PCPCH	Physical common-packet channel
PCRF	Policy charging and rules function
P-CSCF	Proxy CSCF
PCU	Packet control unit
PDA	Personal digital assistant
PDCH	Packet data channel
PDN	Packet data network
PDP	Packet data protocol
PDSCH	Physical downlink shared channel
PDTCH	Packet data traffic channel
PDU	Packet data unit
PHB	Per hop behaviour
PHY	Physical layer
PICH	Paging indication channel
PLMN	Public land mobile network
PMIP	Proxy mobile IP
PMM	Packet mobility management
PN	Pseudonoise
PPCH	Packet paging channel
PPDN	Public packet data network

PPP	Point-to-point protocol
PRACH	Packet random access channel
PRM	Partial response modulation
PSCH	Primary synchronisation channel
PSTN	Public switched telephone network
PTT	Postal telegraph and telephone
PUSC	Partial utilisation of sub-channels
QoS	Quality of service
QPSK	Quadrature phase shift keying
R&TTE	Radio equipment and telecommunications terminal equipment
R99	Release 99
RA	Routeing area
RA100	Rural area @ 100 km/hour
RA250	Rural area @ 250 km/hour
RAB	Radio-access bearer service
RACE	R&D in Advanced Communications networks in Europe
RACH	Random-access channel
RAN	Radio-access network
RAT	Radio-access technology
RF	Radio frequency
RGCH	Relative grant channels
RLC	Radio link control
RNC	Radio network controller
RoT	Rise over thermal noise
RR	Radio resource
RRC	Radio-resource control
RRM	Radio-resource management
RSCP	Received signal-code power
RSSI	Received signal-strength indication
RTG	Receive transmission gap
RTP	Real-time protocol
rtPS	Real-time polling service
RTS	Request to send
RXLEV	Received (signal) level
S/N	Signal-to-noise ratio
SACCH	Slow associated control channel
SAE	System architecture evolution
SAIC	Single antenna interference cancellation
SAR	Segmentation and reassembly
SCCPCH	Secondary common control physical channel
SCH	Synchronisation channel

SC-OFDM	Single carrier OFDM
SDCCH	Stand-alone dedicated control channel
SDF	Service data flow
SDP	Session description protocol
SDU	Service data unit
SF	Spreading factor
SFH	Slow frequency hopping
SFN	Single-frequency network
SGSN	Serving GPRS support node
SGW	Signalling gateway
SHO	Soft handover
SID	Silence descriptor
SIM	Subscriber identity module
SIP	Session initiation protocol
SIR	Signal-to-interference ratio
SLA	Service-level agreement
SM	Session management
SMS	Short-message service
SNDCP	Sub-network dependent convergence protocol
SNR	Signal-to-noise ratio
SRB	Signalling radio bearer
SS	Subscriber station
SSCH	Secondary synchronisation channel
STA	Wireless station (Wi-Fi)
STM-1	Synchronous transfer module 1
SUI	Stanford University Interim (propagation model)
SVN	Software version number (of a mobile)
TA	Timing advance
TAC	Type allocation code (of a mobile)
TBF	Temporary block flow
TCH	Traffic channel
TCH/F	TCH full rate
TCH/H	TCH half rate
TCP	Transmission control protocol
TDD	Time division duplex
TDMA	Time division multiple access
TFCI	Transmit format combination indicator
TID	Tunnelling identifier
TISPAN	Telecoms and Internet converged services and protocols for advanced networks
TMN	Telecommunications management network (ITU protocol)
TPC	Transmit power control
TRAU	Transcoder and rate adaptor unit

TrCH	Transport channel
TRX	Transceiver
TTG	Transmit transmission gap
TTI	Transmit time interval
TU50	Typical urban @ 50 km/hour
TUP	Telephony user part
TUSC	Tile usage of sub-channels
UA	User agent
UBR	Unspecified bit rate
UDP	User datagram protocol
UE	User equipment
UGS	Unsolicited grant service
UMTS	Universal mobile telecommunications service
UPA	Uniform planning area
USIM	UMTS SIM
UTRA	Universal terrestrial radio access
UTRAN	Universal terrestrial radio access network
VAD	Voice-activity detection
VBR	Variable bit rate
VLAN	Virtual LAN
VLR	Visitor location register
VoIP	Voice over IP
WAP	Wireless-application protocol
WCDMA	Wideband CDMA
WG	Working group
Wi-Fi	Wireless fidelity
WiMAX	Wireless interoperability for microwave access
WLAN	Wireless local area network
WWW	World Wide Web
X25	ITU protocol

Index

Printed in the United States
by Baker & Taylor Publisher Services